D1727834

Springer Tracts in Advanced Robotics 90

Editors

For further volumes:
http://www.springer.com/series/5208

STAR (Springer Tracts in Advanced Robotics) has been promoted under the auspices of EURON (European Robotics Research Network)

Sami Haddadin

Towards Safe Robots

Approaching Asimov's 1st Law

Sami Haddadin
German Aerospace Center (DLR)
Robotics and Mechatronics Center
82230 Wessling,
Germany
Sami.Haddadin@dlr.de

ISSN 1610-7438 ISSN 1610-742X (electronic)
ISBN 978-3-642-40307-1 ISBN 978-3-642-40308-8 (eBook)
DOI 10.1007/978-3-642-40308-8
Springer Heidelberg New York Dordrecht London

Library of Congress Control Number: 2013945418

Printed on acid-free paper

Springer is part of Springer Science+Business Media (www.springer.com)

To my dear ladies Julia & Franka

Foreword

Robotics is undergoing a major transformation in scope and dimension. From a largely dominant industrial focus, robotics is rapidly expanding into human environments and vigorously engaged in its new challenges. Interacting with, assisting, serving, and exploring with humans, the future robots will increasingly touch people and their lives.

Beyond its impact on physical robots, the body of knowledge robotics has produced is revealing a much wider range of applications reaching across diverse research areas and scientific disciplines, such as: biomechanics, haptics, neurosciences, virtual simulation, animation, surgery, and sensor networks among others. In return, the challenges of the new emerging areas are proving an abundant source of stimulation and insights for the field of robotics. It is indeed at the intersection of disciplines that the most striking advances happen.

The Springer Tracts in Advanced Robotics (STAR) is devoted to bringing to the research community the latest advances in the robotics field on the basis of their significance and quality. Through a wide and timely dissemination of critical research developments in robotics, our objective with this series is to promote more exchanges and collaborations among the researchers in the community and contribute to further advancements in this rapidly growing field.

Working and interacting in shared workspaces with humans, the new robots are increasingly challenged for achieving the highest levels of dependability and safety. Safety has indeed come to represent one of the major themes in robotics research for those applications that bring robots in contact with humans. These include cooperative material-handling, power extenders and such high-volume markets as rehabilitation, physical training, entertainment, among others. The authoritative monograph by Sami Haddadin focuses on the central problem of controlling the physical interaction between the robot and the human in a safe and dependable manner. Taking a biomechanical approach to assess the level of injury in case of accidental impacts, the analysis is used not only in the design of innovative lightweight arm structures but also for the development of novel control schemes for collision detection, reactive behaviors, and ultimately human-robot coexistence. The extensive experimental validation of the proposed concepts on dummies and even on humans has rapidly risen to the role of a significant milestone in the growing area of soft robotics. By the depth of its analysis and exceptionally salient experimental work, this monograph offers one of the most comprehensive treatments of the safety challenge in our field.

Remarkably, the monograph is based on the author's doctoral thesis, which received the 2012 EURON Georges Giralt PhD Award. A very fine addition to STAR!

Naples, Italy and Stanford, USA
May 2013

Bruno Siciliano and Oussama Khatib
STAR Editors

Foreword

Today's industrial robots still are heavy, fast and strong positioning devices, which are supposed to guarantee precision by high stiffness; but thus in their load/weight relationship (typically 1:5 or less) they are clearly inferior to the human arm, which can hardly guarantee pure positioning precision, but via sensory feedback (vision and tactile) they can perform the most delicate assembly tasks despite of changing environmental conditions. Yet even with a camera and a force sensor in wrist or gripper classical industrial robots are not really competitive with humans in the wide field of assembly and they are strictly separated from humans by fences thus avoiding any kind of touch and danger for humans.

Since a number of years it has become clear that future production assistants and personal assistants for elderly care, cooperating closely with humans, must be of different type. The soft robotics paradigm became popular, e.g. joint torque controlled arms with programmable or self-adaptable stiffness and compliance. The idea of robots which a human can touch every where along the arm and which sensitively react thus avoiding injuries of the human was realized in DLR's light weight robot, the control concepts of which were already inspired by Sami Haddadin's early work. Space flight with its requirements for minimal weight and energy needs had driven these technologies, aiming at a new generation of robonauts cooperating closely with human astronauts, including programming and learning by demonstration e.g. in direct physical contact.

Practically in all future applications of assistance robots the direct interaction of human and artificial arms is a key problem with respect to Asimov's 1st law, which requests that a robot is never allowed to injure a human. But even with the above mentioned soft robotics technologies this request is not easy to fulfill. Even a fully compliant robot with a sharp knife in his hand hitting e.g. a human's eye would easily break this law. No matter how quickly he would retract his arm after the slightest touch. Thus "robot safety" became a "megatopic" in the last years and Sami Haddadin one of the worldwide most experts in the field. He treated the basic problems from the most diverse aspects, not only from the advanced feedback and control techniques but also from the biomechanical side. In close cooperation with medical doctors he investigated the biomechanical quantification of human injuries by incidental collision with robot arms and tools, involving the stabbing of human and animal tissue by knifes and guillotines. His fundamental work for the first time gives clear hints on the influence of diverse robot design parameters (e.g. inertia, speed, kinematics) on the degree of injury for a human in case of a sudden collision. He has performed innumerous simulations, life experiments (including self trials) and crash tests with dummies and thus generated presumably the first systematic data collection and evaluation of robot-human injury. They have laid the basis for his innovative motion control and path planning concepts for the safe assistance robots of the future.

Oberpfaffenhofen, May 2013 Gerd Hirzinger

Preface

Up to now, state of the art industrial robots played the most important role in real-world applications and more advanced, highly sensorized robots were usually kept in lab environments and remained in a prototypical stadium. Various factors like low robustness and the lack of computing power were large hurdles in realizing robotic systems for highly demanding tasks in e.g. domestic environments or as robotic co-workers. The recent increase in technology maturity finally made it possible to realize systems of high integration, advanced sensorial capabilities and enhanced power to cross this barrier and merge living spaces of humans and robot workspaces to at least a certain extent.

In addition, the increasing effort various companies have invested to realize first commercial service robotics products has made it necessary to properly address one of the most fundamental questions of Human-Robot Interaction:

How to ensure safety in human-robot coexistence?

Although the vision of coexistence itself has always been present, very little effort has been made to actually enforce safety requirements, or to define safety standards up to now.

In this monograph, which originates from my PhD thesis to a large extend, the essential question about the necessary requirements for a safe robot is addressed in depth and from various perspectives. The approach taken here focuses on the biomechanical level of injury assessment, addressing the physical evaluation of robot-human impacts and the definition of the major factors that affect injuries during various worst-case scenarios. This assessment is the basis for the design and exploration of various measures to improve the safety in human-robot interaction. They range from control schemes for collision detection, and reaction, to the investigation of novel joint designs. An in-depth analysis of their contribution to safety in human-robot coexistence is carried out.

In addition to this "on-contact" treatment of human-robot interaction, this monograph proposes and discusses real-time collision avoidance methods, i.e. how to design pre-collision strategies to prevent unintended contact. An additional major outcome of this monograph is the development of a concept for a robotic co-worker and its experimental verification in an industrially relevant real-world scenario. In this context, a control architecture that enables a behavior based access to the robot and provides an easy to parameterize interface to the safety capabilities of the robot was developed. In addition, the architecture was applied in various other applications that deal with physical Human-Robot Interaction as e.g. the first continuously brain controlled robot by a tetraplegic person or an EMG[1] controlled robot.

Generally, all aspects discussed in this monograph are fully supported by a variety of experiments and cross-verifications, leading to strong conclusions in this sensitive and immanently important topic. Several surprising and gratifying results, which were registered in the robotics community to great interest, were obtained.

In addition to the scientific output, the outcome of this monograph attracted also significant public attention, confirming the importance of the topic for robotics research.

The major parts and contributions of this monograph are described hereafter in more detail.

Structure of the Book

Chapter 1 gives an introduction into the general context of the thesis, discusses the important open problems in physical Human-Robot Interaction, and describes the major contributions of the thesis.

State of the Art

Human-Robot Interaction (HRI) is one of the grand challenges of robotics research. In HRI, contributions in such diverse fields as robot design, control, manipulation, or human-robot communication have been carried out. Chapter 2 reviews work relevant to ensuring safety to the human. Furthermore, the major cornerstones in the development of safe and dependable robotic systems, standardization efforts taken, and the major existing contributions to "safe robotics" are outlined. The earliest ones focused on strictly separating the workspace of human and robot and therefore minimized the risk of any possibly dangerous situation to occur. In this sense, no cooperation takes place and most accidents happen during maintenance or an operating error. Other work initiated risk analysis from a more formal and classical point of view. Further contributions concentrated on realizing collision avoidance schemes or minimizing the potential risk during collisions based on appropriate measures. Although several criteria, countermeasures, and control schemes for safe physical Human-Robot Interaction (pHRI) were proposed in the literature, the main objective of actually quantifying and evaluating them on a biomechanical basis was

[1] EMG: electromyography.

only marginally addressed. Since the underlying biomechanical and forensic literature has neither been reviewed nor fully introduced into the robotics literature, the most important facts are summarized in Chapter 4.

Approach Taken in This Book

The key to this monograph is to make the human the central entity of the evaluation of safety in robotics, i.e. to analyze injuries a human can actually suffer from direct contact with a robot. It is argued that if the physical properties, i.e. the biomechanics of the human are not taken into consideration, a realistic prediction of the resulting human injury or of the benefit of a particular countermeasure is not possible. In recent work, the awareness and importance of this problem appears to have been realized and some interesting results were obtained.

In the present monograph, the existence of two contradictory paradigms for handling safety in the context of human-robot interaction is proposed.

1. Human-centered Robotics
2. Competitive Robotics

Safety in the context of Human-centered Robotics basically claims to completely prevent any harm to humans. This is especially demanded in a typical domestic environment or in a scenario incorporating a robotic co-worker. Injuries have to be absolutely avoided despite the desired active physical contact.

Competitive Robotics take a different philosophy: Robots shall reach human-like performance and in principle be able to compete e.g. in the realm of sports. The most prominent example is the RoboCup competition whose federation proclaimed the ambitious long term goal that "By mid-21st century, a team of fully autonomous humanoid robot soccer players shall win the soccer game, in compliance with the official rule of the FIFA, against the winner of the most recent World Cup." For such a soccer match between humans and robots to take place implies physical human-robot interaction, including tackles and fouls between humans and robots. In the domain of Human-centered Robotics, robots are designed to cause absolutely no harm to a human. Presumably, a team of such robots would be placed at a significant disadvantage to win a soccer mach. The assumption for Competitive Robotics stated in this monograph is that a human-robot match must not be more dangerous than an ordinary soccer match.

In order to be able to estimate the resulting injury of a human in the context of both approaches, a careful injury assessment is carried out. However, before describing this methodology, a survey on the the novel contributions in the fields of collision detection, reaction, and avoidance made in this monograph is given.

Countermeasures during Physical Contact

A robot sharing its workspace with humans should be able to quickly detect collisions and safely react to limit injuries due to physical contacts. In the absence of external sensing, relative motions between robot and human are not predictable and unexpected collisions may occur at any location along the robot arm. In Chapter 3, various algorithms for coping with this problem are developed and evaluated. Efficient collision detection methods that use only proprioceptive robot sensors and provide also directional information for a safe robot reaction after collisions are introduced and validated. Various reaction schemes to sensed collisions are presented and evaluated on an objective basis, pointing out the resulting benefits. The outcome of these methods is already integrated in the new commercially available KUKA Lightweight Robot and they are considered the key feature to enable safe pHRI with the robot in entirely new types of applications.

Apart from binary detect of collisions and reacting in a very limited, predefined manner, a combination of reaction strategies provides an intuitive and effective response to desired physical interaction or unintended collision/clamping.

Injury Assessment

During unexpected collisions, various injury sources as e.g. fast blunt impacts, dynamic and quasi-static clamping, or cuts by sharp tools are present. In Chapter 5 and 6 various worst-case scenarios in pHRI are discussed and analyzed according to the following scheme

1. Select and/or define and classify the impact type
2. Select the appropriate injury measure(s)
3. Evaluate the potential injury of the human
4. Quantify the influence of the relevant robot parameters
5. Evaluate the effectiveness of countermeasures for injury reduction and prevention

Up to now, attempts to investigate real-world threats via impact tests at standardized crash-test facilities and to use the outcome to analyze safety issues during physical human-robot interaction were carried out for the first time in. In order to quantify the potential danger emanating from the DLR Lightweight-Robot III (LWR-III), impact tests at the Crash-Test Center of the German Automobile Club (ADAC) were conducted and evaluated. The outcome of these dummy crash-tests indicated a very low injury risk with respect to evaluated injury criteria for rigid impacts with the LWR-III, which was confirmed for various body parts with a series of impact tests with a human. Furthermore, it shows that a robot, even with an arbitrary mass, moving not faster than 2 m/s is not able to be dangerous to a *non-clamped* human head with respect to typical severity indices[2]. These strong statements were confirmed by crash-tests with several industrial robots. After evaluating free impacts between humans and robots, dynamic and constrained impacts at high robot speeds are analyzed, which are a major source of potential injuries especially for massive robots. Apart from such dynamic clamping impacts certain situations were identified in which low-inertia robots such as the LWR-III could become seriously dangerous as well. They are related to clamping close to singularities where the robot is able to exert very large external forces.

In addition to the already described experiments, Chapter 5 interprets a large experimental campaign of standardized crash-tests with robots of different weight classes for varying impact situations. It also provides unique data that leads to fundamental insights into the characteristics of robot-human impacts.

Chapter 6 gives an analysis of soft-tissue injuries caused by sharp tools which are mounted on/grasped by a robot. An analysis of soft-tissue injuries was conducted based on available biomechanical and forensic data and presumably for the first time in robotics various experimental results with biological tissue which validate the analysis are presented. Furthermore, the beneficial effect of the collision detection and reaction schemes as possible countermeasures to prevent or at least reduce soft-tissue injury are also analyzed. Again, the obtained results are confirmed with an experimental human volunteer session.

Countermeasures during Task Execution

In addition to exhaustively discussing and evaluating worst-case scenarios of human-robot impacts and investigating the design of collision reaction and detection schemes, this monograph proposes and examines two pre-collision strategies in Chapter 7. They serve as a mechanism to avoid undesired human-robot collisions while the desired task is retained if possible. The first method is strictly task preserving in the sense that the geometric desired trajectory is kept and the robot avoids obstacles or contacts while sliding back and forth along this path. The second method allows, similar as for industrial robots, the generation of predefined motion commands by means of desired velocity and acceleration, as e.g. trapezoidal or sinusoidal path velocity, if no disturbance is present. In case of virtual or physical contact the algorithm does not suffer the usual problem of unpredictable avoidance velocities caused by non-deterministic disturbances, which is a common problem of other methods. On the contrary, it is possible to strictly determine the evading/avoiding behavior in real-time. A key feature of all presented methods is the unified treatment of virtual and physical forces, which allows the systematic fusion of avoiding with collision retraction behavior.

---

[2] Severity indices are injury measures used in the automobile industry. Head injury assessing criteria mostly focus on the evaluation of head acceleration.

Towards the Robotic Co-worker

In addition to generating the fundamental scientific basis for robot safety, a full concept and implementation of a robotic co-worker in an industrially relevant scenario was built up and is described in Chapter 8. Major focus was placed on understanding how to bring a set of sophisticated control features to such a degree of integration and versatility that they can be used effectively in a complex application that incorporates seamless switching between autonomy without human presence, physical Human-Robot Interaction (pHRI), and autonomy under human presence. For this, new concepts on various architectural levels of robot design were developed and implemented on experimental test beds. The resulting state-based human-friendly robot control architecture that unifies the interaction and motion control methods developed in this thesis was e.g. used for the first robotic hand-arm system continuously brain controlled by a tetraplegic person.

The Role of Joint Stiffness

Based on the evaluation for Human-centered Robotics, Chapter 9 discusses Competitive Robotics in the context of RoboCup by extending the preceding analysis that was carried out for physical Human-Robot Interaction. For this, two matches from the (2006) FIFA World Cup in Germany serve as examples and are analyzed with respect to scenes incorporating physical interaction. These interactions are related to the results in Chapter 5,6 and sports science by imagining what would have happened if one of the opponents in the scene was a robot. Additionally, important mechanical considerations on how the robot can endure such interactions and meet the performance needed for competing with humans are discussed. In particular, the joint torque and velocity data of human athletes is compared to a state of the art robot arm. One of the major differences from Human-centered Robotics is the much higher velocity at which impacts would occur, therefore dramatically increasing the resulting injury level.

Although the protection of the human body has the absolute main priority, the protection of the robotic structure requires special focus as well since this directly affects the prospects to effectively react to collisions. Therefore, this thesis also discusses and presents results on how novel joint designs contribute to increasing the robustness and capabilities of a robotic joint. On the other hand, this part also brings new and unexpected results concerning the safety benefit obtained by deliberately introducing joint compliance. It is shown that the possibility to increase safety is definitely present. However, the injury potential can be much higher compared to a stiff robot. It may be argued that designing a more compliant robot makes it inherently less safe due to the intrinsic possibility of energy storage and release in the joint elasticity. Various theoretical and experimental results based on optimal-control theory are outlined to confirm these claims. They are part for the development of the the new DLR hand-arm system, a fully integrated intrinsically compliant manipulator system. First results on the control and modeling of these devices were presented in. Also, human inspired actuation paradigm may be used for implementing human like adaptation of force and impedance.

Standards

Chapter 11 discusses how the outcome of this thesis can be utilized for future standardization efforts in pHRI. The major deficits of current regulations concerning direct physical HRI are pointed out and several concepts for the classification of contact types, injury scaling, and systematic crash-testing are proposed. In particular, an overview of possible injuries, a classification attempt, and related injury severity measures with the goal of assembling a full image of injury mechanisms in Human-centered Robotics which is missing completely in the literature up to now is given. Furthermore, a definition of injury severity tailored to the needs of robotics is proposed. At the end of the chapter, a clear start for establishing standardized situations, representing the most important cases that have to be treated for a full blunt impact evaluation is provided.

Finally, the main conclusions of the thesis are drawn and an outlook on future research directions is provided in Chapter 12.

Oberpfaffenhofen,

Sami Haddadin
October 2012

Acknowledgements

First of all I would like to thank Professor Dr.-Ing. Gerhard Hirzinger and Professor Dr.-Ing. Jürgen Roßmann who gave me the invaluable chance to realize my Ph.D. Thesis, which is the basis for the present monograph, at the Institute of Robotics and Mechatronics of the German Aerospace Center (DLR) and the Institute of Man-Machine-Interaction at the Rheinisch-Westfälische Technische Hochschule Aachen (RWTH Aachen). Being able to scientifically grow up at DLR, starting in a moderately large research group and then being expanded over time to one of the largest robotics centers in the world, was an experience that had significant impact on my career. It certainly gave me the chance to act already very early duirng my PhD as a free researcher, giving me the possibility to select and work on entirely new problems of robotics research. Being bound by imagination (and reality) only, I had the possibility to define my own research directions more or less right from the beginning, which to a large extent was made possible by Prof. Dr.-Ing. Alin Albu-Schäffer. Certainly, many other people at DLR had influence on my work, but first of all I would like to express my gratitude to Alin for being such an excellent supporter and encourager of my work, as well as a great discussion partner. His attitude and passion for his work are an inspiration and encouraged me to follow suit. Over the time I was even able to call him a close friend of mine, who I trust and value highly. Unfortunately, occasions for intense discussions over night are getting very rare, which presumably is the way things go.

Furthermore, I would like to thank Prof. Alessandro De Luca for his collaboration and great support during the time we worked together. Prof. Dr.-Ing. Udo Frese was involving me into his concept of sport robotics, which became an unexpectedly large interest of mine and for that I am very thankful.

Since parts of this work are truly interdiciplinary, unifying Robotics with Biomechanics and Forensics, I would like to thank Professor Dr. rer. nat. Dimitrios Kallieris for his valuable advice in the latter, which was virgin soil for me as an engineer and computer scientist.

During my time at DLR many young as well as already established researchers have been an invaluable source of inspiration and wonderful discussion partners. Dipl.-Ing. Oliver Eiberger, Dr.-Ing. Michael Suppa, Dipl.-Ing. Sebastian Wolf,

Dipl.-Ing. Manfred Schedl, Dr.-Ing. Michael Strohmayr, Dipl.-Ing. Mirko Frommberger, Dipl.-Ing. Mathias Nickl, Dr.-Ing. Markus Grebenstein, Prof. Dr. Patrick van der Smagt, Dipl.-Inf. Holger Urbanek, Dipl.-Inf. Sven Parusel, Dr.-Ing. Tim Bodenmüller, Dr.-Ing. Thomas Wimböck, Dr.-Ing. Christian Ott, Dipl.-Ing. Andreas Stemmer, Prof. Dr.-Ing. Darius Burschka, Dipl.-Phys. Berthold Bäuml, Dr. Gerhard Grunwald, Dipl.-Ing. Georg Plank, Dr.-Ing. Stefan Fuchs, Dipl.-Ing. Alexander Bayer, Dipl.-Ing. Klaus Jöhl, Tim Rokahr, and Dipl.Ing. Felix Huber are the most influential ones I would like to mention at this point without loss of generality. My special thanks go to Dr. Neal Y. Lii for carefully reading the initial manuscript. Furthermore, I would like to express my gratitude to Christopher Schindlbeck, who supported me in finalizing the present monograph and carefully read through the manuscript.

Apart from my direct collaborators during the time I wrote my thesis, I had the extraordinary chance and luck to get into close contact with a number of outstanding roboticists of our time. Prof. Oussama Khatib, Prof. Antonio Bicchi, and Prof. Bruno Siciliano let me experience their deep passion and enthusiasm for robotics and research in general, which gave me great inspiration and also influenced my more recent work. Furthermore, I would like to thank Dr. Steve Cousins for letting me be part of the Willow Garage adventure, which has been a wild ride that I will not forget.

Certainly, many persons are left out here but I hope they know I enjoyed our conversations and deeply appreciate their valuable input to this work.

For my father Fahed and my mother Eija I am truly grateful. They raise me with sincere love and tolerance, passed me a strong will and dedication to what I do, and supported my ideas and desires during all my life. Due to their choice to become strangers in a new country, they gifted me three beating hearts, an invaluable thing to posess.

I would also like to thank my siblings Simon and Jasmin for the great times we spent together and that our bonding was proven to be a very special one. I am proud of you and wish our growing families the same luck and joy we have in our life.

My deepest gratitude and love I send to my wife Julia for being so understanding and supportive over the last years. We certainly had to cut back a lot, but in the end I had the incredible luck of having a loving person at my side that understands and accepts the deep passion for my “work”. In the end, she decided to bring Franka into our life, which was the most beautiful moment of my life.

Contents

Chapter 1
Introduction

For more than half a century it was predicted that robots will eventually interact and work closely with humans in diverse everyday environments as well as support them in industrial scenarios. However, despite large efforts in all major robotic fields, only recently have robots gained capabilities in both sensing and actuation, which may enable operation in the proximity of humans. Direct high performance

Fig. 1.1 Physical Human-Robot Interaction: The upper left image shows a scene from the robot story *Robbie* written by Isaac Asimov. In the lower left a human is equipped with an exoskeleton developed by *Sarcos Inc.* [5]. The center image depicts a human interacting with the DLR humanoid upper body system *Justin* (middle), the upper right a human attached to a bimanual haptic interface, and the lower right a human interacting with the LWR-III.

S. Haddadin: *Towards Safe Robots*,
Springer Tracts in Advanced Robotics 90,
DOI: 10.1007/978-3-642-40308-8_1,

physical interaction became possible without the loss of speed and payload. Recently, some significant contributions in control, design, motion planning, and safety were achieved to provide a solid basis for **physical Human-Robot Interaction (pHRI)**. These innovations are expected to lead to entirely new application domains that will require highly flexible and autonomous robotic systems. Especially

- automation of common daily tasks,
- support of humans in heavy industrial jobs,
- elderly care in elderly-dominated societies,
- tasks fulfillment in hospitals and medical care,
- rehabilitation robotics,
- tele-presence systems during lack or high cost of local human expertise,
- entertainment robotics,
- and unmanned warfare with human augmentation

are most likely to form large markets and cause significant impact on the society. Apart from terrestrial applications, the use of robots in space applications, which is the main activity field of the DLR Institute of Robotics and Mechatronics, intends to relieve astronauts from both physical and mental burden during long and exhaustive tasks. Especially during field work in space, humans carry out complex and possibly dangerous missions. The use of robotic technology may significantly improve the efficiency and reliability of the entire process.

However, despite intense efforts in robotics research, numerous "grand challenges" remain. In order to finally bring robots and humans spatially together as exemplified in Fig. 1.1 especially the fundamental concern of how to ensure the safety to the human by all means has to be treated. This major challenge of robotics was already noted in literature by Isaac Asimov in 1942 [1].

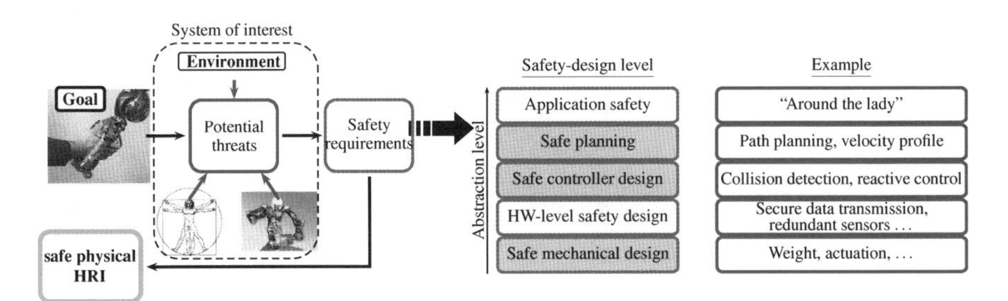

Fig. 1.2 Why is the analysis of safety fundamental to robotics? A problem classification and multi-level safety analysis & design.

Providing safety during Human-Robot Interaction is a multi-faceted challenge and requires an analysis on various levels of abstraction. Physical Human-Robot Interaction aims at the coexistence of humans and robots in a common workspace and at extending their communication modes by physical means. This spatial proximity leads to a variety of *potential threats*, determined by the current state of the system of interest, which consists of the human(s), the robot(s) and their surrounding environment, see Fig. 1.2. In order to adequately rate the situation, *safety requirements* based on a careful analysis of possible injuries a human can suffer are necessary. These *safety requirements* have to be integrated into the layers of a multi-level safety design, treating safety issues on multiple levels of abstraction. Such safety-levels range from formulating requirements on mechanical design, including weight, shape, stiffness, and actuation, up to high-level commands related to the actions of the robot. This leads to the need for

safe physical Human-Robot Interaction.

In recent years, there has been some effort to solve particular problems of robot safety, especially in motion planning and design. There have been initial investigations of robot-human collisions and their related impact characteristics. The resulting contact forces during the impact phase may be reduced by pursuing a lightweight robot design [3], by adding soft visco-elastic covering to the links [6], or by introducing compliance in the driving system so as to mechanically decouple the heavy motor inertia from the link inertia [2, 7]. However, these pioneering works can only be regarded as preliminary efforts towards achieving safe systems. On the standardization side there exist guidelines as e.g. [4] which are also only first steps and hardly applicable to real world pHRI applications.

In order to truly enable future robots to interact and work closely with humans in everyday environments, as well as to support them in industrial applications, e.g. during complex assembly tasks, there are still numerous open problems. One of the most fundamental challenges regarding safety is to ensure it even under *worst-case conditions*. This means that one is first interested in the intrinsic properties of robot-human collisions and hereafter quantify, based on the accordingly gained insight, the potential benefit obtained by control and motion schemes.

In general, safety in pHRI is still a very new and open topic of research. In this sense, the monograph lays the ground work for the knowledge in this field, e.g. by analyzing robot-human collisions and investigating numerous aspects about the worst-case injury humans would suffer from such impacts. Among other things, numerous interaction control schemes and motion generation algorithms were developed in order to achieve not only insight useful for safer robot design, but also for safe robot motion control.

Next, a short survey of the major contributions of the monograph is provided.

Contribution of the Monograph

The present monograph provides new insights for safe pHRI from various perspectives, forming a holistic and unifying approach to the entire problem. Many of the addressed issues are analyzed and discussed for the first time in robotics and can be regarded as fundamental contributions to robotics. The human is put in the center

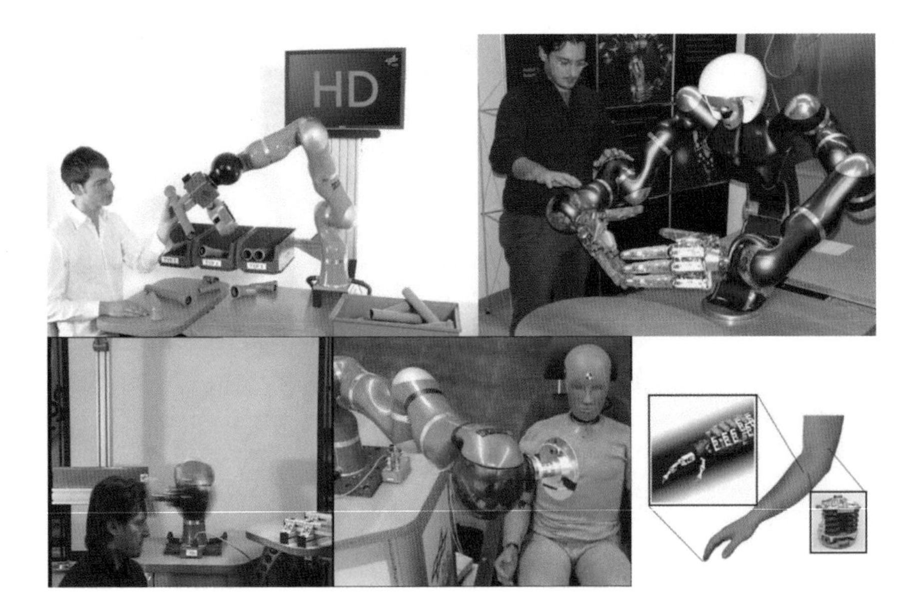

Fig. 1.3 Impressions from the monograph

of the investigations, quantifying injury in its enormous variety by means of biomechanical and forensic analysis. This monograph gives also major theoretical and practical input to establish *Safe Robotics* by thoroughly understanding the injury mechanisms behind potentially dangerous or even lethal situations. For example, the first standardized crash-tests were carried out with robots by using crash-test dummies or investigate soft-tissue injury with pig tissue. In addition, a basis for classifying and evaluating the relevant kinds of injuries in robotics is put together. Such work has been missing until now in the robotics community and the presented approaches will open doors in many ways for physical-human robot interaction as part of future service robots and production assistants. The proposed taxonomy and injury classification, together with the rich basis of experimental verification, already gave inputs to the definition of future norms for human-centered robotic systems. This gives the ability to formulate recommendations for future robot standards based on thorough insights. These findings have already attracted large interest from the industrial and from the standardization bodies as a basis to realize concrete applications incorporating pHRI in real-world scenarios.

A further important contribution of this monograph is the development of new control schemes for collision detection and reaction algorithms. Their performance is showcased in numerous experiments with several DLR robots. They can especially be used to rapidly detect and react to a possibly harmful collision or safely cope with intentional physical contact.

Furthermore, novel methods for real-time collision avoidance are elaborated, which take into consideration virtual and physical forces to circumvent real and virtual objects. The combination of the aforementioned methods provides the necessary components for enabling pre-collision, contact, and post-contact strategies. This complex, together with classical control schemes as e.g. Cartesian impedance control, represents all necessary steps to interface a reactive global motion planner that coarsely provides milestones. It leaves the particular local behavior in dynamic and partially unknown environments to the real-time avoidance/collision detection and reaction module.

Finally, all methods are merged into one framework to provide the architectural basics for building robotic systems for physical Human-Robot Interaction, exemplarily demonstrated in a complex robotic co-worker scenario. An efficient hybrid state based concept is outlined that uses multi-sensor inputs to generate flexible robot behavior. The underlying concept is to provide task completion under the absolute premise of guaranteeing safety for the human.

Apart from the immanent and highly topical contributions of this monograph, the presented biomechanical results are also used for taking a closer look at a vision far ahead:

Competitive Robotics

This term is introduced to distinguish research fields whose applications require an absolute no-harm guarantee to the human (co-workers, service robots, ...) from those, whose implementations aim at designing robots that can compete with human capabilities in physical games as e.g. the long-term vision of *RoboCup*:

Win against the current soccer world champion with a team of soccer robots by the year 2050.

This demands careful analysis of the implications on safety such types of robots would cause. Since they require high physical capabilities to compete with humans, they are potentially more dangerous than machines designed for human supportive tasks. One of the major results is that intrinsic joint compliance turns out to be very beneficial to achieve robustness and performance comparable to humans. Therefore, it is focused on this aspect and various insights into the influence of intrinsic compliance and how it can help to achieve Competitive Robotics are provided.

On the other hand, intrinsic elasticity is also an active research area nowadays. It is mainly considered as a promising actuation technology for intrinsic safety. Usually, it is stated that a clear safety benefit is automatically achieved with passive compliance. In this monograph, it is argued that this is only true for particular cases and sometimes passive compliance can even increase the potential danger a robot is able to cause. Novel aspects about their design and control are elaborated and the methods for collision detection and reaction are extended to this new kind of robot design. Significant results concerning dynamic energy storage and release processes are given, pointing out ways to use intrinsic elasticity for enhancing the performance of robots based on optimal control theory. Furthermore, aspects related to their intrinsic safety properties during robot-human impacts are analyzed, which give a more differentiated look into the one-sided discussion in the robotics literature.

Throughout this monograph, several demonstrators were built, ranging from simple test beds to full scale demonstrators, emphasizing the performance of the taken approaches.

Many results of this monograph have already shown their potential impact, not only in the scientific community, but also on general society. There has been public interest as well as continuous international press coverage in some of the most important newspapers, TV shows, magazines, and online portals. This emphasizes the relevance of *Safe Robotics* to the society by addressing fundamental concerns of the general public regarding robots. At the same time, the results of this monograph were received with great interest by the international research community, as well as by the industrial side, and are considered as core contributions to finally realize pHRI outside restricted lab environments. Figure 1.3 depicts some impressions from the work performed in this monograph.

References

[1] Asimov, I.: The Caves Of Steel, A Robot Novel. Doubleday, Garden City, New York (1954)

[2] Bicchi, A., Tonietti, G.: Fast and soft arm tactics: Dealing with the safety-performance trade-off in robot arms design and control. IEEE Robotics and Automation Mag. 11, 22–33 (2004)

[3] Hirzinger, G., Sporer, N., Albu-Schäffer, A., Krenn, R., Pascucci, A., Schedl, M.: DLR's torque-controlled light weight robot III - are we reaching the technological limits now? In: IEEE Int. Conf. on Robotics and Automation (ICRA 2002), Washington, DC, USA, pp. 1710–1716 (2002)

[4] ISO10218: Robots for industrial environments - Safety requirements - Part 1: Robot (2006)

[5] SARCOS: Sarcos web site, `http://www.sarcos.com`

[6] Yamada, Y., Hirasawa, Y., Huand, S., Umetani, Y.: Fail-safe human/robot contact in the safety space. In: IEEE Int. Workshop on Robot and Human Communication, pp. 59–64 (1996)

[7] Zinn, M., Khatib, O., Roth, B.: A new actuation approach for human friendly robot design. The Int. J. of Robotics Research 23, 379–398 (2004)

Chapter 2
State of the Art

Human-Robot Interaction has become an important and intensive topic of research in the robotics community and is commonly divided into two major branches [1]:

1. cognitive and social Human-Robot Interaction (cHRI)
2. physical Human-Robot Interaction (pHRI)

The former combines such diverse disciplines as psychology, cognitive science, human-computer interfaces, human factors, and artificial intelligence with robotics. Cognitive Human-Robot Interaction intends to understand the social and psychological aspects of possible interaction between humans and robots and seeks for uncovering its fundamental aspects. The latter deals to a large extent with the physical problems of interaction, especially from the view of robot design and control. It focuses on the realization of so called human-friendly robots by combining in a bottom-up approach suitable actuation technologies with advanced control algorithms, reactive motion generators, and path planning algorithms for achieving safe, intuitive, and high performance physical interaction schemes. In pHRI human safety is of primary concern, since continuous physical interaction is desired.

However, robot safety was until now, basically an exclusive topic for applications involving heavy machinery with no physical Human-Robot Interaction. Robots that would have been capable of direct interaction were still suffering from technological immaturity. Consequently, current standards are tailored to exclude the human from the robot workspace and solve the safety problem by segregation. However, due to several breakthroughs in robot design and control, first efforts were undertaken recently to shift focus in industrial environments and consider the close cooperation between human and robot. This necessitates fundamentally different approaches and forces the standardization bodies to specify new standards suitable for regulating Human-Robot Interaction (HRI) . A first step in this direction was taken with the introduction of the *ISO-10218* [23].

In this chapter, the advances in HRI are reviewed and a short survey on existing safety standards is given. It is organized as follows. In Section 2.1 a brief introduction on cHRI is given. In Section 2.2 a more in-depth survey on pHRI outlines the

S. Haddadin: *Towards Safe Robots*,
Springer Tracts in Advanced Robotics 90,
DOI: 10.1007/978-3-642-40308-8_2,

major achievements in this field. Section 2.3 provides an overview on robot safety in industrial settings to elaborate the focus of industrial standards at the current stage[1].

2.1 Social and Cognitive Human-Robot Interaction

Robots designed for social interaction with people in human-centric terms have diverse outward appearance, ranging from humanoid to even animal-like one, see Fig. 2.1.

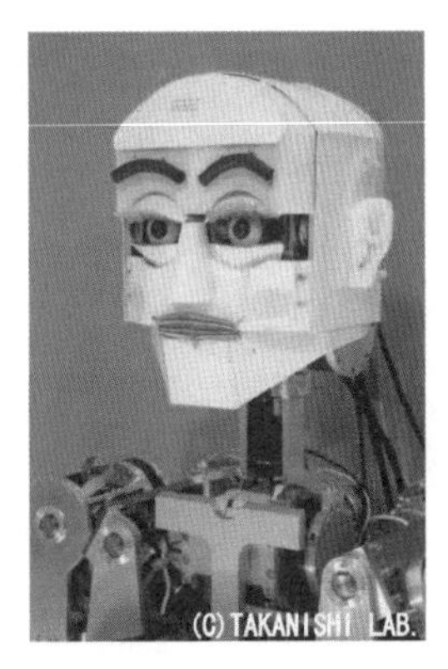

Fig. 2.1 MIT Kismet [6] (left) and Waseda Emotion Expression Humanoid Robot No. 4 Refined II [24] (right)

In contrast to pHRI, their characteristic is that social robots engage people in an interpersonal manner. They communicate with humans via different channels as verbal, nonverbal, or affective modalities. For such robots it is important to communicate and express social-emotional behavior on different modal levels. For being able to finally close the high-level human-robot loop, the perception and interpretation of as well as response to human cues, is of fundamental interest for high intuitiveness. In the following, three major sectors of extensive research within cHRI are shortly introduced to explain some conceptional foundations [58]:

1. Multimodal communication
2. Expressive emotion-based interaction
3. Social-cognitive skills

[1] Please note this chapter gives only an overview and the missing references are completed in the according chapters.

2.1.1 Multimodal Communication

During natural conversations of human beings with each other, nonverbal information supports the primary linguistic information. These *paralinguistic* cues support the smoothening and regulation of interpersonal communication. Currently, several robotic systems have already gained quite advanced paralinguistic and linguistic communication capabilities [12, 42] enhancing the ability of early humanoid robot systems as WABOT and WABOT-II. These first generation designs were already able to carry out simple conversations [57, 31].

2.1.2 Expressive Emotion-Based Interaction

In order to be capable of emotion-based interactions, social robots must be capable of recognizing and correctly interpreting affective signals from humans. Furthermore, they have to possess their own internal models of emotion and need to be able to communicate this to others.

Two of the most successful designs are Kismet [6] from MIT (Fig. 2.1 (left)) and the Waseda Emotion Expression Humanoid Robot No. 4 Refined II [24] from Waseda University (Fig. 2.1 (right)).

2.1.3 Social-Cognitive Skills

Understanding and interacting with *animate entities*, characterized by having a mind and body, is one of the most fundamental requirements of social robots. Their ability is to recognize, understand, and predict human behavior in terms of underlying mental states (beliefs, intents, desires, etc.). In psychology this ability is called *theory of mind*. Estimating the human state is a large research topic and its role in robotics has e.g. been addressed in [32].

After this short introduction to cHRI a more detailed overview on pHRI is given to help familiarize with the fundamental differences of these complementary branches.

2.2 Physical Human-Robot Interaction

In this section, an overview of control, mechanical design, planning, and safety in physical Human-Robot Interaction is given. The central cornerstones achieved in these fields are outlined. For brevity, however, the efforts that were made in the field of dependability are omitted (see e.g. [14, 15]).

2.2.1 Control

The goal of robots and humans coexisting in the same physical domain poses various fundamental problems for the entire robotic design. Unlike their classical counterparts, these robots take into account for the hardware design, control, and planning that the environment is partially *unknown*. Such a robot cannot simply move along computed trajectories without concern for external forces, but must react to unexpected contact with the environment. Therefore, it is usually equipped with proprioceptive sensors, such as Cartesian force-/torque and joint torque sensors [19, 8] and/or tactile arrays resembling a sensitive skin (especially for hands [27]). Alternatively, backdrivable motors can be used to passively react to external forces [62].

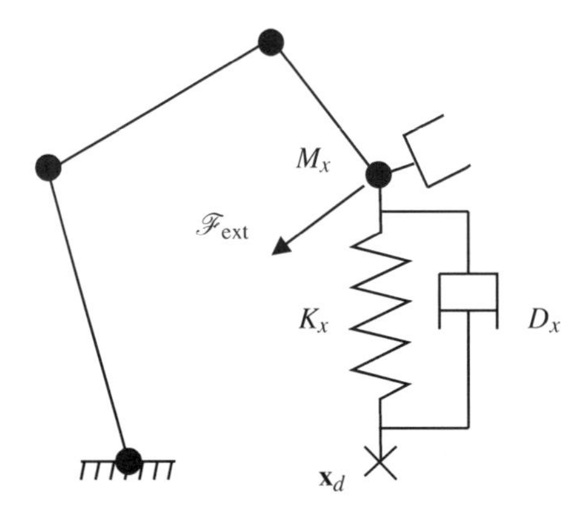

Fig. 2.2 Desired mechanical behavior expressed by mass-spring-damper

In order to incorporate reactions to external forces, the concept of force control has been an active topic of research with initial work in [68, 69], leading to schemes for hybrid position/force control by Craig and Raibert [9]. Paul and Shimano introduced compliance control [48] and Salisbury the conceptually equivalent stiffness control in [54]. However, the most widely used control approach to physically interact with robots is probably impedance control and its related schemes, introduced in the pioneering work of Neville Hogan [20] and extended to flexible joint robots in [13, 3, 74, 4, 46]. This type of controller imposes a desired physical behavior with respect to external forces on the robot. For instance the robot is controlled to behave like a Cartesian second order mass-spring-damper system (see Fig. 2.2) by

$$\mathscr{F}_{\text{ext}} = M_x(\ddot{\mathbf{x}} - \ddot{\mathbf{x}}_d) + D_x(\dot{\mathbf{x}} - \dot{\mathbf{x}}_d) + K_x(\mathbf{x} - \mathbf{x}_d), \tag{2.1}$$

where $\mathbf{x}, \mathbf{x}_d \in \mathbb{R}^6$ are the current robot and desired tip position, $\mathscr{F}_{\text{ext}} \in \mathbb{R}^6$ is the external wrench and $M_x, K_x, D_x \in \mathbb{R}^{6\times6}$ are the desired Cartesian inertia, stiffness, and damping tensors. Consequently, impedance control allows to realize compliance of the robot by means of active control.

Interaction with an impedance controlled robot is robust and intuitive, since in addition to the commanded trajectory, a disturbance response is defined. A major advantage of impedance control (with impedance causality) is that discontinuities like contact-non-contact do not create such stability problems as they occur with for example hybrid force control [9]. However, many open questions still have to be tackled from a control point of view, such as how to adjust the impedance parameters according to the current task.

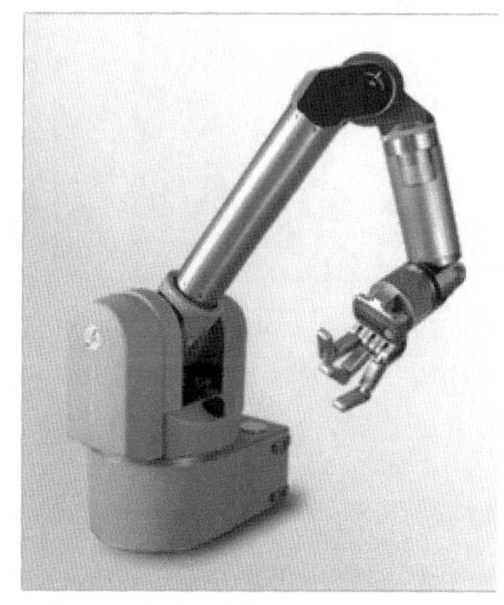

Fig. 2.3 The DLR Lightweight Robot III [19] (left) and the Barrett WAM Arm (right) [62]

2.2.2 *Human-Friendly Mechanical Design*

Apart from the aforementioned control problems, mechanical design plays a fundamental role in safety. Especially, lightweight design is a major requirement for human-friendly robots. Generally, two major design approaches for lightweight robots can be isolated today [18]:

1. A modular mechatronic approach and
2. a tendon actuation approach.

They have following main characteristics in common:

- *Lightweight structures:* Lightweight, high strength metals or composite materials are used for the robot links. Moreover, the design of the entire system (controllers, power supply) is optimized for weight reduction to enable mobility.

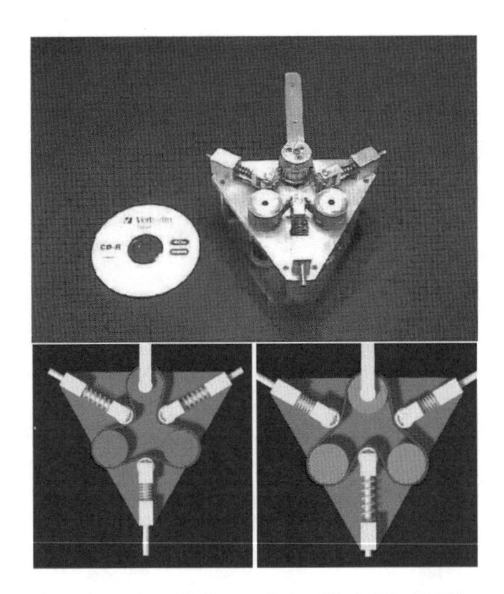

Fig. 2.4 The humanoid Wendy of Waseda University [41] (left) and the Variable Stiffness Actuator of University of Pisa [5] (right)

- *Low power consumption:* This is achieved by small moving inertias and is relevant for both safety and efficiency.

In case of modular mechatronic robots the electronics are usually integrated into the joint structure for allowing highly modular units. Such a design enables the assembly of different kinematics with increasing complexity. This property is also highly desirable for mobile robotic applications. From the actuator side high power/torque at moderate velocity motors and high transmission ratio gears are used. Apart from these design and actuation characteristics the modular mechatronic design is also using additional sensors as e.g. joint torque, force, and current sensing in addition to position sensing only.

The class of tendon-actuated lightweight robots has three main characteristics. First, the actuators are located in the robot base. This reduces the weight of moving parts for a fixed base manipulator. In order to actuate the joints remotely from the base a cable-pulley system is used. Finally, low reduction ratios are used for keeping the system backdriveable. Generally, the benefits of lightweight robots are obtained at the price of higher elasticities in the joints and the structure leading to a more complex dynamic behavior [10]. In this sense, joint elasticity has long been addressed for lightweight robot systems, however more as an undesired consequence that the control has to handle [19, 2]. This requires advanced control techniques in order to obtain accurate, performant motion. The most successful design approaches include the DLR Lightweight Robot III [19, 2] and the Barrett WAM Arm [62], see Fig. 2.3. Series Elastic Actuation (SEA) [49] is an actuator design that represents a different approach. An elastic element with constant stiffness is deliberately introduced between motor and link. It is result of a trade-off between position high control bandwidth and stable high performance force control. Furthermore, the intrinsic elasticity is used for shock absorption. An interesting and promising paradigm currently regaining attention in robotics design is antagonism [61, 65], or more generally Variable Stiffness/Impedance Actuation (VSA/VIA), see Fig. 2.4. Early implementations were carried out in [35, 30, 41]. The idea is to realize joint compliance not by means of control but via adjustable intrinsically compliant joints, inspired by the unquestionably successful design of human and animal muscles. The design and control of such systems were addressed in numerous publications [41, 5, 39, 65, 47, 70, 25].

Due to increased mechanical system complexity for such lightweight and/or compliant systems, higher sensor and component requirements are given compared to industrial robots.

2.2.3 Motion Planning and Obstacle Avoidance

Conventional robot motion planning is typically an off-line process that determines a feasible path (and a dynamically feasible timing), if one exists, connecting an initial and a final arbitrary robot configuration while avoiding obstacles. Generally, complete knowledge of the geometry of the static environment is assumed. For high-dimensional configuration spaces (robots with many degrees of freedom) in crowded environments, the search for a feasible path is very complex and time-consuming. Recently, probabilistic and randomized approaches have been developed to tackle this curse of dimensionality [26, 21, 36].

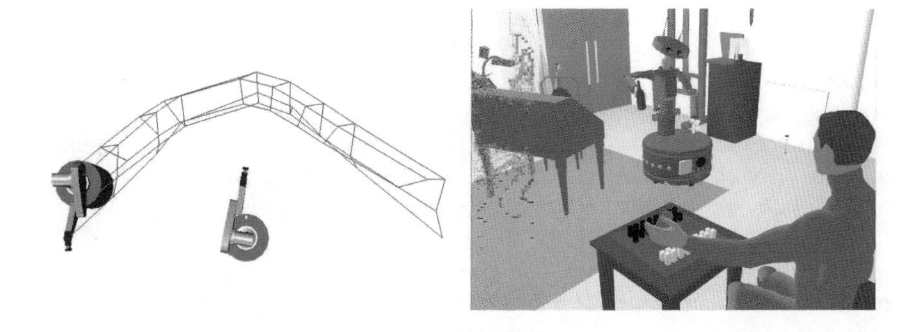

Fig. 2.5 The elastic strips framework of Stanford University [7] (left) and the human-aware motion planner of LAAS [60] (right)

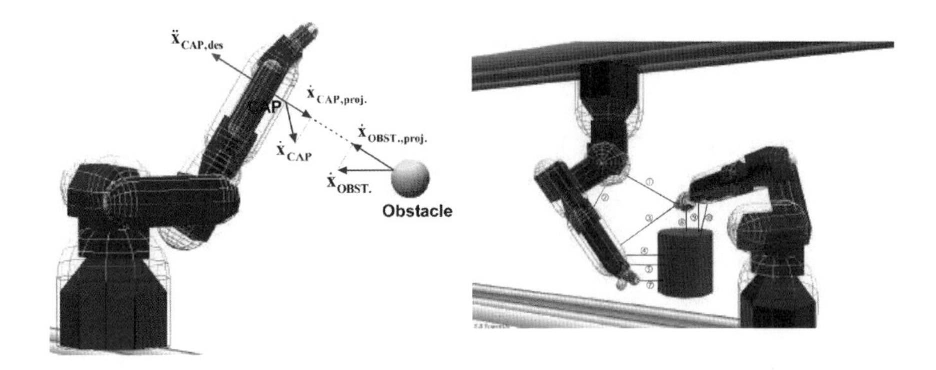

Fig. 2.6 CARE, introduced in [52, 53] and enhanced in [11], calculates Cartesian collision avoidance accelerations (left) and is applicable to multi-robot systems (right)

The intrinsic nature of service robotics is to deal with unstructured, time-varying environments, for which a model is not available. This is to a large extent due to the unpredictable motion of human users. Therefore, the integration of a sensor-based on-line reactive component into a global off-line motion plan becomes mandatory. Various sensors can be used to acquire local information about the relative position of the robot with respect to the human user [17] or to other robots. Based on this, the motion planner should locally modify a nominal path so to achieve collision avoidance under the current task goal. Several reactive motion planning approaches exist for this purpose [45, 50, 67, 72]. Nonetheless, path planning with reactive collision avoidance was mostly investigated in the field of mobile manipulators [72], [44]. A typical task to be fulfilled is to avoid obstacles with or without (partial) task consistency. The obstacles are either known beforehand or suddenly appear, thus necessitating quick response times.

Many collision avoidance methods based on artificial potential fields are introduced in [28], and their algorithmic or heuristic variations in [58, 72, 44].

A virtual repulsive potential is assigned to each known obstacle and an attractive potential to the desired goal configuration. This leads to a directed motion towards the goal while avoiding the obstacles in a reactive fashion. In [44], e.g., the method is applied for the translative motion of a mobile base and in [72] for a manipulator mounted on a mobile base alone. Despite one of its major deficits, namely its possibility to get easily stuck in local minima, this method is popular. Its fast calculation time within the low-level controller cycle of the robot is a well known benefit. One possibility to overcome the local minima drawback is presented with the circulatory fields, introduced in [59]. Each obstacle is attached with a circulatory field, similar to that of a electrical charge in a magnetic field. While this field will then drive the path around the obstacle, this method will not be able to find optimal solutions, but it is far less prone to get stuck in local minima. Furthermore, the potential fields can be extended from a virtual point-shaped particle associated with the robot endeffector to various 3D-objects. These are able to change their orientation accordingly to avoid obstacles [40].

Other promising principles of combining a global path planner with a local collision avoidance strategy are the Elastic Strips [7] or the preceding Elastic Bands [51]. A global path planner searches for a path around the known obstacles, and unforeseen hindrance can then deform the planned path as if it was a rubber band, while avoiding known obstacles remains possible. The local motion deviations are performed in a task-consistent manner, leaving primary (Cartesian) task execution possibly unaffected by obstacle avoidance, see Fig. 2.5 (left). If the modification to the trajectories are given in the operational space, there is the need for an appropriate inverse kinematics system to give the reference values for the velocity/force controllers of the manipulator, possibly considering kinematic redundancy and/or dynamic issues. However, in case of an impedance controlled robot this becomes obsolete (not the appropriate redundancy resolution). The elastic strips and elastic bands, however, are computationally more complex, so that they cannot run in the inner most control loop. Therefore, they increase the time lag until a reaction to an obstacle initiates.

Instead of applying potential field methods in the Cartesian space, one could also apply them in the configuration space (C-space) of the robot [67]. Still, since calculations in the C-space is computational complex, this method is practically only applicable for offline planning, and therefore offers no reactive behavior. This method, however, is able to find valid paths, where Cartesian space based potential field methods will fail. Further related approaches are the Harmonic Potential Fields that are discussed e.g. in [55, 56, 29, 38]. Their most important property is that they solve the problem of local minima by having none.

A general overview of classical motion planning techniques for reactive planning is given in [34], where the work of Jean-Claude Latombe and co-authors on discrete potential fields is reviewed as well.

A concept proposed in [52, 53] is *CARE*, which uses a measure (gauge) for collision danger that is based on inherent robot properties (maximum joint accelerations) and calculates the avoidance acceleration to circumvent objects based on closest distances and relative velocities. Via the required Cartesian avoidance acceleration $\ddot{\mathbf{x}}_{CAP}$ of the Collision Avoidance Point (CAP)[2] the corresponding joint acceleration to be commanded to the robot are obtained. The basic scheme was extended in [11].

In [60] a human aware mobile robot motion planner is considered, which incorporates humans accessibility, their vision field, and their preferences in terms of relative human-robot placement. However, human dynamics are basically excluded from the analysis, see Fig. 2.5 (right).

A different approach focussing on increasing safety is given in [64]. Given a collision free path, a so called proxy acts as an attractor which slides along the path, yet having its own dynamics, therefore smoothing out discontinuities of the

[2] This is the point on the robot structure that is endangered the most by the obstacle.

given path. The robot is then connected to the proxy by a PID like controller. This combination allows for a safer, gentle path-following robot, avoiding sudden jerky motions.

2.2.4 Quantifying Human Safety

The first work to investigate the measurable influence of robot-human impacts was [71]. The authors evaluated the human pain tolerance on the basis of human experiments. In this work, the somatic pain was considered as a suitable criterion for determining a safety limit against mechanical stimuli.

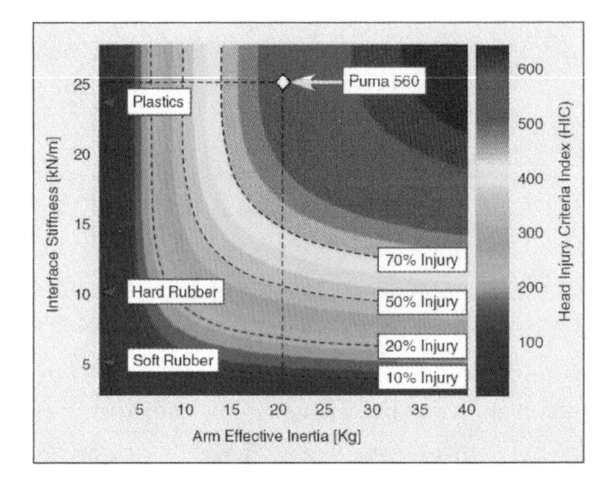

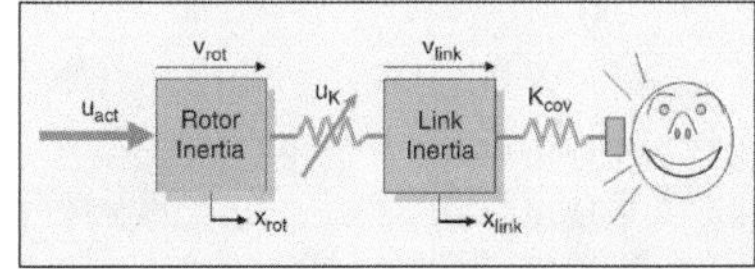

Fig. 2.7 The prediction of HIC for blunt robot-human as originally published in [73] (left). The concept of Variable Stiffness Actuation (VSA) as proposed in [5], which allows the joint elasticity to be a controlled variable during execution of the task (right).

Pioneering work on human-robot impacts under certain worst-case conditions and resulting injuries was carried out in the two independent works [5] and [73], evaluating free rigid impacts at a robot speed of $1-2$ m/s, see Fig. 2.7. Both contributions introduced new compliant joint design concepts and made the first attempt to use the so called Head Injury Criterion (HIC) [66] to quantify the injury potential during occurring collisions. The HIC is defined as

$$\begin{aligned} \mathrm{HIC}_{36} &= \max_{\Delta t}\left\{\Delta t\left(\frac{1}{\Delta t}\int_{t_1}^{t_2}\ddot{\mathbf{x}}_H \mathrm{d}t\right)^{\left(\frac{5}{2}\right)}\right\} \\ \Delta t_{\max} &= \max\{t_2-t_1\} \leq 36\,\mathrm{m}s, \end{aligned} \tag{2.2}$$

where $\ddot{\mathbf{x}}_H \in \mathbb{R}^3$ is the human head acceleration[3].

[3] More information on the HIC and other severity indices is given in Chapter 4.

Table 2.1 Safety design strategies and danger-indices [22]

Design strategy	Impact force	Danger index
Weight reduction	$F = ma$	$\frac{ma}{F_c}$
Soft material	$F = \frac{mv - mv'}{\mathrm{d}t}$ $\mathrm{d}t = \frac{\tan^{-1}\frac{\zeta\omega_n}{\omega_d} + \frac{\pi}{2}}{\omega_d}$ $\omega_n = \sqrt{\frac{k}{m}}$ $\omega_d = \omega_n\sqrt{1-\zeta^2}$ $\zeta = \frac{c}{2}\sqrt{mk}$	$\frac{mv}{F_c\mathrm{d}t}$
Joint elasticity	$F = \frac{I\omega - I\omega'}{l\mathrm{d}t}$ $\mathrm{d}t = \frac{\tan^{-1}\frac{\zeta\omega_n}{\omega_d} + \frac{\pi}{2}}{\omega_d}$ $\omega_n = \sqrt{\frac{K}{I}}$ $\omega_d = \omega_n\sqrt{1-\zeta^2}$ $\zeta = \frac{c}{2}\sqrt{mk}$	$\frac{I\dot{\theta}}{F_c\mathrm{d}t}$
Shape	$F = ma$	$\frac{ma}{F_c}$
Soft material	$\sigma = \frac{F}{A}$ $\lambda = \frac{A_2}{A_1}$ $= \frac{\int_{-l/2}^{l/2}(\mathrm{human}(x) - \mathrm{robot}(x))\mathrm{d}x}{\int_{-l/2}^{l/2}(l/2 - \mathrm{human}(x))\mathrm{d}x}$	$\frac{\lambda}{\lambda_c}$
Surface friction	$F = \mu N$	$\frac{\mu N}{F_c}$

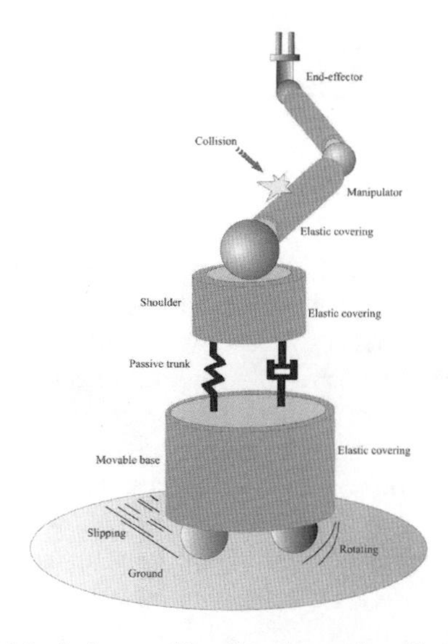

Fig. 2.8 Conceptual model of a human-friendly robot as described in [37]

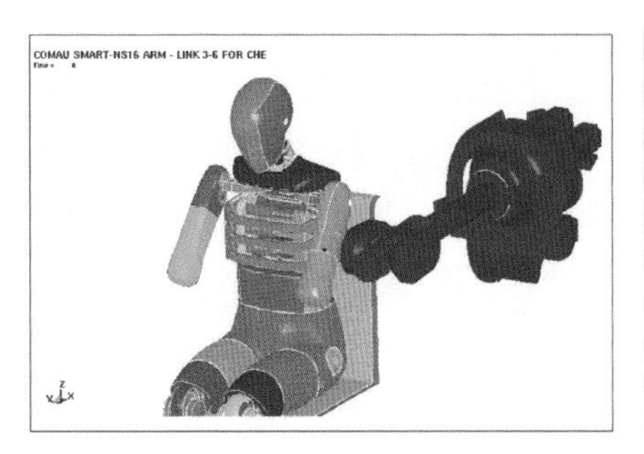

Fig. 2.9 FEM simulations for crash testing evaluations [43]

Further aspects concerning safety in Human-Robot Interaction were introduced in [22], see Tab. 2.1. In this work, several danger indices were proposed based on the design properties of the robot. In [16] a control scheme was developed to limit the impact force of a robot by restricting the torque commands. [37] introduced various design aspects for a mobile robot with physical compliance in its trunk and a passively movable base, see Fig. 2.8. [33] developed an integrated Human-Robot Interaction strategy incorporating a definition of danger by means of reflected inertia, relative velocity, and the distance between human and robot. Recently, [43] utilized FEM simulations for an evaluation of robot-dummy impacts with an industrial robot, see Fig. 2.9. Earlier work presented in [63] focused on a more abstract injury classification. The authors provided a first classification and hazard analysis. In particular, they assumed the human to be exhaustively protected for the given concept.

Next, safety standards in industrial robotics are shortly summarized.

2.3 Robot Safety in Industrial Robotics

For both the traditional industrial and newer service robot applications safety is one of the biggest concerns. Earlier in industrial settings the robots had to be separated from the humans by fences. Recently, so called collaborative modes have been added to the relevant standards allowing, for example, hand guiding in automatic mode. However, the possibilities of collaboration for industrial use remain constrained and service robots are not covered adequately by current standards.

Manufacturers and users of robot technology have to obey numerous laws and standards. Generally, every machine has to follow the machinery directive *2006/42/EC*. Standards derived from this law help the robot developer to adhere to the directive.

The *ISO-10218* defines the current standards for industrial robot safety, including lists of causes of hazards. Safety related elements are explained as e.g. workspace boundaries, cladding, and control mechanisms as emergency stops, electrical connections, or confirmation buttons. Furthermore, it states that one of the following conditions always has to be fulfilled for allowing Human-Robot Interaction: The Tool Center Point (TCP)/flange velocity must be ≤ 0.25 m/s, the maximum dynamic power ≤ 80 W, or the maximum static force ≤ 150 N.

ISO-13849-1 addresses all safety relevant parts of the control system, helps identify safety requirements by providing safety categories and specifies how to fulfill the requirements. According to their dependability, safety related elements of the controller are classified into five categories, describing the fault tolerance level of the system. The particular grading depends on injury severity, occurrence and/or duration of hazard exposure, and the possibility to avoid danger.

A prerequisite for operating a machine according to *2006/42/EC* is to perform a risk analysis and a risk assessment, for which *ISO-14121* provides guidelines. Apart from these standards, there exist others, which can be used for particular cases.

Although *ISO-10218* gives some initial guidelines for Human-Robot Interaction, there are still numerous open issues to be addressed. In this sense, some recent efforts are made to close the gap and define the necessary limits regarding biomechanical injury limits or safety requirements for sophisticated control methods as e.g. in the *ISO/TC 184 / SC 2* committees/working groups for

- personal care safety and
- service robots.

As a result, a committee draft has been compiled on *Robots and Robotic Devices - Safety Requirements - Non-Medical Personal Care Robot.*

References

[1] Albu-Schäffer, A., Bicchi, A., Boccadamo, G., Luca, R.C.A., Santis, A.D., Giralt, G., Hirzinger, G., Lippiello, V., Mattone, R., Schiavi, R., Siciliano, B., Tonietti, G., Villani, L.: Physical human-robot interaction in anthropic domains: Safety and dependability. In: IARP International Workshop on Technical challenges and for Dependable Robots in Human Environments (IARP 2005), Nagoya, Japan (2005)

[2] Albu-Schäffer, A., Haddadin, S., Ott, C., Stemmer, A., Wimböck, T., Hirzinger, G.: The DLR lightweight robot - lightweight design and soft robotics control concepts for robots in human environments. Industrial Robot Journal 34(5), 376–385 (2007)

[3] Albu-Schäffer, A., Ott, C., Hirzinger, G.: A passivity based cartesian impedance controller for flexible joint robots - Part II: Full state feedback, impedance design and experiments. In: Int. Conf. on Robotics and Automation (ICRA 2004), New Orleans, USA, pp. 2666–2673 (2004)

[4] Albu-Schäffer, A., Ott, C., Hirzinger, G.: A unified passivity-based control framework for position, torque and impedance control of flexible joint robots. The Int. J. of Robotics Research 26, 23–39 (2007)

[5] Bicchi, A., Tonietti, G.: Fast and soft arm tactics: Dealing with the safety-performance trade-off in robot arms design and control. IEEE Robotics and Automation Mag. 11, 22–33 (2004)

[6] Breazeal, C.: Towards sociable robots. Robotics and Autonomous Systems 42, 167–175 (2002)
[7] Brock, O., Khatib, O.: Elastic strips: A framework for motion generation in human environments. The Int. J. Robotics Research 21(12), 1031–1052 (2002)
[8] Cheng, G., Hyon, S., Morimoto, J., Ude, A., Hale, J., Colvin, G., Scroggin, W., Jacobsen, S.: CB: a humanoid research platform for exploring neuroscience. Advanced Robotics 21(10), 1097–1114 (2007)
[9] Craig, J., Raibert, M.: A systematic method for hybrid position/force control of a manipulator. In: IEEE Computer Software Applications Conf., pp. 446–451 (1979)
[10] De Luca, A., Book, W.: Robots with flexible elements. In: Springer Handbook of Robotics, pp. 287–319 (2008)
[11] Freund, E., Schluse, M., Roßmann, J.: Dynamic collision avoidance for redundant multi-robot systems. In: International Conference on IEEE/RSJ Intelligent Robots and Systems (IROS 2001), Maui, Hawaii, pp. 1201–1206 (2001)
[12] Fujita, Y.: Personal robot PaPeRo. Journal of Robotics and Mechatronics (14), 60–63 (2002)
[13] Goldsmith, P.B., Francis, B., Goldenberg, A.: Stability of hybrid position/force control applied to manipulators with flexible joints. Int. Journal of Robotics and Automation (14), 146–159 (1999)
[14] Guiochet, J., Motet, G., Baron, C., Boy, G.: Toward a human-centered UML for risk analysis. In: WCC 18th IFIP World Computer Congress, Human Error Safety and System Development (2004)
[15] Guiochet, J., Powell, D., Baudin, É., Blanquart, J.P.: Online safety monitoring using safety modes. In: IARP International Workshop on Technical challenges and for dependable robots in Human environments (IARP 2008), Pasadena, USA (2008)
[16] Heinzmann, J., Zelinsky, A.: Quantitative safety guarantees for physical human-robot interaction. The Int. J. of Robotics Research 22(7-8), 479–504 (2003)
[17] Henrich, D., Kuhn, S.: Modeling intuitive behavior for safe human/robot coexistence cooperation. In: IEEE Int. Conf. on Robotics and Automation (ICRA 2006), Orlando, USA, pp. 3929–3934 (2006)
[18] Hirzinger, G., Albu-Schäffer, A.: Lightweight robots. Scholarpedia 3, 3889 (2008)
[19] Hirzinger, G., Sporer, N., Albu-Schäffer, A., Krenn, R., Pascucci, A., Schedl, M.: DLR's torque-controlled light weight robot III - are we reaching the technological limits now? In: IEEE Int. Conf. on Robotics and Automation (ICRA 2002), Washington, DC, USA, pp. 1710–1716 (2002)
[20] Hogan, N.: Impedance control: An approach to manipulation: Part I - theory, Part II - implementation, Part III - applications. Journal of Dynamic Systems, Measurement and Control 107, 1–24 (1985)
[21] Hsu, D., Kavraki, L., Latombe, J.C., Motwani, R.: Capturing the connectivity of high-dimensional geometric spaces by parallelizable random sampling techniques. In: Pardalos, P.M., Rajasekaran, S. (eds.) Advances in Randomized Parallel Computing. Combinatorial Optimization Series, pp. 159–182. Kluwer Academic Publishers (1998)
[22] Ikuta, K., Ishii, H., Nokata, M.: Safety evaluation method of design and control for human-care robots. The Int. J. of Robotics Research 22(5), 281–298 (2003)
[23] ISO10218: Robots for industrial environments - Safety requirements - Part 1: Robot (2006)

[24] Itoh, K., Miwa, H., Matsumoto, M., Zecca, M., Takanobu, H., Roccella, S., Carrozza, M., Dario, P., Takanishi, A.: Behavior model of humanoid robots based on operant conditioning. In: IEEE-RAS International Conference on Humanoid Robots (HUMANOIDS 2005), Tsukuba, Japan, pp. 220–225 (2005)

[25] Iwata, H., Sugano, S.: Design of human symbiotic robot TWENDY-ONE. In: IEEE Int. Conf. on Robotics and Automation (ICRA 2009), Kobe, Japan, pp. 580–586 (2009)

[26] Kavraki, L.E., Svestka, P., Latombe, J.C., Overmars, M.H.: Probabilistic roadmaps for path planning in high-dimensional configuration spaces. IEEE Transactions on Robotics and Automation 12(4), 566–580 (1996)

[27] Kawasaki, H., Komatsu, T., Uchiyama, K.: Dexterous anthropomorphic robot hand with distributed tactile sensor: Gifu hand II. IEEE/ASME Transactions Mechatronics 7, 296–303 (2002)

[28] Khatib, O.: Real-time obstacle avoidance for manipulators and mobile robots. The Int. J. of Robotics Research 5, 90–98 (1985)

[29] Kim, J., Khosla, P.: Real-time obstacle avoidance using harmonic potential functions. IEEE Transactions on Robotics 8, 338–349 (1992)

[30] Kobayashi, H., Hyodo, K., Ogane, D.: On tendon-driven robotic mechanisms with redundant tendons. The Int. J. of Robotics Research 17(5), 561–571 (1998)

[31] Kobayashi, T., Komori, Y., Hashimoto, N., Iwata, K., Fukazawa, Y., Yazawa, J., Shirai, K.: Speech conversation system of the musician robot. In: International Conference on Advanced Robotics (ICAR 1985), pp. 483–488 (1985)

[32] Kulic, D., Croft, E.: Affective state estimation for human-robot interaction. IEEE Transactions on Robotics 23(5), 991–1000 (2007)

[33] Kulic, D., Croft, E.: Pre-collision strategies for human robot interaction. Autonomous Robots 22(2), 149–164 (2007)

[34] Latombe, J.C.: Robot Motion Planning. Kluwer Academic Publishers, Norwell (1991)

[35] Laurin-Kovitz, K., Colgate, J.E., Carnes, S.D.R.: Design of programmable passive impedance. In: IEEE Int. Conf. on Robotics and Automation (ICRA1991), Sacramento, USA, pp. 1476–1481 (1991)

[36] LaValle, S., Yakey, J., Kavraki, L.: A probabilistic roadmap approach for systems with closed kinematic chains (1999)

[37] Lim, H.O., Tanie, K.: Human safety mechanisms of human-friendly robots: Passive viscoelastic trunk and passively movable Base. The Int. J. of Robotics Research 19(4), 307–335 (2000)

[38] Masoud, A.: Kinodynamic motion planning: A novel type of nonlinear, passive damping forces and advantages. IEEE Robotics and Automation Mag. 17(1), 85–99 (2010)

[39] Migliore, S., Brown, E., DeWeerth, S.: Biologically inspired joint stiffness control. In: IEEE Int. Conf. on Robotics and Automation (ICRA 2005), Barcelona, Spain (2005)

[40] Minoura, H., Kijima, R., Ojika, T.: Collision avoidance using a virtual electric charge in the electrostatic potential field. In: International Conference on Virtual Systems and MultiMedia (VSMM 1996), pp. 289–294 (1996)

[41] Morita, T., Iwata, H., Sugano, S.: Development of human symbiotic robot: WENDY. In: IEEE Int. Conf. on Robotics and Automation (ICRA1999), Detroit, USA, pp. 3183–3188 (1999)

[42] Nisimura, R., Uchida, T., Lee, A., Saruwatari, H., Shikano, K., Matsumoto, Y.: ASKA: receptionist robot with speech dialogue system. In: IEEE/RSJ Int. Conf. on Intelligent Robots and Systems (IROS 2002), Lausanne, Switzerland, pp. 1314–1319 (2007)

[43] Oberer, S., Schraft, R.D.: Robot-dummy crash tests for robot safety assessment. In: IEEE Int. Conf. on Robotics and Automation (ICRA 2007), Rome, Italy, pp. 2934–2939 (2007)
[44] Ögren, P., Egerstedt, N., Hu, X.: Reactive mobile manipulation using dynamic trajectory tracking. In: IEEE Int. Conf. on Robotics and Automation (ICRA 2000), San Francisco, USA, pp. 3473–3478 (2000)
[45] Ogren, P., Petersson, L., Egerstedt, M., Hu, X.: Reactive mobile manipulation using dynamic trajectory tracking: Design and implementation (2000)
[46] Ott, C.: Cartesian Impedance Control of Redundant and Flexible-Joint Robots. Springer Publishing Company, Incorporated (2008)
[47] Palli, G., Melchiorri, C., Wimböck, T., Grebenstein, M., Hirzinger, G.: Feedback linearization and simultaneous stiffness-position control of robots with antagonistic actuated joints. In: IEEE Int. Conf. on Robotics and Automation (ICRA 2007), Rome, Italy, pp. 2928–2933 (2007)
[48] Paul, R., Shimano, B.: Compliance and control. In: Joint Automatic Control Conf. (JACC 1976), San Francisco, USA, pp. 694–699 (1976)
[49] Pratt, G., Williamson, M.: Series elastics actuators. In: IEEE/RSJ Int. Conf. on Intelligent Robots and Systems (IROS 1995), Victoria, Canada, pp. 399–406 (1995)
[50] Quinlan, S.: Efficient distance computation between non-convex objects. In: IEEE International Conference on Robotics and Automation (ICRA 1994), San Diego, USA, pp. 3324–3329 (1994)
[51] Quinlan, S., Khatib, O.: Elastic bands: connecting path planning and control. In: IEEE Int. Conf. on Robotics and Automation (ICRA 1993), Atlanta, USA, pp. 802–807 (1993)
[52] Roßmann, J.: Echtzeitfähige kollisionsvermeidende Bahnplanung für Mehrrobotersysteme. Ph.D. thesis, University of Dortmund (1993) (German)
[53] Roßmann, J.: On-line collision avoidance for multi-robot systems: A new solution considering the robots' dynamics. In: Int. Conf. on Multisensor Fusion and Integration for Intelligent Systems (MFI 1996), Washington D.C., USA, pp. 249–256 (1996)
[54] Salisbuy, J.: Active stiffness control of manipulator in Cartesian coordinates. In: Int. Conf. on Decision and Control (CDC 1980), Albuquerque, USA, pp. 95–100 (1980)
[55] Sato, K.: Collision avoidance in multi-dimensional space using Laplace potential. In: 15th Conf. Robotics Society Japan, pp. 155–156 (1987)
[56] Sato, K.: Deadlock-free motion planning using the Laplace potential field. Advanced Robotics 7(5), 449–461 (1992)
[57] Shirai, K., Fujisawa, H.: An algorithm for spoken sentence recognition and its application to the speech input-output system. IEEE Transactions on Systems, Man, and Cybernetics (4), 475–479 (1974)
[58] Siciliano, B., Khatib, O. (eds.): Springer Handbook of Robotics. Springer (2008)
[59] Singh, L., Stephanou, H., Wen, J.: Real-time motion control with circulatory fields. In: IEEE Int. Conf. on Robotics and Automation (ICRA1996), Minneapolis, USA, pp. 2737–2742 (1996)
[60] Sisbot, E., Marin-Urias, L., Alami, R., Simeon, T.: A human aware mobile robot motion planner. IEEE Transactions on Robotics 23(5), 874–883 (2007)
[61] Tonietti, G., Schiavi, R., Bicchi, A.: Design and control of a variable stiffness actuator for safe and fast physical human/robot interaction. In: IEEE Int. Conf. on Robotics and Automation (ICRA 2005), Barcelona, Spain, pp. 528–533 (2005)

[62] Townsend, W., Salisbury, J.: Mechanical design for whole-arm manipulation. In: Dario, P., Sandini, G., Aebischer, P. (eds.) Robots and Biological Systems: Towards a New Bionics?, pp. 153–164. Springer (1993)
[63] Ulrich, K., Tuttle, T., Donoghue, J., Townsend, W.: Intrinsically safer robots. Tech. rep., 139 Main Street, Kendall Square (1995), http://www.barrett.com/robot/
[64] Van Damme, M., Vanderborght, B., Verrelst, B., Van Ham, R., Daerden, F., Lefeber, D.: Proxy-based sliding mode control of a planar pneumatic manipulator. The Int. Journal of Robotics Research, 266–284 (2009)
[65] Vanderborght, B., Verrelst, B., Ham, R.V., Damme, M.V., Lefeber, D., Duran, B., Beyl, P.: Exploiting natural dynamics to reduce energy consumption by controlling the compliance of soft actuators. The Int. J. of Robotics Research 25(4), 343–358 (2006)
[66] Versace, J.: A review of the severity index. SAE Paper No.710881, Proc. 15th Stapp Car Crash Conf., pp. 771–796 (1971)
[67] Warren, C.: Global path planning using artificial potential fields. In: IEEE Int. Conf. on Robotics and Automation (ICRA1989), Scottsdale, USA, pp. 316–321 (1989)
[68] Whitney, D.: Force feedback control of manipulator fine motions. ASME J. of Dynamic Systems, Measurement, and Control 99, 91–97 (1977)
[69] Whitney, D.: Quasi-static assembly of compliantly supported rigid parts. ASME J. of Dynamic Systems, Measurement, and Control 99, 65–77 (1982)
[70] Wolf, S., Hirzinger, G.: A new variable stiffness design: Matching requirements of the next robot generation. In: IEEE Int. Conf. on Robotics and Automation (ICRA 2008), Pasadena, USA, pp. 1741–1746 (2008)
[71] Yamada, Y., Hirasawa, Y., Huand, S., Umetani, Y.: Fail-safe human/robot contact in the safety space. In: IEEE Int. Workshop on Robot and Human Communication, pp. 59–64 (1996)
[72] Yamamoto, Y., Yun, X.: Coordinated obstacle avoidance of a mobile manipulator. In: IEEE Int. Conf. on Robotics and Automation (ICRA1995), Nagoya, Aichi, Japan, vol. 3, pp. 2255–2260 (1995)
[73] Zinn, M., Khatib, O., Roth, B.: A new actuation approach for human friendly robot design. The Int. J. of Robotics Research 23, 379–398 (2004)
[74] Zollo, L., Siciliano, B., De Luca, A., Guglielmelli, E., Dario, P.: Compliance control for an anthropomorphic robot with elastic joints: Theory and experiments. ASME Journal of Dynamic Systems, Measurements and Control (127), 321–328 (2005)

Chapter 3
Soft-Robotics Control

Accidental collisions that may harm humans should be avoided by anticipating dangerous situations. The effects of physical collisions should be mitigated by having the robot react promptly so as to recover a safe operative condition. In the *pre-impact phase*, collision avoidance is the primary goal and requires (at least, local) knowledge of the current environment geometry and computationally expensive motion planning techniques. Anticipating initiating collisions or recognizing them in real-time is typically based on the use of additional external sensors, such as sensitive skins [23], on-board vision [13], strain gauges and force load cells.

In the *post-impact phase*, the first task is to detect the collision occurrence, which may have happened at any location along the robot arm. The controller should then promptly react with an appropriate reaction strategy. The simplest is to stop the robot. Less expensive solutions are able to detect a collision without the need of additional sensors. A rather intuitive scheme is to compare the commanded torque (or, the current in an electrical drive) with the nominal model based command (i.e., the torque expected in the absence of collision) and to look for fast transients due to possible collision [34, 30, 29]. This approach has been refined by including adaptive compliance control [24, 19]. However, tuning of collision detection thresholds in these schemes is difficult because of the highly varying dynamic characteristics of the commanded torques. Moreover, their common drawback, even with fully identified robot dynamics, is that the inverse dynamics computation for torque comparison is based on acceleration estimates. This introduces inherent noise (due to numerical differentiation of velocity or position data) and/or an intrinsic delay in a digital implementation. A detection scheme for similar conditions that avoids the above drawbacks has been recently proposed in [11]. Collisions are viewed as faulty behaviors of the robot actuating system, while the design of a detector takes advantage of the decoupling property of the robot generalized momentum [9, 20]. Moreover, this detection scheme is particularly convenient for switching control strategies, since it is independent from the control methods used to generate the commanded motor torques.

In this chapter, new collision detection and reaction schemes are developed, which are compared with respect to their ability to quickly and robustly detect

S. Haddadin: *Towards Safe Robots*,
Springer Tracts in Advanced Robotics 90,
DOI: 10.1007/978-3-642-40308-8_3, © Springer-Verlag Berlin Heidelberg 2014

collisions and adequately react to them. The algorithms range from basic schemes to nonlinear observer structures. They are evaluated in simulation and experiment for gathering an insight into their capabilities and drawbacks. Furthermore, a method based on the estimation of external torques, which makes it possible to push a robot back and fourth along a desired trajectory was designed. Prior to introducing and evaluating these schemes, some preliminaries of robot dynamics modeling, the mechanical design properties of the LWR-III, and its basic control schemes are surveyed.

The chapter is organized as follows. Section 3.1 outlines the formulation of robot dynamics for the rigid body and flexible joint case, while Sec. 3.2 describes the control schemes of the LWR-III. Section 3.3 and Sec. 3.4 give the theoretical background for the proposed collision detection and reaction schemes, whereas Sec. 3.5 provides extensive evaluation of the proposed methods.

3.1 Robot Dynamics and Modeling

In the following, the modeling of a rigid robot and the incorporation of collisions with the environment into the formulation are described. The description is then extended to the case of flexible joints, characterized by a constant but non-negligible elasticity between motor and link inertia.

3.1.1 Rigid Joint Model

First, robot manipulators are considered as open kinematic chains of rigid bodies, having n (rotational) rigid joints.

3.1.1.1 Contact-Free Dynamics

The generalized coordinates $\mathbf{q} \in \mathbb{R}^n$ can be associated to the position of the links. The dynamic model is [25]

$$M(\mathbf{q})\ddot{\mathbf{q}} + C(\mathbf{q},\dot{\mathbf{q}})\dot{\mathbf{q}} + \mathbf{g}(\mathbf{q}) = \tau_m + \tau_F =: \tau_{\text{tot}}, \tag{3.1}$$

with $\mathbf{q} \in \mathbb{R}^n$ being the link position, $M(\mathbf{q}) \in \mathbb{R}^{n \times n}$ the symmetric and positive definite inertia matrix, $C(\mathbf{q},\dot{\mathbf{q}})\dot{\mathbf{q}} \in \mathbb{R}^n$ the centripetal and Coriolis vector, and $\mathbf{g}(\mathbf{q}) \in \mathbb{R}^n$ the gravity vector. $\tau_m \in \mathbb{R}^n$ is the motor torque, $\tau_F \in \mathbb{R}^n$ the motor friction torque, and $\tau_{\text{tot}} \in \mathbb{R}^n$ the sum of active and dissipative torques.

From the skew-symmetry of matrix $N(\mathbf{q},\dot{\mathbf{q}}) = \dot{M}(\mathbf{q}) - 2C(\mathbf{q},\dot{\mathbf{q}})$ it follows that

$$\dot{M}(\mathbf{q}) = C(\mathbf{q},\dot{\mathbf{q}}) + C(\mathbf{q},\dot{\mathbf{q}})^T. \tag{3.2}$$

This property is directly derived from $\dot{M}(\mathbf{q}) - 2C(\mathbf{q},\dot{\mathbf{q}}) = -(\dot{M}(\mathbf{q}) - 2C(\mathbf{q},\dot{\mathbf{q}}))^T$ (definition of skew symmetry). A proof of the skew-symmetry of $\dot{M}(\mathbf{q}) - 2C(\mathbf{q},\dot{\mathbf{q}})$ can be found in most standard robotics textbooks as [25]. The total energy E of the robot is the sum of its kinetic energy T and potential energy U due to gravity:

$$E = T + U = \frac{1}{2}\dot{\mathbf{q}}^T M(\mathbf{q})\dot{\mathbf{q}} + U_g(\mathbf{q}), \tag{3.3}$$

with $\mathbf{g}(\mathbf{q}) = (\partial U_g(\mathbf{q})/\partial \mathbf{q})^T$. From (3.1) and (3.2)

$$\dot{E} = \dot{\mathbf{q}}^T \tau_{\mathrm{tot}} \tag{3.4}$$

can be derived, which represents the power balance in the system. This property can be shown by elaborating (3.3).

$$\begin{aligned}
\dot{E} &= \frac{1}{2}\ddot{\mathbf{q}}^T M(\mathbf{q})\dot{\mathbf{q}} + \frac{1}{2}\dot{\mathbf{q}}^T \dot{M}(\mathbf{q})\dot{\mathbf{q}} + \frac{1}{2}\dot{\mathbf{q}}^T M(\mathbf{q})\ddot{\mathbf{q}} + \frac{\partial U_g(\mathbf{q})}{\partial \mathbf{q}}\dot{\mathbf{q}} && (3.5)\\
&= \frac{1}{2}\dot{\mathbf{q}}^T M(\mathbf{q})\ddot{\mathbf{q}} + \frac{1}{2}\dot{\mathbf{q}}^T \dot{M}(\mathbf{q})\dot{\mathbf{q}} + \frac{1}{2}\dot{\mathbf{q}}^T M(\mathbf{q})\ddot{\mathbf{q}} + \dot{\mathbf{q}}^T \mathbf{g}(\mathbf{q}) && (3.6)\\
&= \dot{\mathbf{q}}^T M(\mathbf{q})\ddot{\mathbf{q}} + \frac{1}{2}\dot{\mathbf{q}}^T \dot{M}(\mathbf{q})\dot{\mathbf{q}} + \dot{\mathbf{q}}^T \mathbf{g}(\mathbf{q}) && (3.7)
\end{aligned}$$

Since $\dot{M} - 2C(\mathbf{q},\dot{\mathbf{q}}) = N(\mathbf{q},\dot{\mathbf{q}})$, $\dot{M}$ may be expressed as $N(\mathbf{q},\dot{\mathbf{q}}) + 2C(\mathbf{q},\dot{\mathbf{q}})$.Inserting this into (3.7) gives

$$\dot{E} = \dot{\mathbf{q}}^T M(\mathbf{q})\ddot{\mathbf{q}} + \frac{1}{2}\dot{\mathbf{q}}^T (N(\mathbf{q},\dot{\mathbf{q}}) + 2C(\mathbf{q},\dot{\mathbf{q}}))\dot{\mathbf{q}} + \dot{\mathbf{q}}^T \mathbf{g}(\mathbf{q}). \tag{3.8}$$

Furthermore, due its skew-symmetry following property holds for N.

$$\mathbf{w}^T N(\mathbf{q},\dot{\mathbf{q}})\mathbf{w} = 0 \tag{3.9}$$

for any $(n \times 1)$ vector $\mathbf{w}$. This may be derived from the definition of a skew-symmetric matrix: A is skew symmetric if $A = -A^T$. This can be written as $\mathbf{w}^T A\mathbf{w} = -\mathbf{w}^T A^T \mathbf{w}$, which can be transformed into the form of (3.9). In particular, this condition leads to $\dot{\mathbf{q}}^T N(\mathbf{q},\dot{\mathbf{q}})\dot{\mathbf{q}} = 0$. In turn $\dot{E}$ is

$$\begin{aligned}
\dot{E} &= +\dot{\mathbf{q}}^T M(\mathbf{q})\ddot{\mathbf{q}} + \frac{1}{2}\dot{\mathbf{q}}^T (N(\mathbf{q},\dot{\mathbf{q}}) + 2C(\mathbf{q},\dot{\mathbf{q}}))\dot{\mathbf{q}} + \dot{\mathbf{q}}^T \mathbf{g}(\mathbf{q}) && (3.10)\\
&= +\dot{\mathbf{q}}^T (M(\mathbf{q})\ddot{\mathbf{q}} + C(\mathbf{q},\dot{\mathbf{q}})\dot{\mathbf{q}} + \mathbf{g}(\mathbf{q})) && (3.11)\\
&= \dot{\mathbf{q}}^T \tau_{\mathrm{tot}}. && (3.12)
\end{aligned}$$

The generalized momentum of the robot is defined as

$$\mathbf{p} = M(\mathbf{q})\dot{\mathbf{q}}. \tag{3.13}$$

Using (3.13), (3.1) and (3.2), the time evolution of $\mathbf{p}$ is given by

$$
\begin{aligned}
\dot{\mathbf{p}} &= M(\mathbf{q})\ddot{\mathbf{q}} + \dot{M}\dot{\mathbf{q}} &(3.14)\\
&= \tau_{\text{tot}} - C(\mathbf{q},\dot{\mathbf{q}})\dot{\mathbf{q}} - \mathbf{g}(\mathbf{q}) + \dot{M}\dot{\mathbf{q}} &(3.15)\\
&= \tau_{\text{tot}} - C(\mathbf{q},\dot{\mathbf{q}})\dot{\mathbf{q}} + C(\mathbf{q},\dot{\mathbf{q}})\dot{\mathbf{q}} + C^T(\mathbf{q},\dot{\mathbf{q}})\dot{\mathbf{q}} - \mathbf{g}(\mathbf{q}) &(3.16)\\
&= \tau_{\text{tot}} + C^T(\mathbf{q},\dot{\mathbf{q}})\dot{\mathbf{q}} - \mathbf{g}(\mathbf{q}). &(3.17)
\end{aligned}
$$

The derivation of the i-th component of $\dot{\mathbf{p}}$ takes advantage of the robot Lagrangian

$$
L = T - U = \frac{1}{2}\dot{\mathbf{q}}^T M(\mathbf{q})\dot{\mathbf{q}} - U_g(\mathbf{q}). \tag{3.18}
$$

Define the manipulator Hamiltonian as

$$
H := \mathbf{p}^T\dot{\mathbf{q}} - L. \tag{3.19}
$$

Hamilton's equations of motion are [21]

$$
\dot{\mathbf{q}} = \frac{\partial H}{\partial \mathbf{p}} \tag{3.20}
$$

$$
-\dot{\mathbf{p}} = \frac{\partial H}{\partial \mathbf{q}} - \tau_{\text{tot}}. \tag{3.21}
$$

This yields

$$
\begin{aligned}
\dot{\mathbf{p}} &= -\frac{1}{2}\frac{\partial}{\partial \mathbf{q}}(\mathbf{p}^T M^{-1}(\mathbf{q})\mathbf{p}) - \mathbf{g}(\mathbf{q}) + \tau_{\text{tot}} &(3.22)\\
&= -\frac{1}{2}\frac{\partial}{\partial \mathbf{q}}\left((M(\mathbf{q})\dot{\mathbf{q}})^T M^{-1}(\mathbf{q})M(\mathbf{q})\right) - \mathbf{g}(\mathbf{q}) + \tau_{\text{tot}} &(3.23)\\
&= -\frac{1}{2}\frac{\partial}{\partial \mathbf{q}}\left(\dot{\mathbf{q}}^T M(\mathbf{q})\dot{\mathbf{q}}\right) - \mathbf{g}(\mathbf{q}) + \tau_{\text{tot}}. &(3.24)
\end{aligned}
$$

Due to the fact that $\dot{\mathbf{q}}^T M(\mathbf{q})\dot{\mathbf{q}}$ is scalar one may write for the $i-$th row

$$
\begin{aligned}
\dot{p}_i &= \tau_{\text{tot},i} - \frac{1}{2}\frac{\partial}{\partial q_i}\dot{\mathbf{q}}^T M(\mathbf{q})\dot{\mathbf{q}} - g_i(\mathbf{q}) &(3.25)\\
&= \tau_{\text{tot},i} - \frac{1}{2}\dot{\mathbf{q}}^T\frac{\partial M(\mathbf{q})}{\partial q_i}\dot{\mathbf{q}} - g_i(\mathbf{q}), &(3.26)
\end{aligned}
$$

for $i = 1,\ldots,n$. The components of the generalized momentum are thus decoupled with respect to the torques acting on the right-hand side of (3.1).

Next, the case of contact along the robot structure is considered.

3.1.1.2 Cartesian Collision

During normal operation, the robot arm may collide with a standing or moving person/obstacle in its workspace. For simplicity, it is assumed that there is at most a single link involved in the collision. Let

$$\dot{\mathbf{x}}_c = \begin{bmatrix} \mathbf{v}_c \\ \omega_c \end{bmatrix} = \begin{bmatrix} J_{c,\text{lin}} \\ J_{c,\text{ang}} \end{bmatrix} \dot{\mathbf{q}} = J_c(\mathbf{q})\dot{\mathbf{q}} \in \mathbb{R}^6 \tag{3.27}$$

be the stacked vector of the linear velocity at the contact point and the angular velocity of the associated robot link (the screw). $\dot{\mathbf{x}}_c$ and the (geometric) contact Jacobian $J_c(\mathbf{q})$ are unknown in advance. Accordingly, the Cartesian collision forces and moments are denoted by

$$\mathscr{F}_{\text{ext}} = \begin{bmatrix} \mathbf{f}_{\text{ext}} \\ \mathbf{m}_{\text{ext}} \end{bmatrix} \in \mathbb{R}^6. \tag{3.28}$$

When a collision occurs, the robot dynamics (3.1) become

$$M(\mathbf{q})\ddot{\mathbf{q}} + C(\mathbf{q},\dot{\mathbf{q}})\dot{\mathbf{q}} + \mathbf{g}(\mathbf{q}) = \tau_m + \tau_F + \tau_{\text{ext}} = \tau_{\text{tot}}, \tag{3.29}$$

with $\tau_{\text{ext}} \in \mathbb{R}^n$ being the external torque[1], which is associated to the (generalized) Cartesian collision force $\mathscr{F}_{\text{ext}}$ by

$$\tau_{\text{ext}} = J_c^T(\mathbf{q})\mathscr{F}_{\text{ext}}. \tag{3.30}$$

In the next subsection the rigid body robot model is extended to the flexible joint case, taking into account elasticity in the joints, which could be due to compliance in the gears or in a joint torque sensor, etc.

3.1.2 Flexible Joint Model

For lightweight design manipulators such as the LWR-III[2] it is not sufficient to model the robot by a second-order rigid body system as described in the preceding

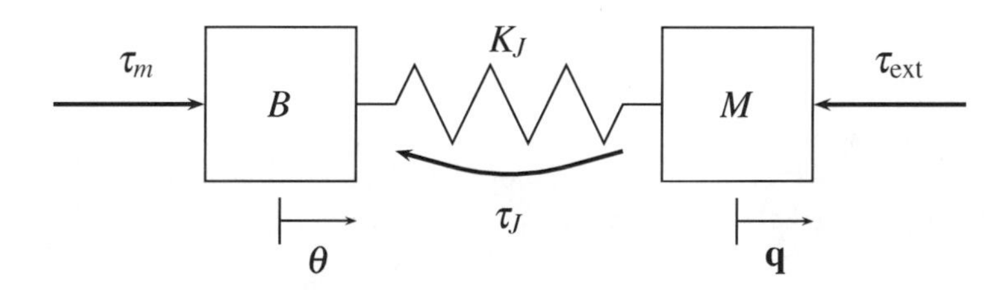

Fig. 3.1 1-DoF model of a flexible joint robot

[1] Whether the external torques appear on the left or right side varies in the chapter. The choice is made for convenience in the particular context.

[2] For details on the robot, please see Sec. 3.2.

subsection. In case of the LWR-III, e.g., the non-negligible joint elasticity between motor and link inertia due to the Harmonic Drive gears and the joint torque sensor has to be taken into account into the model equation. The 1 Degree of Freedom (DoF) case for such a flexible joint is depicted in Fig. 3.1. For a flexible joint robot the following model can be assumed [28]:

$$M(\mathbf{q})\ddot{\mathbf{q}} + C(\mathbf{q},\dot{\mathbf{q}})\dot{\mathbf{q}} + \mathbf{g}(\mathbf{q}) = \tau_J + D_J K_J^{-1}\dot{\tau} + \tau_{\text{ext}} = \tau_{\text{tot}} \tag{3.31}$$

$$B\ddot{\theta} + \tau_J + D_J K_J^{-1}\dot{\tau}_J = \tau_m - \tau_F \tag{3.32}$$

$$\tau_J = K_J(\theta - \mathbf{q}), \tag{3.33}$$

with $\theta \in \mathbb{R}^n$ being motor side position, $\tau_J \in \mathbb{R}^n$ the joint torque, $K_J = \text{diag}\{k_{J,i}\} \in \mathbb{R}^{n\times n}$ the diagonal positive definite joint stiffness matrix, $D_J = \text{diag}\{d_{J,i}\} \in \mathbb{R}^{n\times n}$ the diagonal positive definite joint damping matrix, and $B = \text{diag}\{b_i\} \in \mathbb{R}^{n\times n}$ the diagonal positive definite motor inertia matrix.

Equation (3.31) has basically the same properties as (3.1) and is coupled via $\tau_a := K_J(\theta - \mathbf{q}) + D_J(\dot{\theta} - \dot{\mathbf{q}}) = \tau_J + D_J K_J^{-1}\dot{\tau}_J$, which is the elastic force transmitted through the joints, to the motor dynamics. For a more complete overview on the properties of flexible joint robots please refer to [8].

3.2 Unified Control for the LWR-III

Since the LWR-III is extensively used as a reference platform in this monograph, some technical details of the LWR-III and the underlying unified control algorithms for joint and Cartesian space are presented in the following. These make use of the previously discussed flexible joint formulation.

3.2.1 The DLR Lightweight Robot III

Figure 3.2 shows the history of the DLR LWR, resulting in its commercialized version: the KUKA LWR. Apart from minor modifications, this manipulator has exactly the same design as the 3rd generation of the DLR LWR-III. The DLR LWRs are kinematically redundant, 7-DoF, joint torque controlled flexible joint robots. The current version is the result of 15 years of research that produced three consecutive generations. Since the LWR-III weighs 13.5 kg and is able to handle loads up to 15 kg, an approximate load-to-weight ratio of 1 is achieved[3]. The robot is a

[3] Please note that the nominal payload for the KUKA LWR is 7 kg, but it is able to handle up to 15 kg for research purposes.

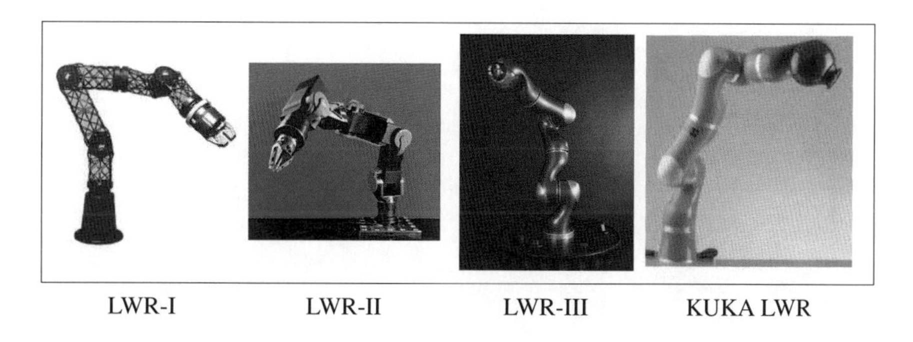

Fig. 3.2 The generations of DLR lightweight robots (LWR-I, LWR-II, and LWR-III) and the commercialized version (KUKA LWR)

modular system and the joints are linked via carbon-fiber structures. The electronic parts, including power converting elements are integrated into the structure of the arm. Each joint is equipped with a motor position and a joint torque sensor. Additionally, a 6-DoF force sensor is embedded in the wrist. In Figure 3.2 the robot is depicted in a configuration with no tools being attached. Each joint is electrically isolated from the following one and they communicate via a fiber optical bus system with each other. All electronics, motors, and gears are integrated into the arm, which makes the robot very compact and portable.

3.2.2 Control Architecture

Prior to introducing some theoretical basics in the next subsection, the hierarchical control architecture of the robot, which was introduced in [3], is depicted in Fig. 3.3. The motor is assumed to be an ideal torque-source and thus torque is the interface to subsequent controllers. The motor control cycle is 25 μs. Further control components are on a Cartesian or joint level, whereby the controllers on joint level are subdivided into local or central ones. Local joint control runs on one Digital Signal Processing (DSP) board per joint at a cycle time of 330 μs. Central joint control with 1 ms cycle time can be used as well. Thus, new controllers which include the full model of the robot can be implemented easily. Cartesian control is also running at a cycle time of 1 ms in the most recent implementation. In general, the following types of controllers are implemented on the robot:

- Impedance control
- Admittance control
- Stiffness control

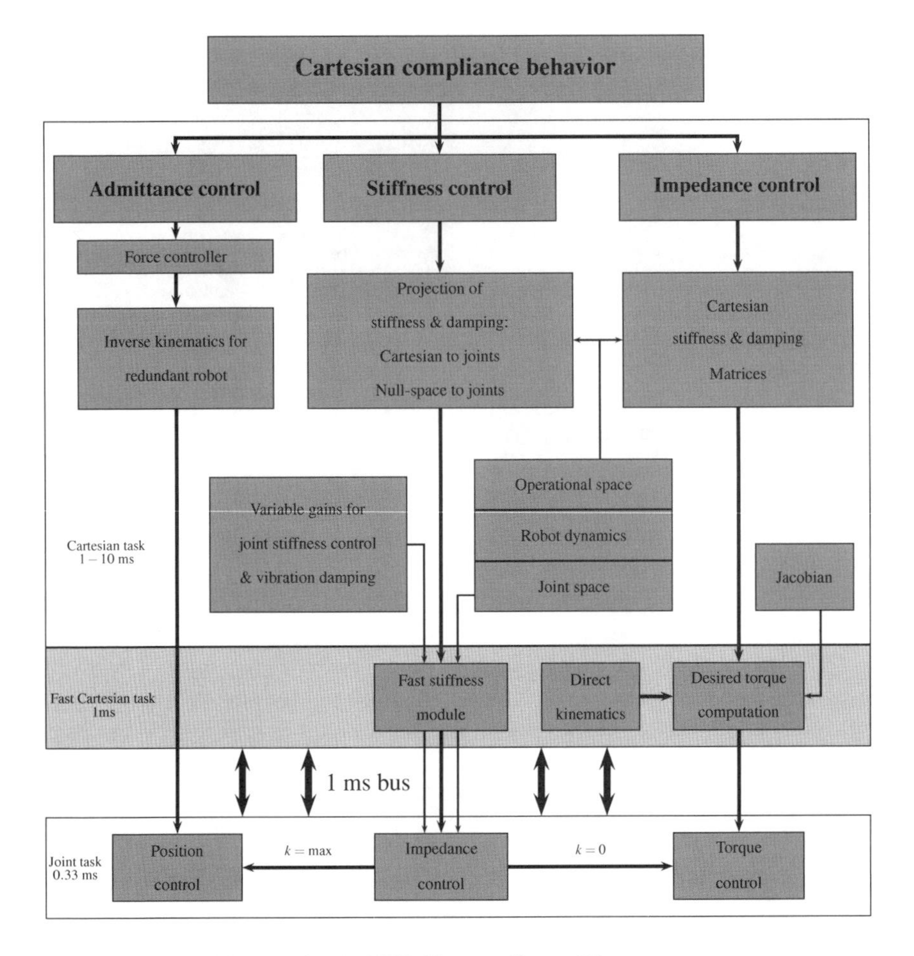

Fig. 3.3 Control architecture for the LWR-III according to [4]

- Position control
- Torque control

Forward kinematics, robot dynamics, and inverse kinematics are also calculated at 1 ms cycle.

After this short introduction of the control architecture, some theoretical background of the torque control strategy of the LWR-III is surveyed.

3.2.3 Implementation in Joint Space

With the additional measurement of the joint torque τ_J in the LWR-III a low-level torque control loop can be implemented, as described in [1]. The overall goal is a full state feedback controller acting on position, velocity, joint torque, and joint torque derivative

$$\tau_m = -K_P\tilde{\theta} - K_D\dot{\theta} + K_T(\mathbf{g}(\mathbf{q}_d) - \tau_J) - K_S\dot{\tau}_J + \mathbf{g}(\mathbf{q}_d), \tag{3.34}$$

with $\theta_d \in \mathbb{R}^n$ being the desired motor position, $\mathbf{q}_d \in \mathbb{R}^n$ the desired link position, $\mathbf{g}(\mathbf{q}_d) \in \mathbb{R}^n$ a gravity compensation term, and K_P, K_D, K_T, K_S positive definite diagonal gain matrices. The motor position error $\tilde{\theta} = \theta - \theta_d$ is obtained from the desired joint position by calculating the desired motor position

$$\theta_d = \mathbf{q}_d + K_J^{-1}\mathbf{g}(\mathbf{q}_d). \tag{3.35}$$

However, the selection of stable controller gains that at the same time provide sufficient performance is not straightforward. In [5] a combination of joint torque control with a superimposed position control structure was proposed, which provides the desired capabilities. The chosen equations for the joint torque feedback and position PD-controller are

$$\tau_m = BB_\theta^{-1}\mathbf{u} + (I - BB_\theta^{-1})(\tau_J + D_J K_J^{-1}\dot{\tau}_J) \tag{3.36}$$

$$\mathbf{u} = -K_\theta(\theta - \theta_d) - D_\theta\dot{\theta} + \mathbf{g}(\mathbf{q}_d), \tag{3.37}$$

with $B_\theta = \mathrm{diag}\{b_{\theta,i}\} \in \mathbb{R}^{n\times n}$ being the diagonal positive definite desired motor inertia matrix with $b_{\theta,i} < b_i$. $K_\theta \in \mathbb{R}^{n\times n}$ is the diagonal positive definite desired stiffness matrix, $D_\theta \in \mathbb{R}^{n\times n}$ the diagonal positive definite desired damping matrix, and $\mathbf{u} \in \mathbb{R}^n$ is a new input variable, which is given by the desired joint position controller. Together with (3.32) the torque controller (3.36) can be written as (friction is neglected)

$$\mathbf{u} = B_\theta\ddot{\theta} + \tau_J + D_J K_J^{-1}\dot{\tau}_J. \tag{3.38}$$

The physical interpretation of (3.38) is that seen by a new input $\mathbf{u}$ the principal structure of the motor dynamics remain, while the motor inertia is scaled down from B to B_θ. In other words, the actuator kinetic energy is shaped. In order to achieve good joint torque damping a slightly modified version of (3.36) can be used

$$\tau_m = BB_\theta^{-1}\mathbf{u} + \tau_J + D_J K_J^{-1}\dot{\tau}_J - BB_\theta^{-1}(\tau_J + D_s K_J^{-1}\dot{\tau}_J), \tag{3.39}$$

where D_s is the independent diagonal joint torque derivative gain matrix D_s. Similar to (3.38) the physical interpretation may be written as

$$\mathbf{u} = B_\theta\ddot{\theta} + \tau_J + D_s K_J^{-1}\dot{\tau}_J. \tag{3.40}$$

Limited by sensor noise the motor inertia ratio BB_θ^{-1} can be chosen between 4 and 6 for the LWR-III. In order to achieve the desired full state feedback controller (3.34), (3.37) are inserted into (3.36). This leads to

$$\tau_m = -BB_\theta^{-1}K_\theta\tilde{\theta} - BB_\theta^{-1}D_\theta\dot{\theta} + BB_\theta^{-1}\mathbf{g}(\mathbf{q}_d) \tag{3.41}$$

$$-(BB_\theta^{-1} - I)\tau_J + \mathbf{g}(\mathbf{q}_d) - \mathbf{g}(\mathbf{q}_d), \tag{3.42}$$

which can be written as

$$\tau_m = -BB_\theta^{-1}K_\theta\tilde{\theta} - BB_\theta^{-1}D_\theta\dot{\theta} + (BB_\theta^{-1} - I)(\mathbf{g}(\mathbf{q}_d) - \tau_J) \tag{3.43}$$

$$-(BB_\theta^{-1} - I)D_J K_J^{-1}\dot{\tau}_J + \mathbf{g}(\mathbf{q}_d), \tag{3.44}$$

with

$$\begin{aligned} K_P &= BB_\theta^{-1}K_\theta \\ K_D &= BB_\theta^{-1}D_\theta \\ K_T &= BB_\theta^{-1} - I \\ K_S &= (BB_\theta^{-1} - I)D_J K_J^{-1}. \end{aligned} \tag{3.45}$$

For the torque control loop given by (3.39) K_S changes to $K_S = (BB_\theta^{-1}D_s - D)K^{-1}$. With this interface different types of controllers can be implemented on joint level. Depending on the selection of the controller matrices (3.45) stiff position control or joint impedance control can be obtained (alternatively torque control with gravity compensation). However, while this structure performs well for joint position control, there are two drawbacks for impedance control. First, minimal values for K_θ, respectively K_P, need to be ensured as the gravity compensation depends on the desired link position. This would prevent to implement zero stiffness. Furthermore, controllers of type (3.37) are only able to realize a desired stiffness relation locally. For details on this, the passivity-based stability proof[4], and certain extensions of the controller that increase performance via gain scheduling please refer to [5].

Next, the Cartesian impedance controller of the LWR-III is described.

3.2.4 Implementation in Cartesian Space

The desired impedance behavior of a robot is usually defined with respect to Cartesian coordinates $\mathbf{x}(\mathbf{q}) \in \mathbb{R}^m$, which describe the position and orientation of the end-effector of the robot. The obvious choice would be the classical Cartesian impedance controller proposed by Hogan [17].

$$\mathbf{u} = -J(\mathbf{q})^T(K_x\tilde{\mathbf{x}}(\mathbf{q}) + D_x\dot{\mathbf{x}}(\mathbf{q}) + \mathbf{g}(\mathbf{q}) \tag{3.46}$$

$$\tilde{\mathbf{x}}(\mathbf{q}) = \mathbf{x}(\mathbf{q}) - \mathbf{x}_d \tag{3.47}$$

$$\dot{\mathbf{x}}(\mathbf{q}) = J(\mathbf{q})\dot{\mathbf{q}}, \tag{3.48}$$

[4] Passivity is the underlying concept of all controllers for the LWR-III. Passivity-based controller design guarantees stability and high robustness.

with $K_{\mathbf{x}} \in \mathbb{R}^{m \times m}$ being the diagonal positive definite desired stiffness matrix, $D_{\mathbf{x}} \in \mathbb{R}^{m \times m}$ the diagonal positive definite desired damping matrix, $\mathbf{x}_d \in \mathbb{R}^m$ the desired tip pose in Cartesian coordinates, $\mathbf{x}(\mathbf{q}) = f(\mathbf{q})$ the position and orientation of the tip computed by the direct kinematics map f. $J(\mathbf{q}) = \frac{\partial f(\mathbf{q})}{\partial \mathbf{q}}$ is the Jacobian of the manipulator, and $\mathbf{g}(\mathbf{q}) \in \mathbb{R}^n$ the gravity compensation. While this controller is passive with respect to the input-output pair $\{\tau_a + \tau_{\text{ext}}, \dot{\mathbf{q}}\}$ for the stiff robot case, it lacks this property with respect to the input-output pair $\{\dot{\mathbf{q}}, -\tau_a\}$ in the flexible joint case[5].

In order to overcome this drawback $\mathbf{u}$ needs to be chosen as a function of $\theta, \dot{\theta}$ only [32, 2, 26, 35]. A solution proposed in [5] uses the static equilibrium $\bar{\mathbf{q}} = h^{-1}(\theta)$ of $\mathbf{q}$ instead of $\mathbf{q}$ itself. h is defined as $h(\mathbf{q}) = \mathbf{q} + K^{-1}\tau_J$. $\bar{\mathbf{q}}$ depends only on the motor position[6]. Now the controller (3.46) can be modified to

$$\mathbf{u} = -J(\bar{\mathbf{q}})^T (K_x \tilde{\mathbf{x}}(\bar{\mathbf{q}}) + D_x \dot{\mathbf{x}}(\bar{\mathbf{q}}) + \mathbf{g}(\bar{\mathbf{q}}) \tag{3.49}$$

$$\tilde{\mathbf{x}}(\bar{\mathbf{q}}) = \mathbf{x}(\bar{\mathbf{q}}) - \mathbf{x}_d \tag{3.50}$$

$$\dot{\mathbf{x}}(\bar{\mathbf{q}}) = J(\bar{\mathbf{q}})\dot{\theta}, \tag{3.51}$$

which provides the desired exact Cartesian impedance behavior. The gravity compensation term $\mathbf{g}(\bar{\mathbf{q}})$ from [26] is a function of the motor position only and is designed so that it provides exact gravity compensation (for $\tau_{\text{ext}} = \mathbf{0}$) in all stationary/static points from the set $\Omega := \{(\mathbf{q}, \theta) | K_J(\theta - \mathbf{q}) = \mathbf{g}(\mathbf{q})\}$. This property can be written as

$$\mathbf{g}(\mathbf{q}) - \mathbf{g}(\bar{\mathbf{q}}) = \mathbf{0} \quad \forall (\mathbf{q}, \theta) \in \Omega. \tag{3.52}$$

With (3.31), (3.38), and (3.49) the closed loop system is

$$M(\mathbf{q})\ddot{\mathbf{q}} + C(\mathbf{q}, \dot{\mathbf{q}})\dot{\mathbf{q}} + \mathbf{g}(\mathbf{q}) = \tau_a + \tau_{\text{ext}} \tag{3.53}$$

$$B_\theta \ddot{\theta} + J(\bar{\mathbf{q}})^T K_x \tilde{\mathbf{x}}(\bar{\mathbf{q}}) - \mathbf{g}(\bar{\mathbf{q}}) + J(\bar{\mathbf{q}})^T D_x J(\bar{\mathbf{q}})\dot{\theta} + \tau_a = \mathbf{0}. \tag{3.54}$$

For the passivity analysis of the Cartesian impedance, the explanation of $h^{-1}(\theta)$, and the details on joint impedance control within the passivity-based framework refer to [5].

After this introduction to the control architecture and control schemes of the LWR-III, the developed collision detection and reaction schemes are described next.

3.3 Collision Detection Schemes

In this section, the theoretical basis for different collision detection methods is derived. The methods range from simple energy-based collision detectors to more advanced methods, which are able to give an accurate estimation of the external torques.

---

[5] This pair is chosen as one is interested in the passivity of the complete flexible joint system.

[6] It was shown that under some mild conditions there exists a unique mapping between the equilibria θ_0 and $\mathbf{q}_0$.

3.3.1 Energy-Based Detection

In order to only recognize the occurrence of a collision (the detection problem) for a rigid robot, an energy argument is sufficient[7].

Define the following scalar quantity

$$\hat{r}(t) = k_O \left[E(t) - \int_0^t (\dot{\mathbf{q}}^T \tau + \hat{r}) \mathrm{d}s - E(0) \right], \tag{3.55}$$

with initial value $\hat{r}(0) = 0$, $k_O > 0$, and where $E(t)$ is the total robot energy at time $t \geq 0$, as defined in (3.3). $\tau \in \mathbb{R}^n$ denotes the link driving torque, which is τ_m for rigid robots and τ_J for flexible joint robots. Based on the integration of the input power and an unknown disturbance $\hat{r}$ the integrand essentially monitors the energy dynamics of the robot. Note that $\hat{r}$ can be computed using the measured joint position $\mathbf{q}$, the joint velocity $\dot{\mathbf{q}}$ (possibly obtained through numerical differentiation), and the commanded motor torque. The latter may be the result of any type of control action. No acceleration measurement is needed. Using eqs. (3.4) and (3.29), the resulting dynamics of $\hat{r}$ is

$$\dot{\hat{r}} = -k_O \hat{r} + k_O \dot{\mathbf{q}}^T \tau_{\mathrm{ext}}, \tag{3.56}$$

i.e., that of a first-order stable linear filter driven by the work performed by the joint torques due to collision. During free motion, $\hat{r} = 0$ up to measurement noise and unmodeled disturbances. In response to a generic collision, $\hat{r}$ raises exponentially with a time constant $1/k_O$ and detection occurs as soon as $|\hat{r}| > \hat{r}_{\mathrm{CD}}$. The scalar $\hat{r}_{\mathrm{CD}}$ is a suitable threshold whose actual value depends on the noise characteristics in the system. Dynamic thresholding can be used to avoid false detection due to spurious spikes in noisy signals, as shown in [10]. When contact is lost, $\hat{r}$ rapidly returns to zero. Because of these properties, $\hat{r}$ is called a **collision detection signal**. Not all possible collision situations are detected by this scheme. With the robot at rest ($\dot{\mathbf{q}} = \mathbf{0}$), the instantaneous value of τ_{ext} does not affect $\hat{r}$, but begins as soon as the robot starts moving. As a consequence, with the robot at rest, true impulse collision forces/torques cannot be detected with this scheme. On the other hand, when the robot is in motion, collision can be detected if the Cartesian collision force $\mathscr{F}_{\mathrm{ext}}$ produces motion at the contact. Using eq. (3.30) the following relationship can be formed.

$$\dot{\mathbf{q}}^T \tau_{\mathrm{ext}} = \dot{\mathbf{q}}^T J_c^T(\mathbf{q}) \mathscr{F}_{\mathrm{ext}} = \dot{\mathbf{x}}_c \mathscr{F}_{\mathrm{ext}} = 0 \Longleftrightarrow \dot{\mathbf{x}}_c \perp \mathscr{F}_{\mathrm{ext}} \tag{3.57}$$

For example, a lateral (horizontal) force due to a human colliding against a $2R$ planar arm in motion in the vertical plane will not be detected, being fully compensated by the reaction forces of the manipulator structure. When evaluated in terms of reactive motions that the robot may take in response to this collision, such behavior of the detection scheme is rather natural. In fact, no possible robot motion would be able to reduce the force loading in this case. Suppose now to add a vertical joint axis at the base (obtaining a $3R$ elbow-type robot) and let the second and third links

[7] The extension to the flexible joint case is straightforward by substituting τ_m with τ_J.

be in motion in the vertical plane as before (i.e., with $\dot{q}_1 = 0$). The same previous lateral force will be felt initially only at the first joint $\tau_{\text{ext},1} \neq 0$), which is at rest. Therefore, $\dot{\mathbf{q}}^T \tau_{\text{ext}} = 0$ and thus $\dot{\hat{r}}(0) = 0$. Provided that the joint position controller is soft enough, the first joint will start moving in response to the collision with low contact stiffness before the contact force has been removed and detection may then occur.

From here on the methods are discussed for the flexible joint case. However, please note that they are directly applicable to the rigid case by substituting the joint torque τ_J with the motor torque τ_m.

The next methods aim at obtaining the full external torque vector τ_{ext}.

3.3.2 Direct Derivation Method

Using (3.31) the external torque τ_{ext} can be expressed as[8]

$$\tau_{\text{ext}} = \tau_J - M(\mathbf{q})\ddot{\mathbf{q}} - C(\mathbf{q},\dot{\mathbf{q}})\dot{\mathbf{q}} - \mathbf{g}(\mathbf{q}), \tag{3.58}$$

where $\dot{\tau}_J$ and τ_F are not taken into account for sake of clarity. The most simple estimation of the external torques is thus obtained by using joint torque and link position. (3.58) would be a straightforward derivation of the external torque. However, this approach is not applicable in practice because in reality $\mathbf{q}$ cannot be differentiated two times due to the presence of non-negligible noise. A scheme to circumvent this problem can be derived for smooth desired trajectories. As described next, this method can be used for a robot that is being controlled by a high performance position controller.

3.3.3 Derivation from Desired Dynamics

For a commanded motor trajectory with smooth derivatives of higher order and a well parameterized position controller, it can be assumed that

$$\mathbf{q}_d \approx \mathbf{q}, \tag{3.59}$$

where $\mathbf{q}_d \in \mathbb{R}^n$ is the desired joint position. Thus, $\mathbf{q}$ and its derivatives can be approximated by $\mathbf{q}_d$ and its derivatives. An estimate of the external joint torque due to collision is given by combining the expected joint torque computed with (3.31) for $\tau_{\text{ext}} = \mathbf{0}$.

$$\hat{\tau}_J(\mathbf{q}_d,\dot{\mathbf{q}}_d,\ddot{\mathbf{q}}_d) = M(\mathbf{q}_d)\ddot{\mathbf{q}}_d + C(\mathbf{q}_d,\dot{\mathbf{q}}_d)\dot{\mathbf{q}}_d + \mathbf{g}(\mathbf{q}_d) \tag{3.60}$$

with the measurement of the joint torque τ_J. This leads to an estimation of the external torque

[8] Note that from here on, a different sign convention for τ_{ext} will be used.

$$\hat{\tau}_{\text{ext}} = \tau_J - \hat{\tau}_J \approx \tau_{\text{ext}}. \tag{3.61}$$

Even though this method suffices for stiff position control and smooth desired motions, the scheme is not a general estimation of external torques that is independent from the controller and desired trajectories (i.e. it is not a kind of virtual sensor). In the next two subsections two different disturbance observers that neither suffer from the problem of requiring $\ddot{\mathbf{q}}$ nor from the demanded a priori knowledge of $\mathbf{q}_d$ are proposed. Both methods may be interpreted as a general sensor for external torques along the robot structure [7, 15, 16].

3.3.4 Observing Joint Velocity

The underlying idea of the present method is to observe the joint velocity $\dot{\mathbf{q}}$ with a reduced state and disturbance observer. A reduced state observer can be used in order to make the algorithm react faster to changes of the system since the derivative reflects the timely evolution itself[9]. First $\ddot{\mathbf{q}}$ has to be calculated:

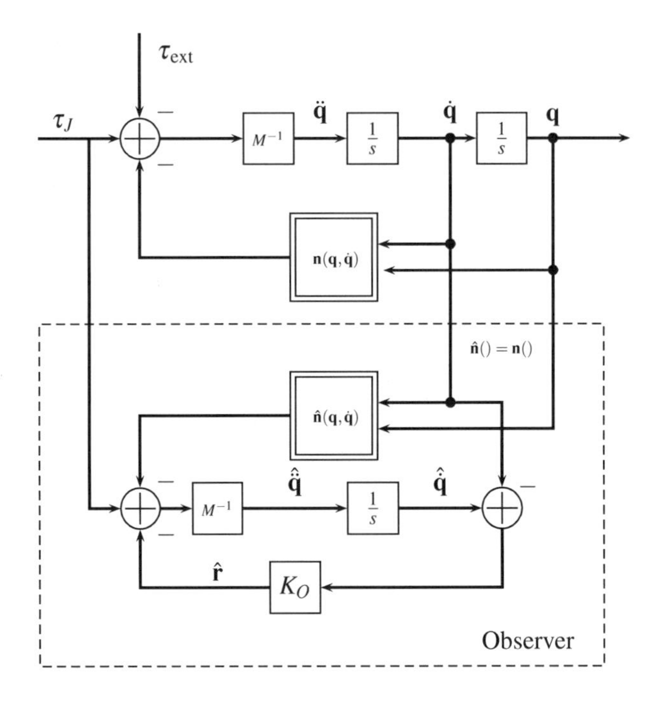

Fig. 3.4 Block diagram of a reduced state observer for $\dot{\mathbf{q}}$ to estimate the external torque τ_{ext}

[9] The same argument states for the scheme proposed in Sec. 3.3.5.

$$\ddot{\mathbf{q}} = \frac{\mathrm{d}\dot{\mathbf{q}}}{\mathrm{d}t} = M^{-1}\left(\tau_J - \mathbf{n}(\mathbf{q},\dot{\mathbf{q}}) - \tau_{\text{ext}}\right), \tag{3.62}$$

where

$$\mathbf{n}(\mathbf{q},\dot{\mathbf{q}}) := C(\mathbf{q},\dot{\mathbf{q}})\dot{\mathbf{q}} + \mathbf{g}(\mathbf{q}). \tag{3.63}$$

The assumed model to observe the external torque τ_{ext} is, according to [14], chosen as

$$\hat{\mathbf{r}} = \hat{\tau}_{\text{ext}} \tag{3.64}$$

$$\dot{\hat{\mathbf{r}}} = \mathbf{0}. \tag{3.65}$$

Such a model is used if no further information on the expected behavior is available. In Figure 3.4 the block diagram of the link side of the robot and the proposed observer are depicted. The observed disturbance $\hat{\mathbf{r}}$ can be expressed as follows:

$$\hat{\ddot{\mathbf{q}}} = M^{-1}\left(\tau_J - \mathbf{n}(\mathbf{q},\dot{\mathbf{q}}) - \hat{\mathbf{r}}\right) \tag{3.66}$$

$$\dot{\hat{\mathbf{r}}} := K_O(\hat{\ddot{\mathbf{q}}} - \ddot{\mathbf{q}}) \tag{3.67}$$

$$\hat{\mathbf{r}} = K_O \int_0^T \left[M^{-1}(\tau_J - \mathbf{n}(\mathbf{q},\dot{\mathbf{q}}) - \hat{\mathbf{r}}) - \frac{\mathrm{d}}{\mathrm{d}t}(\dot{\mathbf{q}})\right] \mathrm{d}t \tag{3.68}$$

$$= K_O\left(\int_0^T \left[M^{-1}(\tau_J - \mathbf{n}(\mathbf{q},\dot{\mathbf{q}}) - \hat{\mathbf{r}})\right] \mathrm{d}t - \dot{\mathbf{q}}\right), \tag{3.69}$$

where $K_O = \text{diag}\{k_O^i\}$ is the observer gain. The transfer function from the external torque τ_{ext} to the observed disturbance $\hat{\mathbf{r}}$ is obtained from (3.62), (3.66), and (3.67).

$$\dot{\hat{\mathbf{r}}} = K_O(\hat{\ddot{\mathbf{q}}} - \ddot{\mathbf{q}}) \tag{3.70}$$

$$= K_O M^{-1}(\tau_{\text{ext}} - \hat{\mathbf{r}}) \tag{3.71}$$

$$\dot{\hat{\mathbf{r}}} + K_O M^{-1}\hat{\mathbf{r}} = K_O M^{-1}\tau_{\text{ext}} \tag{3.72}$$

$$\hat{\mathbf{r}} = (sI + K_O M^{-1})^{-1} K_O M^{-1}\tau_{\text{ext}} \tag{3.73}$$

In other words, a filtered version of τ_{ext} with a variable filter frequency is obtained.

In the following, a concept for estimating the external torques based on the observation of the generalized momentum $\mathbf{p}$ is introduced.

3.3.5 Observing Generalized Momentum

The method developed now has a similar structure to the previous one. However, in contrast to the velocity observer its basic concept is to observe the generalized angular momentum

$$\mathbf{p} = M(\mathbf{q})\dot{\mathbf{q}}, \tag{3.74}$$

as proposed in [10, 11] with a disturbance and a reduced state observer. First, the physical relationship between $\mathbf{p}$ and $\mathbf{q}$ of the robot equation is obtained. Together with (3.74) the link side dynamics (3.31) can be rewritten to

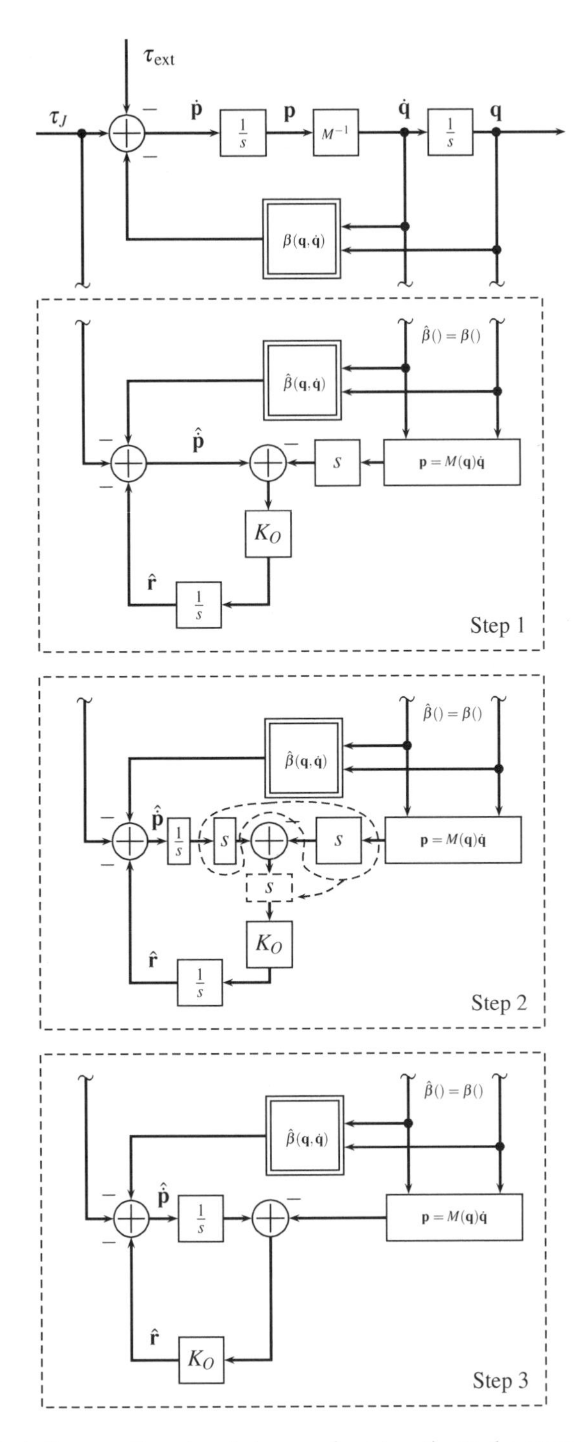

Fig. 3.5 Block diagram of a reduced observer for $\dot{\mathbf{p}}$ to estimate the external torque τ_{ext}

$$\dot{\mathbf{p}} = \frac{\mathrm{d}}{\mathrm{d}t}(M(\mathbf{q})\dot{\mathbf{q}}) = \dot{M}(\mathbf{q})\dot{\mathbf{q}} + M(\mathbf{q})\ddot{\mathbf{q}} \tag{3.75}$$

$$\dot{\mathbf{p}} = \tau_J - \beta(\mathbf{q},\dot{\mathbf{q}}) - \tau_{\text{ext}}, \tag{3.76}$$

where

$$\beta(\mathbf{q},\dot{\mathbf{q}}) := \mathbf{n}(\mathbf{q},\dot{\mathbf{q}}) - \dot{M}(\mathbf{q})\dot{\mathbf{q}} = C(\mathbf{q},\dot{\mathbf{q}}) + \mathbf{g}(\mathbf{q}) - \dot{M}(\mathbf{q})\dot{\mathbf{q}}. \tag{3.77}$$

The model to observe the external torque τ_{ext} is then

$$\hat{\mathbf{r}} = \hat{\tau}_{\text{ext}} \tag{3.78}$$

$$\dot{\hat{\mathbf{r}}} = \mathbf{0}. \tag{3.79}$$

In Figure 3.5 the block diagram of the detection scheme is shown. The actual variable being observed is the derivation of the angular momentum $\dot{\mathbf{p}}$. An obvious drawback is the necessary differentiation of $\mathbf{p}$. Fortunately, this can be avoided by a simple restructuring, see Fig. 3.5. Upon further examination of Fig. 3.5, the equations for $(\dot{\hat{\mathbf{p}}},\hat{\mathbf{r}},\dot{\hat{\mathbf{r}}})$ can be written as

$$\dot{\hat{\mathbf{p}}} = \tau_J - \beta(\mathbf{q},\dot{\mathbf{q}}) - \hat{\mathbf{r}} \tag{3.80}$$

$$\dot{\hat{\mathbf{r}}} := K_O(\dot{\hat{\mathbf{p}}} - \dot{\mathbf{p}}) \tag{3.81}$$

$$\hat{\mathbf{r}} = K_O \int_0^T (\dot{\hat{\mathbf{p}}} - \dot{\mathbf{p}})\mathrm{d}t \tag{3.82}$$

$$= K_O \left(\int_0^T [\tau_J - \beta(\mathbf{q},\dot{\mathbf{q}}) - \hat{\mathbf{r}}]\, \mathrm{d}t - M(\mathbf{q})\dot{\mathbf{q}} \right) \tag{3.83}$$

$$= K_O(\hat{\mathbf{p}} - \mathbf{p}). \tag{3.84}$$

Combining (3.76) and (3.80) it can be shown that the observed disturbance is a component-wise filtered version of the real external torque τ_{ext}:

$$\dot{\hat{\mathbf{p}}} - \dot{\mathbf{p}} = \tau_J - \beta(\mathbf{q},\dot{\mathbf{q}}) - \hat{\mathbf{r}} - (\tau_J - \beta(\mathbf{q},\dot{\mathbf{q}}) - \tau_{\text{ext}}) \tag{3.85}$$

$$\Leftrightarrow K_O^{-1}\dot{\hat{\mathbf{r}}} = -\hat{\mathbf{r}} + \tau_{\text{ext}} \tag{3.86}$$

The dynamics of $\hat{\mathbf{r}}$ is therefore

$$\dot{\hat{\mathbf{r}}} = -K_O\hat{\mathbf{r}} + K_O\tau_{\text{ext}} \tag{3.87}$$

and its components can be written as

$$\hat{r}^i = \frac{1}{sT_O^i + 1}\tau_{\text{ext}}^i = \frac{K_O^i}{s + K_O^i}\tau_{\text{ext}}^i \approx \tau_{\text{ext}}^i \quad \forall i \in \{1, ..., n\} \tag{3.88}$$

$$\hat{\mathbf{r}} = (\hat{r}^1 \cdots \hat{r}^n). \tag{3.89}$$

Each K_O^i can also be interpreted as a filter constant $T_O^i = 1/K_O^i$ of the $i-$th external joint torque signal component. In ideal conditions,

$$K_O \to \infty \Rightarrow \hat{\mathbf{r}} \approx \tau_{\text{ext}},$$

which means in practice that the gains should be as large as possible. Moreover, $\hat{\mathbf{r}}$ is sensitive to collisions, even at $\dot{\mathbf{q}} = \mathbf{0}$. When the contact occurs on the i-th link of the robot kinematic chain, $\hat{\mathbf{r}}$ takes the form

$$\hat{\mathbf{r}} = [\hat{r}_1 \quad \ldots \quad \hat{r}_{i-1} \; 0 \quad 0 \quad \ldots \quad 0]. \tag{3.90}$$

Assuming $\hat{\mathbf{r}} = \hat{\tau}_{\text{ext}} = J_c^T(\mathbf{q})\mathscr{F}_{\text{ext}}$, this follows from the fact that, for a collision on link i, the last $N-i$ columns of the Jacobian $J_c(\mathbf{q})$ are identically zero. The first i components of vector $\hat{\mathbf{r}}$ will be generally different from zero, at least for the time interval of contact, and will start decaying exponentially toward zero as soon as contact is lost. The residual $\hat{\mathbf{r}}$ will be affected only by Cartesian collision forces $\mathscr{F}_{\text{ext}}$ that perform virtual work on admissible robot motion, i.e., those forces that do not belong to the kernel of $J_c^T(\mathbf{q})$. In general, the sensitivity to $\mathscr{F}_{\text{ext}}$ of each of the affected residuals (proximal to the robot base) will vary with the arm configuration. Thanks to the properties of the generalized momentum, this dynamic analysis can be carried out based only on the static transformation matrix $J_c^T(\mathbf{q})$ from Cartesian forces to joint torques. In fact, the residual dynamics in (3.87) is unaffected by robot velocity and acceleration.

3.3.6 Response Behavior of Momentum Observer

In order to ensure fast collision detection a high observer filter frequency is needed, i.e. large K_O. However, noise deteriorates the signal and thus certain low pass filter characteristics are needed. This tradeoff necessitates the analysis of $\hat{\mathbf{r}}$ with respect to the observer gain. To evaluate this, an impact between a full dynamic model of the LWR-III and a spherical object with human head stiffness properties is simulated. Figure 3.6 depicts the mentioned dependency for axis 2 and 4. For a near-ideal collision detection, K_O should be set to 500, assuming the communication delay between hardware components to be $\Delta t_c = 0$ s.

3.3.7 Comparison of Collision Detection Schemes

All methods described in the previous sections are suitable for real-time demands. However, each of them has certain characteristics, which leads to different detection performance.

The method introduced in Sec. 3.3.1 has the advantage of very low computational complexity. However, it does not respond while resting, and does not provide an estimation of the external torques. From a theoretical standpoint, using the direct derivation as described in Sec. 3.3.2 would be most accurate. However, it suffers

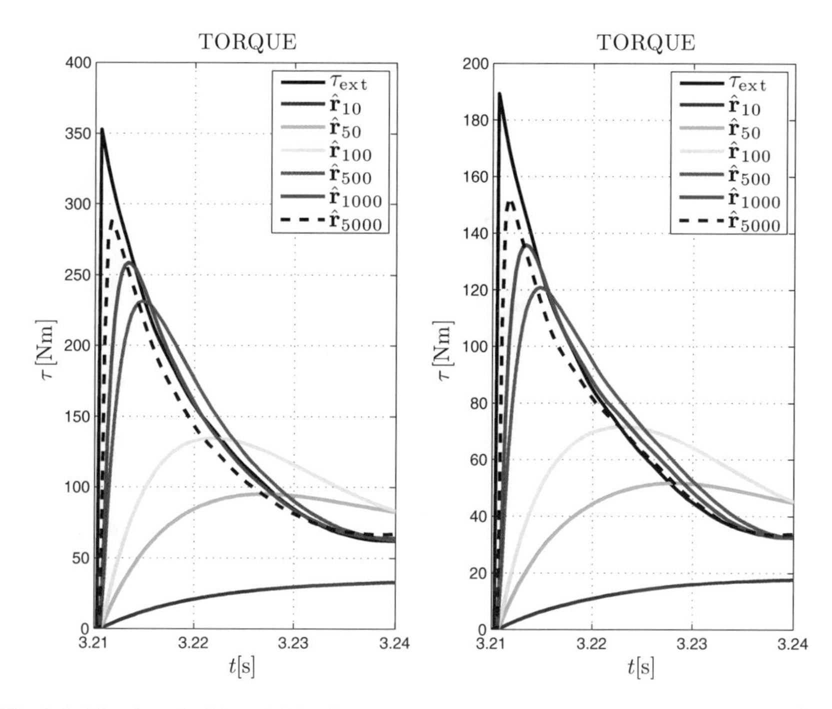

Fig. 3.6 $\hat{\mathbf{r}}$ is plotted with variable observer gain K_O, denoted by the index (i.e. $\hat{\mathbf{r}}_i = \hat{\mathbf{r}}(K_O = \text{diag}\{i\})$). On the left side the graphs for joint 2 and on the right side the ones for joint 4 are shown.

from the severe practical limitation of requiring the joint acceleration. A major benefit from the approach given in Sec. 3.3.3 is that due to not using position sensor signals and its numerical derivatives only the joint torque sensor noise is present. This is in practice considerably lower than the one for velocity. Significant error sources for this scheme are the model error and assumption error (3.59). Therefore, this approach performs better for a stiff position controller and pre-planned trajectories with smooth $\mathbf{q}_d$, for a very soft impedance controller or jerky desired positions (e.g., generated online from a vision system), the detection schemes presented in Sec. 3.3.4 and Sec. 3.3.5 are clearly advantageous. For the joint velocity approach two main disadvantages were identified:

- The computation of $M^{-1}(\mathbf{q})$ is necessary.
- Nonlinear filter dynamics between $\hat{\mathbf{r}}$ and τ_{ext} are introduced. Thus, a variable cut-off frequency due to the coupling in the joints is the consequence.

On the other hand, the angular momentum scheme from Sec. 3.3.5 has a minor practical drawback:

- $\beta(\mathbf{q}, \dot{\mathbf{q}})$ has to be computed instead of $\mathbf{n}(\mathbf{q}, \dot{\mathbf{q}})$.

Therefore, operations being necessary for the mass matrix derivation $\dot{M}(\mathbf{q})$ in $\boldsymbol{\beta}$ could be saved for the velocity observer. However, in the end the momentum based strategy turned out to be the better alternative. The velocity based approach was not working satisfactory even in simulation. This is mainly due to the variable cut-off frequency, which is difficult to determine.

This gives two good methods for collision detection, with the one from Sec. 3.3.3 performing well for a stiff position controlled robot and the momentum observer that can be used for any type of underlying control scheme.

In the following section some practical remarks are given.

3.3.8 Practical Remarks

3.3.8.1 Practical Implementation

The practical implementation of the collision detection method described in Sec. 3.3.5 is depicted in Fig. 3.7. The motor and link side dynamics are interconnected via the joint torque. The motor and friction torque act on the motor inertia, while the external torque is applied to the rigid body dynamics. Due to the joint torque sensing of the robot, the collision detection can be decoupled from friction (left). The link side position sensing of the robot is less accurate than the motor position sensing. Furthermore, a good flexible joint model of the robot is available. Therefore, the motor position and velocity $\boldsymbol{\theta}, \dot{\boldsymbol{\theta}} \in \mathbb{R}^n$ as well as the joint model for estimating the link side position and velocity $\hat{\mathbf{q}}, \dot{\hat{\mathbf{q}}} \in \mathbb{R}^n$ can be utilized (right). As a result, a model-based estimation $\hat{\mathbf{q}}$ can be obtained from (3.31).

$$\hat{\mathbf{q}} = \boldsymbol{\theta} - K_J^{-1}\boldsymbol{\tau}_J \rightarrow (\hat{\mathbf{q}}, \dot{\hat{\mathbf{q}}}) \tag{3.91}$$

This remark is valid for all implemented collision detection methods that make use of the link side position and its derivatives.

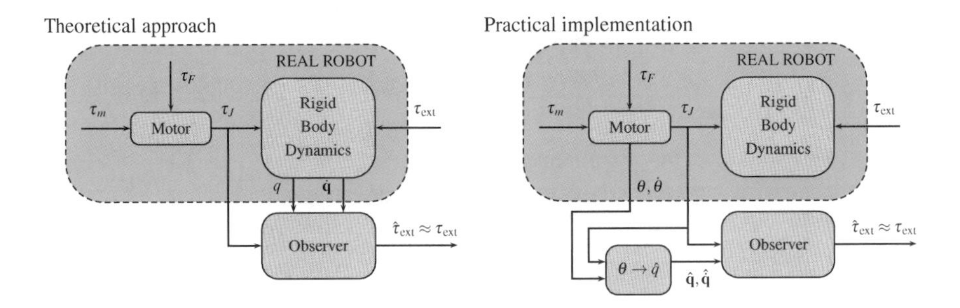

Fig. 3.7 Principle (left) and practical implementation (right) of the disturbance observer

3.3.8.2 Coping with Robot Model Errors

The above Collision Detection mechanisms use a collision threshold, which should be small enough to allow fast firing of a reaction scheme (sensitivity), but also large enough not to be activated by measurement errors and/or dynamic model errors (false alarms):

$$CD = \begin{cases} 1 \text{ if } \exists i: & |\hat{r}^i| > \hat{r}^i_{\text{det}} \\ 0 \text{ else.} \end{cases} \tag{3.92}$$

$\hat{r}^i_{\text{det}} > 0$ is the collision threshold for the ith axis. The main sources of errors limiting the detection threshold for this approach are the model errors and the sensor noise (torque and numerical velocity estimation). By choosing a slower filter constant for $\hat{\mathbf{r}}$ the noise can be reduced at the price of some detection delay. Apart from using these collision thresholds it is possible to use the frequency information of the collision detection signal as well. For the relevant motion velocities, the robot dynamics contains low-frequency signals when compared to the impact torque. One possibility to cope with this robot modelling errors is thus to high-pass filter both detectors (3.88) and (3.61) component-wise. This leads for the momentum observer to

$$\hat{r}^i_{\text{hpf}} = T^i_O s \hat{r}^i = \frac{s}{s + K^i_O} \tau^i_{\text{ext}} \quad \forall i \in \{1, ..., n\} \tag{3.93}$$

if $T^i_O = 1/K^i_O$. For the scheme in Sec. 3.3.3 this means to use

$$\hat{\tau}^{\text{hpf},i}_{\text{ext}} = T^i_O s(\tau^i - \hat{\tau}^i) \quad \forall i \in \{1, ..., n\} \tag{3.94}$$

as a detector. This implies to ignore the very low frequent external torques but they still can be estimated in parallel by (3.88) or (3.61). Therefore, for high frequency torque components, i.e. fast rigid impacts, a more sensitive detector is obtained. Both versions were implemented on the LWR-III and allowed to reduce the detection threshold by 50 % in all joints to $0.05\tau_{\max}$ in comparison to the initial version. Moreover, current improvements in the model allow a threshold of $0.02\tau_{\max}$, which corresponds closely with the resolution of the joint torque sensor.

As an alternative to high-pass filtering of the signals, one could use the difference of the two detection schemes as a collision detector. This will remove the model error entirely and the detector happens to be again a high-pass filtered version of τ_{ext}:

$$\underbrace{\tau^i - \hat{\tau}^i}_{\hat{\tau}^i_{\text{ext}}} - \hat{r}^i \approx \tau^i_{\text{ext}} - \frac{1}{T^i_O s + 1} \tau^i_{\text{ext}} = \frac{T^i_O s}{T^i_O s + 1} \tau^i_{\text{ext}} \tag{3.95}$$

3.3.8.3 Collision Severity Stages

In order to fully exploit the information gained from an accurate estimation of the external joint torques as described in Sec. 3.3.5, it is not only used for differentiation

Table 3.1 Collision severity stages

Severity type	percentage of max. joint torque
Contact	3 %
Grasping failure	15 %
Slight collision	50 %
Severe collision I	90 %
Severe collision II	98 %

between contact and no-contact. Its magnitude is utilized as well. Since the mechanical design of the robot defines constraints on the maximum allowable joint torque, different contact classes are expressed in relation to the maximal joint torque. An example is given in Tab. 3.1, defining collision severity from contact to severe collisions. This classification was chosen for the interactive bin-picking application outlined in Chapter 8.3. The stages are used later for appropriately reacting in an effective and fault tolerant manner, depending on the current state and severity of contact.

3.3.9 Estimating the Contact Wrench

Thus far an accurate estimation of τ_{ext} has been elaborated. However, e.g. for online load estimation or Cartesian collision reaction methods the contact wrench information is of interest. Consider the case of (single point) contact on the tool tip. In order to calculate the corresponding external forces (3.30) is used.

$$J_c^T \mathscr{F}_{\text{ext}} = \tau_{\text{ext}} \tag{3.96}$$

$$J_c J_c^T \mathscr{F}_{\text{ext}} = J_c \tau_{\text{ext}} \tag{3.97}$$

$$\mathscr{F}_{\text{ext}} = (J_c J_c^T)^{-1} J_c \tau_{\text{ext}} = J_c^{\#} \tau_{\text{ext}} \tag{3.98}$$

This relation applies to only non-singular J_c to useful solutions. Since the Jacobian may be interpreted as a sensitivity matrix, forces acting along singular directions cannot be sensed. These forces are simply resisted by the manipulator. However, in the nominal workspace (3.98) may be used instead of a wrist force torque sensor for measuring TCP forces. For this $J_{TCP} = J_c$ is assumed, i.e. the point of contact is known. This is useful for tasks in Operational space, which require tip force control as e.g. in assembly processes. Especially for industrial tasks it is desirable to keep the costs for external sensing low. Furthermore, the number of components is reduced, leading to lower maintenance costs and potential failures.

Estimating the full external contact wrench along the entire structure is for many cases not possible. Singular configurations in the according subspace of dimension i with the associated Jacobian of dimension $\dim(J) = (i,N)$ lead to a transfer of the contact forces into the mechanical structure. Furthermore, due to the possibility of $i < 6$ there may be no torque, for which (3.98) can be satisfied.

The next step to be taken is to develop effective reaction schemes. Even though Cartesian reaction strategies were also developed, the main focus will be on joint level reaction for the sake of brevity.

3.4 Collision Reaction

This section builds upon the previous results and presents the remaining reaction during the post-impact phase. Different strategies are elaborated, leading to significantly different behavior. In particular, the directional information on contact forces provided by the identification scheme is used to safely drive the robot away from the human.

3.4.1 Reflex like Collision Reaction Schemes

3.4.1.1 Stop the Robot

The most obvious idea is to stop the robot as soon as a collision has been detected. This behavior can e.g. be obtained by setting $q_d = q(t_c)$, where t_c is the instant of collision detection.

3.4.1.2 Stop the Robot and Drive Back

Just stopping the robot after a collision has been detected is a basic approach, which can cause uncomfortable or dangerous situations. It could e.g. lead to the human operator being clamped between the robot and a mechanical counterpart. Therefore, the robot should not only stop but also drive back along the previous path. In this approach, it is assumed that no further object is present that could be involved in another collision with the backdriving robot.

Another way to react to a collision is to switch between different control modes. Initially, the robot moves along a desired trajectory with a position reference-based controller (e.g. position or impedance control). In case of a collision the control mode is switched to a compliance-based controller that ignores the previous task trajectory.

3.4.1.3 Control Switch to Torque Control with Gravity Compensation

The first useful control change identified is to switch to torque control with gravity compensation, meaning to set $\tau_d = 0$. The human operator can simply push the robot away while feeling only the reduced inertia of the robot (see Sec. 3.2). For accomplishing this behavior switch with the LWR-III, the weighting matrices in (3.34) for position or torque control only need to be changed to parameterize the reaction behavior within one computation cycle. Please note that this strategy does not explicitly take into account any information or estimation of τ_{ext}.

An admittance type strategy is described in the following.

3.4.1.4 Admittance-Based Strategy

For this scheme no switch e.g. from position to torque control is needed. Since $\hat{\mathbf{r}}$ is the observed external torque the basic idea is to use this information to evade from external torques. This is accomplished by multiplying $\hat{\mathbf{r}}$ with the matrix K_R and let $\dot{\mathbf{q}}_d$ point towards negative $\hat{\mathbf{r}}$ direction. Thus, a desired velocity vector is obtained that lets the robot evade from the external contact producing τ_{ext}.

$$\dot{\mathbf{q}}_d = -K_R\hat{\mathbf{r}} \tag{3.99}$$

$$\mathbf{q}_d = -\int_0^T K_R\hat{\mathbf{r}}\,\mathrm{d}t, \tag{3.100}$$

with $K_R = \{\text{diag}\{K_{R1},...,K_{Rn}\}|\forall i : K_{Ri} \geq 1\}$. With this scheme the robot quickly drives away from the external torque source and decreases the contact forces until they decay to zero[10].

The next strategy extends the torque control based strategy by explicitly incorporating the external torque information.

3.4.1.5 Reflex Torque-Based Strategy

From the full robot model and the torque control loop in Sec. 3.2 the closed-loop system behavior is

$$\tau_J = M(\mathbf{q})\ddot{\mathbf{q}} + C(\mathbf{q},\dot{\mathbf{q}})\dot{\mathbf{q}} + \mathbf{g}(\mathbf{q}) + \tau_{\text{ext}}. \tag{3.101}$$

$$\mathbf{u} = B_\theta\ddot{\theta} + \tau_J \tag{3.102}$$

In order to keep the discussion simpler $D_J K_J^{-1}\dot{\tau}_J$ is omitted. If $\mathbf{u}$ is chosen to be

$$\mathbf{u} := \mathbf{g}(\bar{\mathbf{q}}) + (I - K_v)\tau_{\text{ext}} \tag{3.103}$$

$$K_v = \{\text{diag}\{K_{v1},...,K_{vn}\}|\forall i : K_{vi} \geq 1\} \tag{3.104}$$

it can be shown that this results in a scaling of the robot dynamics by K_v^{-1}. If $\mathbf{g}(\mathbf{q}) \approx \mathbf{g}(\bar{\mathbf{q}})$ (3.102) is used together with (3.101) the relation

$$\begin{aligned} \mathbf{g}(\bar{\mathbf{q}}) + (I - K_v)\tau_{\text{ext}} &= B_\theta\ddot{\theta} + M(\mathbf{q})\ddot{\mathbf{q}} + C(\mathbf{q},\dot{\mathbf{q}})\dot{\mathbf{q}} + \mathbf{g}(\mathbf{q}) + \tau_{\text{ext}} \\ \mathbf{0} &= B_\theta\ddot{\theta} + M(\mathbf{q})\ddot{\mathbf{q}} + C(\mathbf{q},\dot{\mathbf{q}})\dot{\mathbf{q}} + K_v\tau_{\text{ext}} \end{aligned} \tag{3.105}$$

[10] Please note this method also works with a joint impedance controller.

is obtained, leading to

$$\underbrace{K_v^{-1}B_\theta}_{B_{\theta 1}}\ddot{\theta}+\underbrace{K_v^{-1}M(\mathbf{q})}_{M_\theta(\mathbf{q})}\ddot{\mathbf{q}}+K_v^{-1}C(\mathbf{q},\dot{\mathbf{q}})\dot{\mathbf{q}}+\tau_{\text{ext}}=\mathbf{0}. \tag{3.106}$$

The already shaped motor inertia B_θ is further reduced by K_v^{-1}. The inertia matrix $M(\mathbf{q})$ experiences the same scaling as well. Following relations hold component-wise.

$$B > B_\theta > B_{\theta 1} \tag{3.107}$$

$$M(\mathbf{q}) > M_\theta(\mathbf{q}), \tag{3.108}$$

$B_{\theta 1}$ and M_θ are the new motor and link inertia respectively. Finally, the new control law has to be derived. By substituting (3.103) into (3.36), τ_m is obtained:

$$\begin{aligned}\tau_m &= BB_\theta^{-1}(\mathbf{g}(\bar{\mathbf{q}})+(I-K_v)\tau_{\text{ext}})+(I-BB_\theta^{-1})(\tau_J+D_JK_J^{-1}\dot{\tau}_J) && (3.109)\\ &= K_T(\mathbf{g}(\bar{\mathbf{q}})-\tau_J)-K_S\dot{\tau}_J+K_v^*\tau_{\text{ext}}+\mathbf{g}(\bar{\mathbf{q}}) && (3.110)\\ &\approx K_T(\mathbf{g}(\bar{\mathbf{q}})-\tau_J)-K_S\dot{\tau}_J+K_v^*\hat{\mathbf{r}}+\mathbf{g}(\bar{\mathbf{q}}), && (3.111)\end{aligned}$$

where

$$\begin{aligned}K_T &:= BB_\theta^{-1}-I\\ K_S &:= (I-BB_\theta^{-1})D_JK_J^{-1}\\ K_v^* &:= BB_\theta^{-1}(I-K_v)\end{aligned} \tag{3.112}$$

By substituting (3.103) into (3.39) K_S becomes $K_S=(BB_\theta^{-1}D_s-D_J)K_J^{-1}$.

From now on, the different strategies are referred to as follows.

- Strategy 0: Do not react at all
- Strategy 1: Stop the robot
- Strategy 2: Switch to torque control with gravity compensation
- Strategy 3: Reflex torque-based strategy
- Strategy 4: Admittance-based strategy

In the presence of low friction, it may be necessary to limit the excursion of the robot reflex motions as previously described. In such cases a phase of maximum dissipation of kinetic energy is executed in order to rapidly stop the robot. The according control strategy is outlined in the following.

3.4.1.6 Energy Dissipating Strategy

This paragraph refers again to the rigid body case. Let the available motor torques at each joint be bounded by

$$|\tau_{m,i}| \le \tau_{m,\max,i} \qquad i=1,\ldots,n. \tag{3.113}$$

Part of this motor torque is spent for the gravity compensation. By defining the configuration-dependent bounds

$$\tau'_{m,i}(\mathrm{q}) := -(\tau_{m,\max,i} + g_i(\mathbf{q})) < 0 \tag{3.114}$$

$$\tau'_{M,i}(\mathrm{q}) := \tau_{m,\max,i} - g_i(\mathbf{q}) > 0, \tag{3.115}$$

the remaining part of the available torque $\tau' = \tau_m - g(\mathbf{q})$ satisfies

$$\tau'_{m,i}(\mathrm{q}) \leq \tau'_i \leq \tau'_{M,i}(\mathrm{q}) \qquad i = 1, \ldots, n. \tag{3.116}$$

Since the time evolution of the kinetic energy is $\dot{T} = \dot{\mathbf{q}}^T \tau'$, the following control law locally realizes the largest decrease of T:

$$\tau_{m,i} = \left\{ \begin{array}{cl} \tau'_{m,i}(\mathrm{q}) & \text{if } \dot{q}_i \geq \varepsilon_i \\ \tau'_{m,i}(\mathrm{q})\dot{q}_i/\varepsilon_i & \text{if } \varepsilon_i > \dot{q}_i \geq 0 \\ \tau'_{M,i}(\mathrm{q})\dot{q}_i/\varepsilon_i & \text{if } -\varepsilon_i < \dot{q}_i \leq 0 \\ \tau'_{M,i}(\mathrm{q}) & \text{if } \dot{q}_i \geq \varepsilon_i \dot{q}_i \leq -\varepsilon_i \end{array} \right\} + g_i(\mathbf{q}) \tag{3.117}$$

with $i = 1, \ldots, n$. For each velocity $\dot{q}_i$, the insertion of a small ultimate region of amplitude $2\varepsilon_i > 0$ allows a tradeoff between the almost minimum-time solution and a smooth reaching of the final condition $\dot{\mathbf{q}} = \mathbf{0}$.

Thus far only reaction schemes with entire task abortion were presented, i.e. every contact is classified as a collision. However, with the LWR-III it is often the case that physical unintentional interaction takes place. Therefore schemes are necessary for safe interaction without entirely losing the current task of the robot. For this purpose the method of trajectory scaling was developed. It is described in the following.

3.4.2 Trajectory Scaling

The idea of trajectory scaling is to preserve the original motion path and at the same time provide compliant behavior by influencing the time generator of the desired trajectory, see Fig. 3.8. This scheme can be used to enable a position (reference) controlled robot to react compliantly in such a way that it remains on the nominal path. Whereas in case of external disturbances the robot is only able to exert certain maximum forces. Note that this trajectory scaling scheme is driven by the observer output $\hat{\mathbf{r}}$ (for the joint case) or by $\mathscr{F}_{\text{ext}}$ for Cartesian motion generation and can also be combined with any of the previous reaction strategies so as to reduce external torques to zero in case of too dangerous situations.

A desired trajectory is usually parameterized with respect to time, i.e. $\mathbf{q}_d(t) \in \mathbb{R}^n$ in joint space or $\mathbf{x}_d(t) \in SE(3)$ in the Cartesian case, whereas the joint case is described from now on. For the discrete sampling time Δt used in the implementation the current time instant can be written as $t_i = t_{i-1} + \Delta t$. If the increment Δt is now modified in such a way that it is used to respond to external forces, it can be used

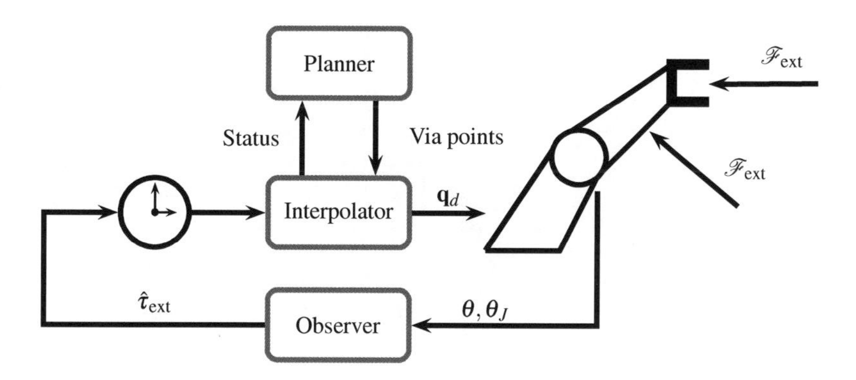

Fig. 3.8 Idea behind the trajectory scaling: "pushing interpolation time back and forth". The estimation of the external torque is fed to the time step generator of the interpolator and directly modulates its behavior.

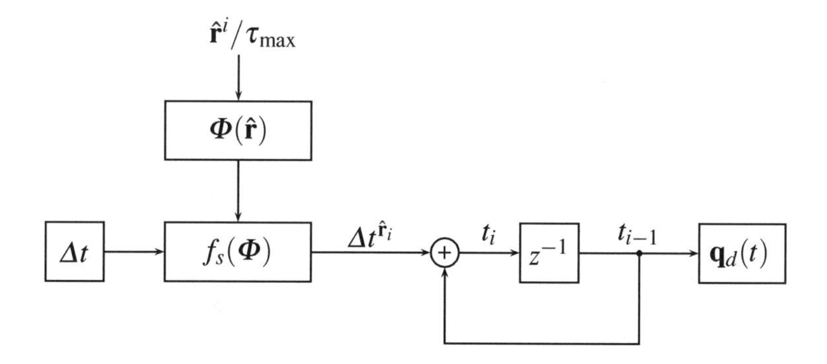

Fig. 3.9 Block diagram of the time generator in the trajectory scaling

to step back and forth along the desired joint path, as a matter of fact by "scaling the trajectory in time", see Fig. 3.9. This can simply be done by re-defining the path parameter as

$$t_i := t_{i-1} + f_s(\Psi(\hat{\mathbf{r}}_i))\Delta t. \tag{3.118}$$

The implementation of the trajectory scaling input based on the estimated external torque was chosen to be

$$\Psi(\hat{\mathbf{r}}_i) = \frac{1}{\alpha}\left(\frac{\hat{\mathbf{r}}_i}{\tau_{\max}} \cdot \frac{\Delta \mathbf{q}_d^i}{||\Delta \mathbf{q}_d^i||}\right)_+ \tag{3.119}$$

where $\Delta \mathbf{q}_d^i = \mathbf{q}_d^{i+1} - \mathbf{q}_d^i$ denotes the difference vector of two consecutive desired via points (e.g. provided by a path planner), $\tau_{\max} \in \mathbb{R}^n$ is the vector of the maximal

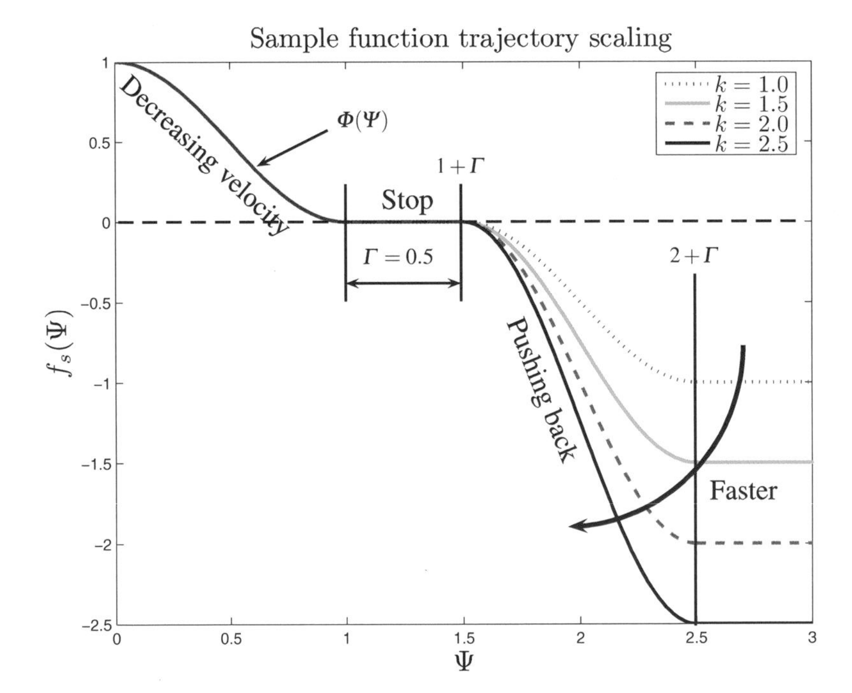

Fig. 3.10 Scaling function of the time increment

nominal joint torques specified for the robot[11], and "+" denotes the restriction of the term in brackets to positive values for each component. In this way, only external forces pushing **against** the natural evolution of the trajectory will have an effect on the behavior of the robot[12]. α is a value for adjusting the overall disturbance sensitivity by specifying the normalized collision torque along the trajectory for which the motion should stop, see Fig. 3.10. The function $f_s(\Psi)$ is given by

$$f_s(\Psi) = \begin{cases} \Phi(\Psi) & 0 \leq \Psi < 1 \\ 0 & 1 \leq \Psi \leq 1+\Gamma \\ k\Phi(\Psi - (1+\Gamma)) - k & 1+\Gamma < \Psi \leq 2+\Gamma \\ -k & 2+\Gamma < \Psi, \end{cases} \tag{3.120}$$

where $k \in \mathbb{R}^+$ is a positive factor that determines the decrement velocity. Γ is an optional dead-zone. Furthermore, $\Phi(.)$ is a monotonically decreasing function

$$\Phi : [0,1] \rightarrow [0,1]. \tag{3.121}$$

---

[11] Dividing $\hat{\mathbf{r}}_i$ by $\tau_{\max}$ weights external torques according to the specified maximum torque for each joint.

[12] However, one could use the signal as well to accelerate the robot if a human pushes it along its desired trajectory.

Depending on the disturbance input $\boldsymbol{\Psi}$ to slow down the robot until zero velocity, and after overcoming a dead-zone $\boldsymbol{\Gamma}$, the piecewise defined function $f_s(\boldsymbol{\Psi})$ enables pushing it back along its original path. A sample function for $f_s(\boldsymbol{\Psi})$ is given in Fig. 3.10. It shows two sinusoidal branches that define the slowing down and back-pushing velocity and an optional dead-zone. The monotonically decreasing function $\boldsymbol{\Phi}$ was implemented as

$$\boldsymbol{\Phi}(\boldsymbol{\Psi}) = \frac{1}{2}(1 + \cos(\pi\boldsymbol{\Psi})). \tag{3.122}$$

This function shows better performance than e.g. linear scaling because noise in the detection signal has reduced influence on the trajectory scaling in the absence of external torques ($\boldsymbol{\Psi} = 0$). A related approach, but for scaling of rhythmic movements was introduced in [33]. Further related work can be found in [22, 31].

For slow trajectories the approach presented up to now is well suited and realizes intuitive behavior. However, for very high desired joint velocities $\dot{\mathbf{q}}_d$ it is desired to make the approach independent of the desired velocities. This inherent dependency is introduced when scaling the interpolation time as described. Since the scaling depends only on the normalized joint torques, the effect of the same residual $\boldsymbol{\Psi}$ is different for each particular desired trajectory. In other words, scaling time implicitly scales the desired velocity, which is task depending. Therefore, it is not an optimal choice to use the same scaling shape for every desired trajectory. However, this drawback can be solved as follows.

$$\mathbf{q}_d(t_n) = \mathbf{q}_d(t_{n-1} + f_s(\boldsymbol{\Psi})\Delta t) \tag{3.123}$$

By using the "first order" Taylor series approximation

$$P_f(x) = f(a) + \frac{f'(a)}{1}(x - a), \tag{3.124}$$

$\mathbf{q}_d(t_n)$ can be written as

$$\mathbf{q}_d(t_n) \approx \mathbf{q}_d(t_{n-1}) + \dot{\mathbf{q}}_d(t_{n-1})(t_n - t_{n-1}) = \tag{3.125}$$

$$= \mathbf{q}_d(t_{n-1}) + \dot{\mathbf{q}}_d(t_{n-1})\underbrace{f_s(\boldsymbol{\Psi})\Delta t}_{\Delta t^*} \tag{3.126}$$

This can be reformulated to

$$\dot{\mathbf{q}}'_d(t_{n-1}) = \frac{\mathbf{q}_d(t_n) - \mathbf{q}_d(t_{n-1})}{\Delta t} = \dot{\mathbf{q}}_d(t_{n-1}) f_s(\boldsymbol{\Psi}). \tag{3.127}$$

Let $f_s(\boldsymbol{\Psi}, \alpha)$ be a monotonically decreasing function $\boldsymbol{\Phi} : [0, 2/\alpha] \to [-1, 1]$ with $(1/\alpha, 0)$ being its origin of symmetry. Therefore, f_s can be written as

$$f_s(\boldsymbol{\Psi}, \alpha) = 1 - \boldsymbol{\Phi}(\boldsymbol{\Psi}, \alpha), \tag{3.128}$$

leading to, together with (3.127):

$$\dot{\mathbf{q}}'_d(t_{n-1}) = \dot{\mathbf{q}}_d(t_{n-1}) - \dot{\mathbf{q}}_d(t_{n-1})\boldsymbol{\Phi}(\boldsymbol{\Psi},\alpha). \tag{3.129}$$

Substituting $t_{n-1} \to t_n$ this leads to

$$\dot{\mathbf{q}}'_d(t_n) = \dot{\mathbf{q}}_d(t_n) - \dot{\mathbf{q}}_d(t_n)\boldsymbol{\Phi}(\boldsymbol{\Psi},\alpha) \tag{3.130}$$

Next, $\boldsymbol{\Phi}(\boldsymbol{\Psi},\alpha)$ needs to be chosen such that α depending on the desired scaling function $\frac{1}{\dot{\mathbf{q}}_d(t_n)}\boldsymbol{\Phi}(\boldsymbol{\Psi},\alpha=1)$ is obtained. This selection makes (3.130) independent from $\dot{\mathbf{q}}_d(t_n)$

$$\boldsymbol{\Phi}(\boldsymbol{\Psi},\alpha) \stackrel{!}{=} \frac{1}{\dot{\mathbf{q}}_d(t_n)}\boldsymbol{\Phi}(\boldsymbol{\Psi},\alpha=1) \tag{3.131}$$

$$\to \alpha = g(\boldsymbol{\Psi},\dot{\mathbf{q}}_d(t_n)) \tag{3.132}$$

For the linear case the function $\boldsymbol{\Phi}$ can be defined as

$$\boldsymbol{\Phi}(\boldsymbol{\Psi},\alpha) = \alpha K \boldsymbol{\Psi} \quad K \in \mathbb{R}. \tag{3.133}$$

Together with

$$\alpha(\dot{\mathbf{q}}_d) := \frac{1}{\dot{\mathbf{q}}_d(t_n)} \tag{3.134}$$

this leads to

$$\dot{\mathbf{q}}'_d(t_n) = \dot{\mathbf{q}}_d(t_n) - K\boldsymbol{\Psi}, \tag{3.135}$$

which no longer depends on the desired velocity.

Combining Trajectory Scaling, Collision Detection and Reaction

Typically, undesired impacts are characterized by high peak forces and joint torques. Therefore, a basic way to distinguish between desired interaction and accidental collisions is to use the magnitude of $\hat{\mathbf{r}}$ (or any other estimation $\hat{\tau}_{\text{ext}}$ of the external torques). Trajectory scaling ensures that during normal operation mode only a certain maximum static force (depending on f_s) can act on a human. If he/she pushes harder, the robot moves back along $\mathbf{q}_d$ and as soon as the pushing force is too high ($||\hat{\mathbf{r}}|| \geq r_{\max}^{\text{switch}} \in \mathbb{R}^+$), the robot switches to one of the other reaction strategies (e.g. in case of Strategy 2 the robot poses due to its compliance no threat anymore), see Fig. 3.11. Thus, a combination of reaction strategies performs an intuitive and effective response to desired physical interaction or unintended collision/clamping. In Figure 3.12 this combined reactive strategy is depicted.

In the next section the benefits gained from the use of the collision detection and reaction algorithms are shown by evaluating impact tests with the LWR-III on different human body parts, a dummy, and a test-bed designed for this purpose.

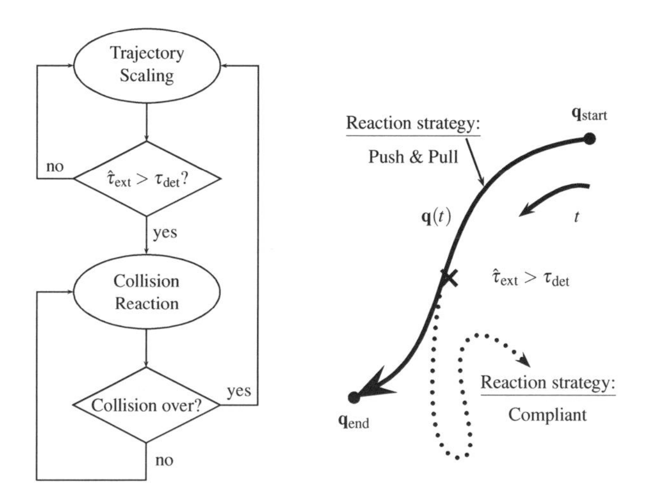

Fig. 3.11 Combining trajectory scaling and other reaction strategies based on the magnitude of the disturbance signal. As long as the torque estimation remains within a certain limit band, trajectory scaling is active. In case this threshold is exceeded the robot switches to one of the other reaction schemes.

Fig. 3.12 Combining safety during the execution of a task and during a real collision: (a.) The robot moves position controlled along its desired trajectory. (b.) The robot slows down (trajectory scaling) and in the end stops after physical contact with the human. If the human would step aside the robot would continue to move along its desired trajectory. (c.) The human pushes harder against the robot and consequently the collision detection is triggered. (d.) The robot compliantly floats away in torque control with gravitation compensation (Strategy 2). (e.) Now, the robot can easily be moved around without being able to cause any harm.

3.5 Experiments

3.5.1 Energy-Based Collision Detection

Figure 3.13 visualizes the behavior of the energy-based collision detection signal defined in Sec. 3.3.1 during a straight line Cartesian motion of the LWR-III in the $(x,y)-$plane. The experiment showcases all aspects outlined in Sec. 3.3.1. The plot depicts the desired and real robot motion (position and velocity), the external contact force, the residual (3.55), and the sensitivity measure $s = \mathbf{f}_{\text{ext}}\dot{\mathbf{x}}$. The indicated phases $A - F$ are characterized by varying respective behavior as follows.

- Phase A : The external force is acting against the motion direction while the robot moves and is well recognized by the detection scheme.
- Phase B : No external forces are applied.

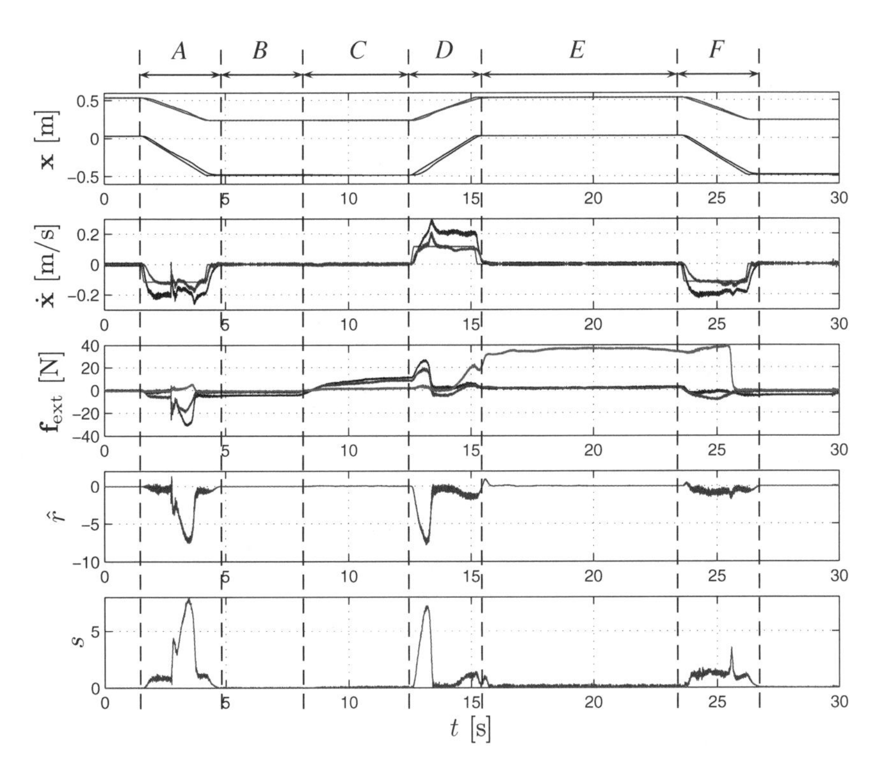

Fig. 3.13 Experimental behavior of the energy-based detection scheme. The Cartesian impedance controlled robot moves on a straight line (first plot, indicated by $\mathbf{x}$). The desired and measured velocity are denoted in the second plot. The external disturbance force $\mathbf{f}_{\text{ext}}$, the residual obtained from (3.55), and the sensitivity of the scheme represented by $s = \mathbf{f}_{\text{ext}}\dot{\mathbf{x}}$ are visualized hereafter. x,y,z are denoted as blue, black, and red.

- Phase C : The external forces in (x,y) plane are gradually increasing during standstill. They are not recognized.
- Phase D : The robot starts moving again, which immediately leads to an increase in sensitivity and consequently residual magnitude.
- Phase E : The robot stands still, while a strong force in z-direction is applied without being recognized.
- Phase F : The robot moves again, while the z-force is still applied. However, as it acts orthogonal to the motion vector, it cannot be recognized.

3.5.2 Balloon Test

In order to show the effect of the developed collision detection from Sec. 3.3.5 and the respective reaction algorithms, initial collision tests with the LWR-III and a balloon were conducted. In these experiments the balloon is fixed on a table. The setup and motion of the robot are shown in Fig. 3.14. The tests were performed using a trapezoidal joint velocity profile with cruise speeds between 10 ^{o}/s and 100 ^{o}/s, as reference trajectory. Start and final configuration are

$$\mathbf{q}_0 = [60\,31\ -78\,23\,158\ -15\ -15]^T\,[^o]$$
$$\mathbf{q}_1 = [60\,65\ -78\,53\,158\ -15\ -15]^T\,[^o],$$

respectively (motion is limited to joints 2 and 4). The robot hits the constrained balloon with its spherical wrist while coming from above, see Fig. 3.14. The detection gain matrix is $K_O = 25\ I$, while the reflex reaction gains are $K_v = 1.4\ I$ (for Strategy 3) or $K_R = 0.05\ I$ (for Strategy 4). The component-wise detection thresholds $r_{\mathrm{low,j}}, j = 1,\ldots,7$ are $0.01\ \tau_{J,\max}$.

Fig. 3.14 Balloon tests with strategy 0 (upper row) and strategy 3 (lower row)

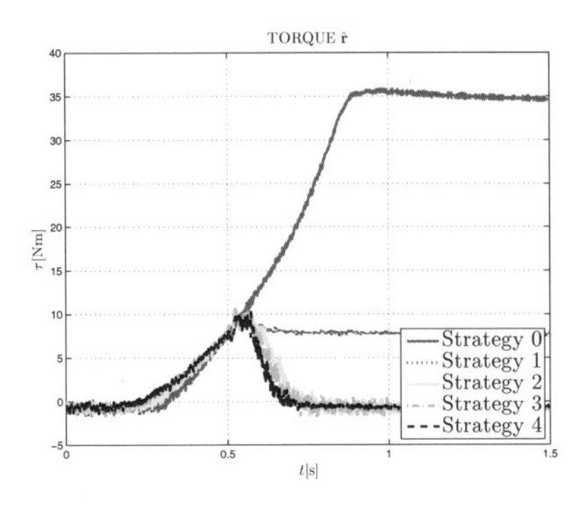

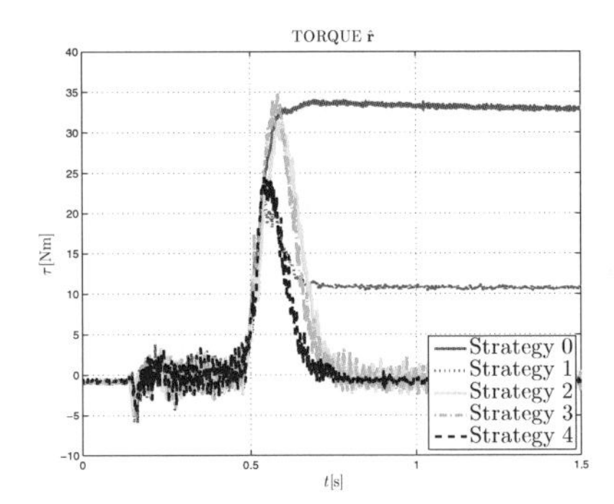

Fig. 3.15 Estimated external torque $\hat{r}_4$

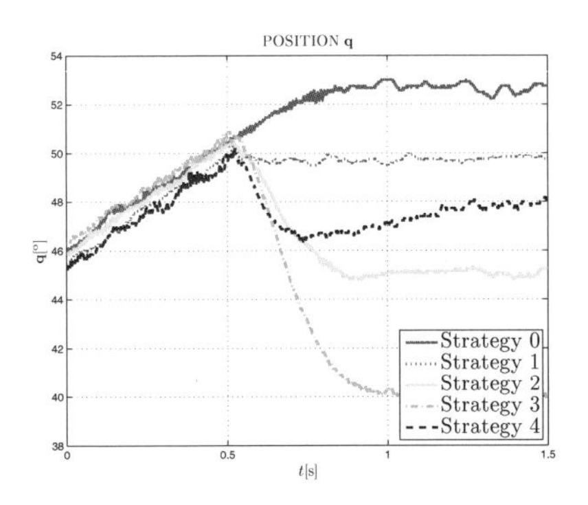

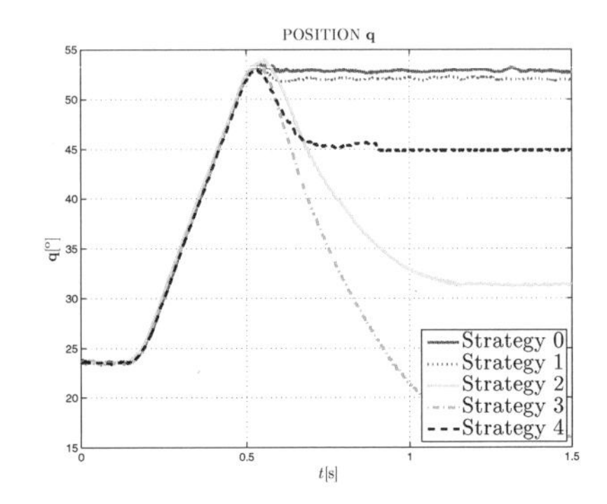

Fig. 3.16 Link Position q_4

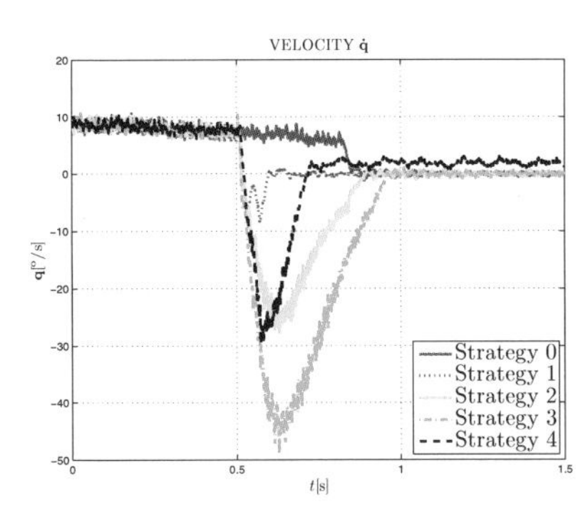

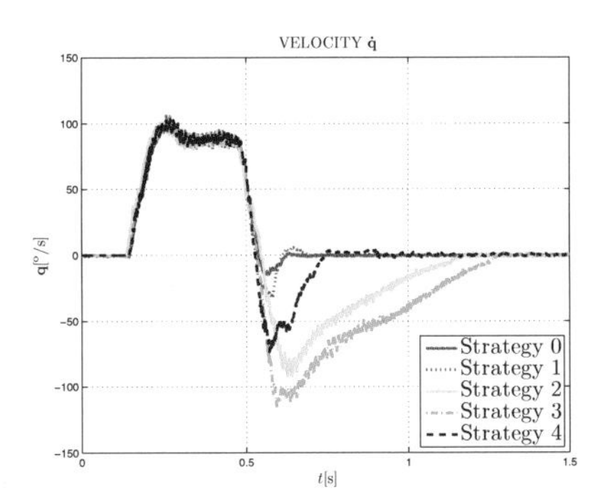

Fig. 3.17 Link Velocity $\dot{q}_4$. The left column shows the experimental results for 10 ^{o}/s, the right one for 100 ^{o}/s.

In Figure 3.14 the resulting motion for strategy 0 (upper row) and strategy 3 (lower row) are given. In Figure 3.15-3.17 the measurements of the experiments are shown for axis 4 at two different impact velocities. On the left side the results for $\dot{q}_4 = 10\,^o/\mathrm{s}$ and on the right side for $\dot{q} = 100\,^o/\mathrm{s}$ are shown.

With strategy 0 the robot moves further along the path commanded by the trajectory generator and therefore the link position increases. Since the robot simply follows its desired trajectory the external torque continuously grows. As soon as the desired trajectory timely ends due to the desired configuration being reached, it remains at 35 Nm. In contrast, all collision reaction strategies stop this growth (strategy 1) or change the direction of motion in order to reduce $\hat{\mathbf{r}}$ and then stop after a while due to friction (2 or 3) or due to the absence of external torques ($\hat{\mathbf{r}} = \mathbf{0}$). With strategy 3 and 4 $\dot{\mathbf{q}}$ changes its sign quickly, thus driving away from the source of external forces. For strategy 3 $\hat{\mathbf{r}}$ decays to zero. Strategy 4 shows the quickest reaction. By comparing strategy 2 and 3, the active part incorporating the estimation of external joint torques shows its influence and makes strategy 3 faster than strategy 2. The link position $\mathbf{q}$ supports this statement. For $100\,^o/\mathrm{s}$ impact velocity (Fig. 3.15-3.17 (right)) similar observations can be made. The external torque rapidly increases after initial contact for all strategies. After the collision is detected, the first and fourth strategy reverse $\dot{\mathbf{q}}$ the quickest. The torque control-based approaches do not decelerate that fast. However, the third scheme outperforms the second one, since it lowers the external torque significantly faster and drives quicker out of the collision area.

These results were expected from tests where the robot drives against an outstretched human arm of different persons. All subjects stopped the robot while it was driving along a desired trajectory. Every subject described to have the feeling of high safety awareness due to the collision detection and reaction.

In the next subsection the first results are discussed on the quantifiable effect of the collision detection based on a collision test bed that represents a simplified model of the human arm.

3.5.3 Human Arm Measurements and Collision Test-Bed

In order to objectively compare collision reaction strategies, a simple collision test-bed was built to emulate robot-human arm impacts. This is a 1-DoF mechanism with adjustable impedance, of which a spring stiffness and a mass can be adapted to fit with impact characteristics of interest, see Fig. 3.18 (left). The impact behavior of the human arm is mimicked in a typical impact configuration, shown in Fig. 3.20 (right), and used as a basis for comparing the presented reaction strategies. The author is aware of the problems in fitting a certain model to a human arm that is potentially nonlinear and of higher order. Furthermore, one could argue that the human reacts with an impedance response to the impact. However, at this point only the rough behavior of the human arm is intended to be replicated for a specific situation. This gives a common ground for comparing impact reaction strategies on

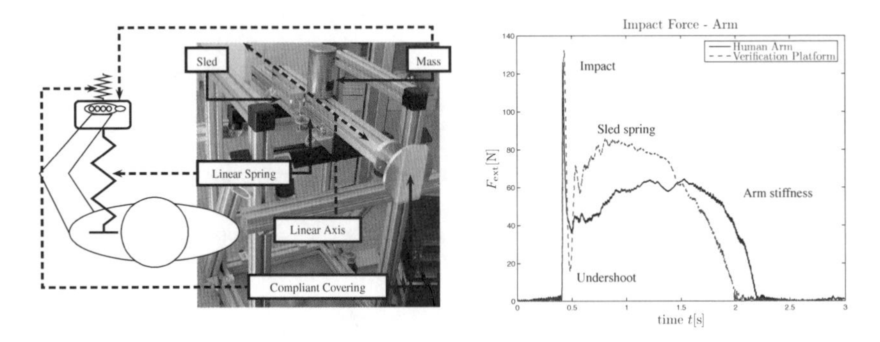

Fig. 3.18 Collision test-bed (left), representing a simplified model of the human arm (left), and the resulting impact forces for a human arm and the collision test-bed at a robot velocity of 0.4 m/s (right). After an initial peak a stiffness profile is observed, representing the human response in this particular experiment.

a fair basis. It was not intended to construct an anthropomorphic model of the human arm.

The force occurring during a typical arm impact is shown in Fig. 3.18 (right) for the reconfiguration trajectory from "elbow up" to "elbow down". The robot was used to measure contact forces[13], kinematic configurations, and velocities. In comparing the test-bed with the human, some differences in the impact characteristics can be observed. Particularly, when the damping in the collision test-bed is considerably lower, an undershoot after the consistent impact force is shown. To partially overcome this deficit, the sled spring was pretensioned, leading to a biased spring and thus to higher forces during the bending process.

3.5.4 Performance Comparison of Reaction Strategies

The results for impacting the LWR-III against the collision test-bed with various reaction schemes are shown in Fig. 3.19. From the instant of impact on, the contact force (upper) and the Cartesian displacement (lower) are shown. Furthermore, in the lower plot the collision detection signal is also reported, indicating how fast the robot actually reacts as soon as a collision is detected. Here, trajectory scaling was not evaluated on purpose, since it is intended to serve as a feature during task execution to allow interaction. This is not as a collision reaction scheme, which shall only be activated during high load impacts. Strategy 1 and 4 show very fast reaction after the first force peak and then lose contact with the accelerated sled. Due to the backlash of the sled a second impact occurs in both cases. Strategy 4 seems to be the fastest to withdraw from the external force in the first 200 ms. However, it

[13] In all remaining experiments in this chapter the contact force was measured with a JR3 force/torque sensor.

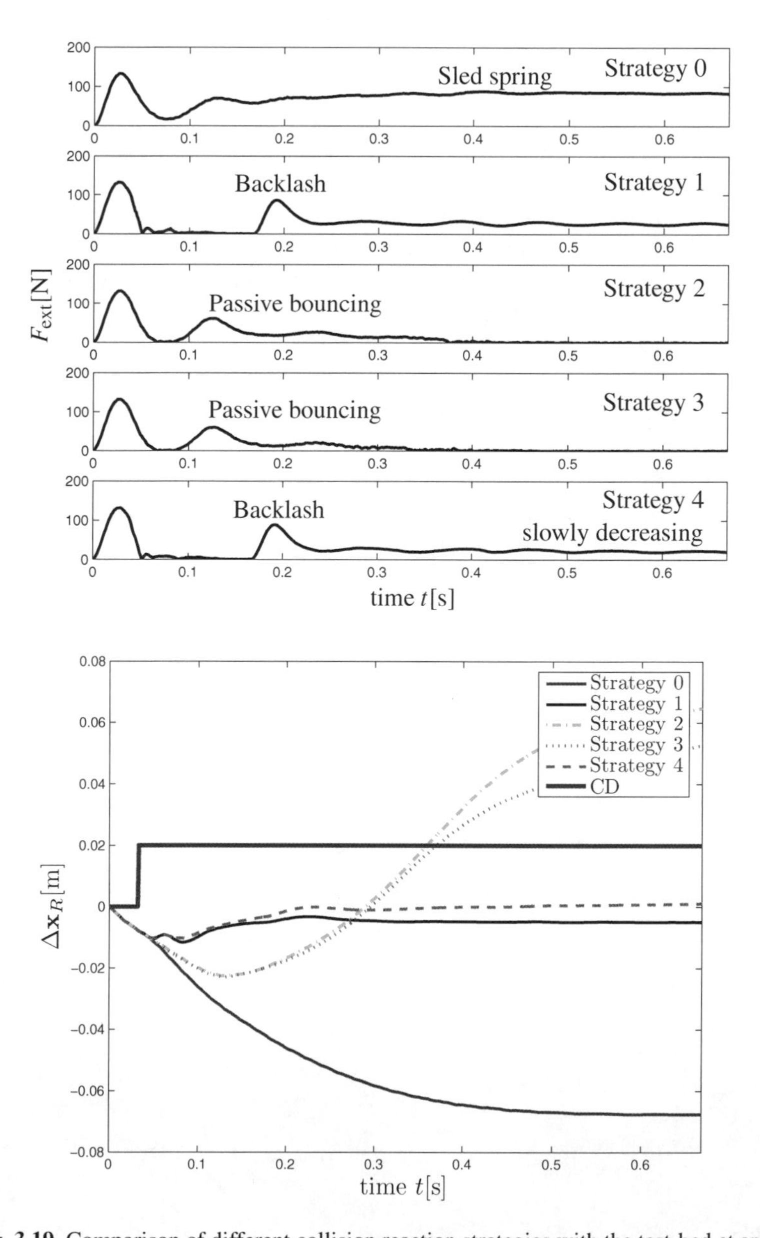

Fig. 3.19 Comparison of different collision reaction strategies with the test-bed at an impact velocity of 0.4 m/s. The point of origin with $t = 0$ indicates the instant of impact. Apparently, the maximum initial peak force, which is passed after less than 25 ms, cannot be reduced for the impact with the test-bed. Although for Strategy 2 and 3 no backlash can be observed, a second "impact" occurs. This is a further bending of the sled spring due to the passive behavior of the robot in these control modes, i.e. similar to Strategy 0, but due to the compliant behavior, in a very alleviated from.

could not be tuned such that the slowly decreasing contact force after the backlash is eliminated. This drawback is probably caused by the time delay in the admittance control loop and the higher Coulomb friction of the robot compared to the one used in Sec. 3.5.2. However, the maximum displacement for both strategies is $\approx 10-12$ mm, showing a much faster reduction than Strategies 2 and 3. Additionally, the influence of the test-bed spring is entirely canceled.

In general, Strategies 2 and 3 show similar behavior, leading to the conclusion that the additional inertia shaping (Strategy 3) does not significantly contribute to an improvement in reaction behavior. Apart from that, these two strategies do not lose contact as abruptly as Strategies 1 and 4 do, but the contact force reduces after < 400 ms to zero due to the compliant behavior. These observations lead to the recommendation to combine the speed of Strategy 4 to avoid the higher displacement and entire influence of the sled spring, with the convenient compliant behavior of Strategy 2 by subsequently switching to this mode.

3.5.5 Collisions with the Human Arm and Chest

In order to show the effectiveness of the collision detection mechanisms, real impact tests were conducted with a non-clamped human chest and an outstretched arm. The human stood relaxed and was not able to see the robot coming. In Figure 3.20 the impact positions are shown. The position for the human arm was chosen such that it is in a comfortable configuration and not pre-stressed. The robot impact velocity ranged up to 2.7 m/s. Since for these tests a difference in the contact forces for the compliant reaction strategies is not measurable due to the large variation caused by the human[14], the focus lies on Strategies $0, 1, 2$. For the chest impacts the

Fig. 3.20 Real collisions with the chest and arm were conducted up to a robot velocity of 2.7 m/s

[14] This is one important reason why the collision test-bed was built.

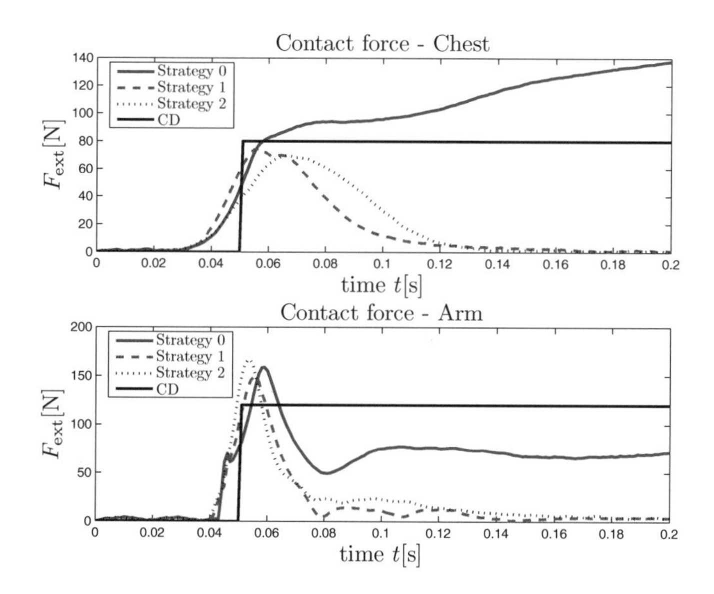

Fig. 3.21 Resulting contact force with and without collision detection and reaction strategy for the human chest (upper) at impact velocities of 0.7 m/s and the arm (lower) at 1.5 m/s. These tests were carried out up to an impact velocity of 2.7 m/s but at such impact velocities it is hard to reproduce testing conditions accurately enough.

detection activates within $\approx$ 14 ms, bringing the contact force down to zero within $<$ 100 ms and limiting it below $\approx$ 75 N for both the active strategies that were evaluated, see Fig. 3.21 (upper). For Strategy 1 the human is accelerated fast enough due to the impact force and thus loses contact in case the robot abruptly stops. Generally, even without collision detection and reaction, the impact forces can be kept far below the tolerance force $F_{\text{ext}}^{x,\text{tol}} \in [1.15, 1.7]$ kN of the chest [27]. Furthermore, the collision reaction limits the contact forces far below the proposed value of 150 N in *ISO-10218* [18] which would be exceeded for chest impacts with Strategy 0. The contact force for the arm is illustrated in Fig. 3.21 (lower), showing a somewhat different behavior. After a short impact, which cannot be prevented or attenuated by the collision detection and reaction, the impact forces reduce to zero for Strategy 1 and 2. For Strategy 0, another safety feature of the LWR-III activates because of the increasing contact force. In fact, a low-level stop is triggered by the exceeding of the measured joint torque. For the human arm very limited biomechanical tolerance data is available. At this point, it becomes apparent that the 150 N proposed by *ISO-10218* are from too conservative for blunt arm impacts due to the fact that a 50% risk of elbow fracture corresponds to forces as large as 1780 N [12].

3.5.6 Trajectory Scaling

Experimental results for the trajectory scaling are depicted in Fig. 3.22. A reference trajectory q_d^1 for the first joint (solid line) is given for nominal free motion. It is a 5th order polynomial from $q^1_{d,\text{start}} = -23°$ to $q^1_{d,\text{end}} = 22°$. During the execution of this trajectory the human pushes against the robot and the resulting scaled desired position q_d^1 (dashed line) shows the slowing down and back-pushing along the trajectory depending on the disturbance input $\boldsymbol{\Psi}(\hat{\mathbf{r}})$ (dashed-dotted line).

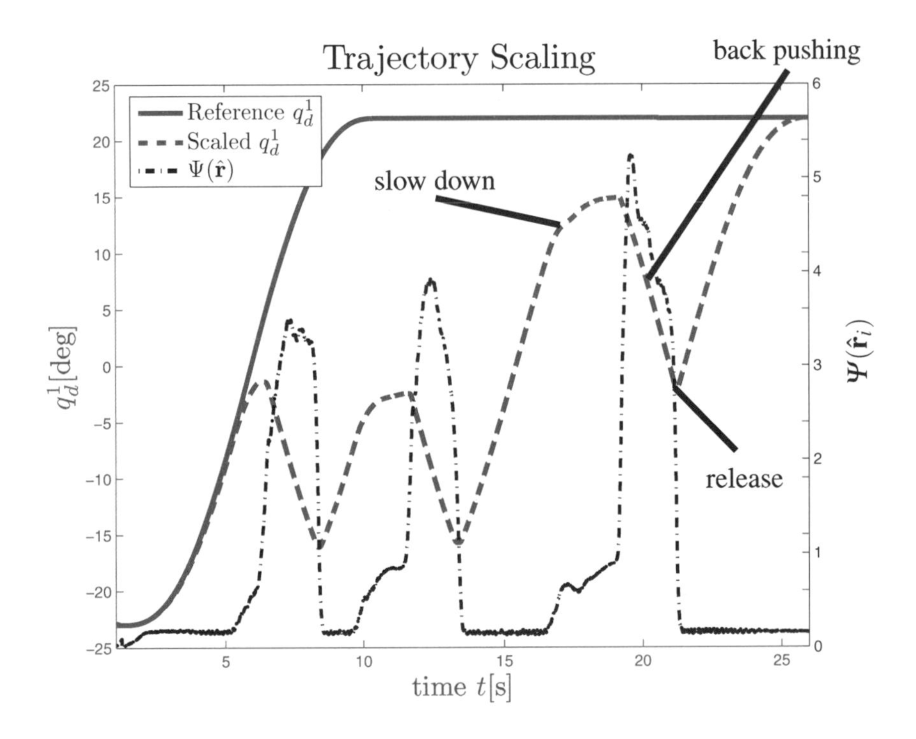

Fig. 3.22 Measured trajectory scaling for a sample trajectory implemented on the LWR-III. The left $y-$axis is relevant for the reference and scaled joint angle q_d^1.

Trajectory scaling is intended for continuous physical interaction without switching the control mode. It is used when the robot is position or impedance controlled, leading to a convenient way to interact with the robot without forcing a global change of its behavior. At the same time, the user has the possibility to almost instantaneously stop the robot by pushing against it.

A major advantage of trajectory scaling is that for a complex robot, such as the DLR dual-arm humanoid *Justin* [6], only one of the sub-robots it consists of has to

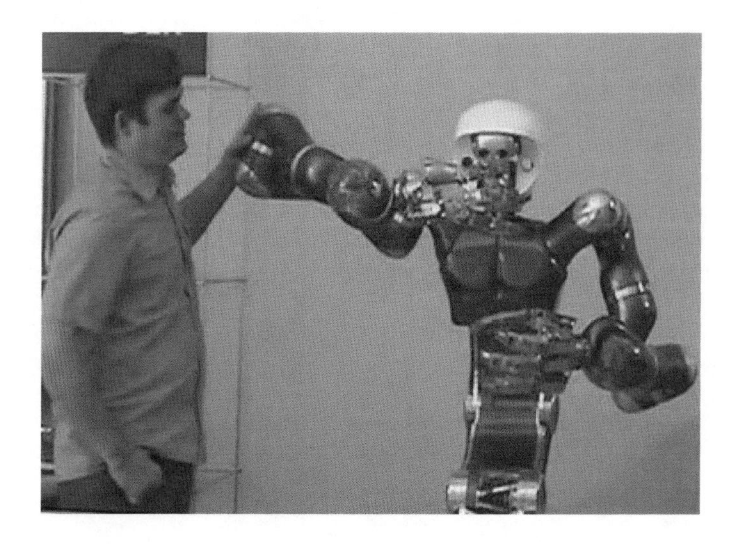

Fig. 3.23 Physical interaction with the DLR humanoid *Justin*

slow down. For instance, pushing against one of the elbows as shown in Fig. 3.23 slows down and finally stops/reverts both arms, both hands, the torso, and the neck and not only the touched arm[15].

3.6 Summary

In this chapter, a complete approach was presented, from detection to reaction, for handling human-robot collisions without the need of external sensing. Collision detection and identification signals can be efficiently generated e.g. by resorting to energy arguments, or based on the robot generalized momentum by using only proprioceptive measurements. After collision has been detected, a reactive control strategy e.g. reduces the effective inertia seen by the Cartesian contact forces. The robot retracts itself safely and rapidly away from the collision area, using the local directional information collected during the impact. The developed methodology covers both the case of rigid manipulators and of robots with elastic joints.

Furthermore, the method of trajectory scaling was introduced, which enables the user to push the robot back and forth along a desired trajectory. This gives a convenient modality to interact with the robot during task execution without forcing the abortion of its current task. In combination with the collision detection and reaction methods it is used to establish a multi-level contact/collision reaction architecture.

[15] At this point one common time basis for all parts of the robot is assumed.

Using the LWR-III, which was especially designed for interactive and cooperative tasks, it was shown how the reactive control strategies can significantly contribute to ensuring safety to the human during physical interaction. Several collision tests were carried out, illustrating the feasibility and effectiveness of the proposed approach. While a subjective "safe" feeling is experienced by users when being able to naturally stop the robot in autonomous motion, a quantitative analysis of different reaction strategies was lacking. In order to compare these strategies on an objective basis, a mechanical verification platform has been built that mimics some impact behavior of the human arm. The proposed collision detection and reactions methods prove to work reliably and are effective in reducing contact forces far below any level which is dangerous to humans. Furthermore, evaluations of impacts between robot and human arm or chest up to a maximum robot velocity of 2.7 m/s are presented.

References

[1] Albu-Schäffer, A.: Regelung von Robotern mit elastischen Gelenken am Beispiel der DLR-Leichtbauarme. Ph.D. thesis, Technical University of Munich (2002) (German)

[2] Albu-Schäffer, A., Hirzinger, G.: A globally stable state-feedback controller for flexible joint robots. J. of Advanced Robots 15(8), 799–814 (2001)

[3] Albu-Schäffer, A., Hirzinger, G.: Cartesian impedance control techniques for torque controlled light-weight robots. In: IEEE Int. Conf. on Robotics and Automation (ICRA 2002), Washington, DC, USA, pp. 657–663 (2002)

[4] Albu-Schäffer, A., Ott, C., Hirzinger, G.: A unified passivity based control framework for position, torque and impedance control of flexible joint robots. In: International Symposium on Robotics Research (ISRR 2005), San Francisco, USA, pp. 5–21 (2005)

[5] Albu-Schäffer, A., Ott, C., Hirzinger, G.: A unified passivity-based control framework for position, torque and impedance control of flexible joint robots. The Int. J. of Robotics Research 26, 23–39 (2007)

[6] Borst, C., Ott, C., Wimböck, T., Brunner, B., Zacharias, F., Bäuml, B., Hillenbrand, U., Haddadin, S., Albu-Schäffer, A., Hirzinger, G.: A humanoid upper body system for two-handed manipulation. In: VIDEO, IEEE Int. Conf. on Robotics and Automation (ICRA 2007), Rome, Italy, pp. 2766–2767 (2007)

[7] De Luca, A., Albu-Schäffer, A., Haddadin, S., Hirzinger, G.: Collision detection and safe reaction with the DLR-III lightweight manipulator arm. In: IEEE/RSJ Int. Conf. on Intelligent Robots and Systems (IROS 2006), Beijing, China, pp. 1623–1630 (2006)

[8] De Luca, A., Book, W.: Robots with flexible elements. In: Springer Handbook of Robotics, pp. 287–319 (2008)

[9] De Luca, A., Mattone, R.: Actuator fault detection and isolation using generalized momenta. In: IEEE Int. Conf. on Robotics and Automation (ICRA 2003), Taipei, Taiwan, pp. 634–639 (2003)

[10] De Luca, A., Mattone, R.: An adapt-and-detect actuator FDI scheme for robot manipulators. In: IEEE Int. Conf. on Robotics and Automation (ICRA 2004), New Orleans, USA, pp. 4975–4980 (2004)

[11] De Luca, A., Mattone, R.: Sensorless robot collision detection and hybrid force/motion control. In: IEEE Int. Conf. on Robotics and Automation (ICRA 2005), Barcelona, Spain, pp. 1011–1016 (2005)
[12] Duma, S., Boggess, B., Crandall, J., MacMahon, C.: Fracture tolerance of the small female elbow joint in compression: the effect of load angle relative to the long axis of the forearm. Stapp Crash Journal 46, 195–210 (2002)
[13] Ebert, D., Henrich, D.: Safe human-robot-cooperation: Image-based collision detection for industrial robots. In: IEEE/RSJ Int. Conf. on Intelligent Robots and Systems (IROS 2002), Lausanne, Switzerland, pp. 239–244 (2002)
[14] Foellinger, O.: Regelungstechnik. Huethig Buch Verlag (1992)(German)
[15] Haddadin, S.: Evaluation criteria and control structures for safe human-robot interaction. Master's thesis, Technical University of Munich (TUM) & German Aerospace Center, DLR (2005)
[16] Haddadin, S., Albu-Schäffer, A., Luca, A.D., Hirzinger, G.: Collision detection & reaction: A contribution to safe physical human-robot interaction. In: IEEE/RSJ Int. Conf. on Intelligent Robots and Systems (IROS 2008), Nice, France, pp. 3356–3363 (2008)
[17] Hogan, N.: Impedance control: An approach to manipulation: Part I - theory, Part II - implementation, Part III - applications. Journal of Dynamic Systems, Measurement and Control 107, 1–24 (1985)
[18] ISO10218: Robots for industrial environments - Safety requirements - Part 1: Robot (2006)
[19] Kosuge, K., Matsumoto, T., Morinaga, S.: Collision detection system for manipulator based on adaptive control scheme. Transactions of the Society of Instrument and Control Engineers 4(39), 552–558 (2003)
[20] Kuntze, H.B., Frey, C., Giesen, K., Milighetti, G.: Fault tolerant supervisory control of human interactive robots. In: IFAC Workshop on Advanced Control and Diagnosis, pp. 55–60 (2003)
[21] Lewis, F.L., Dawson, D.M., Abdallah, C.T.: Robot Manipulator Control Theory and Praxis. Marcel Dekker, New York (2004)
[22] Li, P., Horowitz, R.: Passive velocity field control of mechanical manipulators. In: IEEE Int. Conf. on Robotics and Automation (ICRA 1995), Nagoya/Aichi,Japan, pp. 2764–2770 (1995)
[23] Lumelsky, V., Cheung, E.: Real-time collision avoidance in teleoperated whole-sensitive robotarm manipulators. IEEE Transactions on Systems, Man and Cybernetics 23, 194–203 (1993)
[24] Morinaga, S., Kosuge, K.: Collision cetection system for manipulator based on adaptive impedance control law. In: IEEE Int. Conf. on Robotics and Automation (ICRA 2002), Washington DC, USA, pp. 1080–1085 (2003)
[25] Murray, R., Li, Z., Sastry, S.: A Mathematical Introduction to Robotic Manipulation, 1st edn. CRC (1994)
[26] Ott, C., Albu-Schäffer, A., Hirzinger, G.: A passivity based cartesian impedance controller for flexible joint robots - Part I: Torque feedback and gravity compensation. In: Int. Conf. on Robotics and Automation (ICRA 2004), New Orleans, USA, pp. 2659–2665 (2004)
[27] Patrick, L.: Impact force deflection of the human thorax. In: SAE Paper No.811014, Proc. 25th Stapp Car Crash Conference, pp. 471–496 (1981)
[28] Spong, M.: Modeling and control of elastic joint robots. IEEE Journal of Robotics and Automation, 291–300 (1987)

[29] Suita, K., Yamada, Y., Tsuchida, N., Imai, K., Ikeda, H., Sugimoto, N.: A failure-to-safety "kyozon" system with simple contact detection and stop capabilities for safe human - autonomous robot coexistence. In: IEEE Int. Conf. on Robotics and Automation (ICRA 1995), Nagoya/Aichi, Japan, pp. 3089–3096 (1995)
[30] Takakura, S., Murakami, T., Ohnishi, K.: An approach to collision detection and recovery motion in industrial robot. In: Annual Conference of IEEE Industrial Electronics Society (IECON 1989), Philadelphia, USA, pp. 421–426 (1989)
[31] Tarn, T.J., Xi, N., Bejczy, A.: Path-based approach to integrated planning and control for robotic systems. Automatica 32(12), 1675–1687 (1996)
[32] Tomei, P.: A simple pd controller for robots with elastic joints. IEEE Transactions on Automatic Control 36(10), 1208–1213 (1991)
[33] Urbanek, H., Albu-Schäffer, A., van der Smagt, P.: Learning from demonstration: repetitive movements for autonomous service robotics. In: IEEE/RSJ Int. Conf. on Intelligent Robots and Systems (IROS 2004), Sendai, Japan, pp. 3495–3500 (2004)
[34] Yamada, Y., Hirasawa, Y., Huang, S., Umetani, Y., Suita, K.: Human-robot contact in the safeguarding space. IEEE/ASME Transactions on Mechatronics 2(4), 230–236 (1997)
[35] Zollo, L., Siciliano, B., De Luca, A., Guglielmelli, E., Dario, P.: Compliance control for an anthropomorphic robot with elastic joints: Theory and experiments. ASME Journal of Dynamic Systems, Measurements and Control (127), 321–328 (2005)

Chapter 4
Biomechanics and Forensics

In this monograph, various injury measures from biomechanics and forensics are used for analyzing human injury in robotics. In order to give the full picture, an overview on the most important existing injury classification metrics and biomechanical injury measures is given in this chapter.

As suggested by the New FMVSS[1], safety-measured regions can be divided into following complexes:

1. head
2. neck
3. chest
4. lower extremities

In the following, numerous injury indicators and measures for the head, neck, and chest are described. In this chapter, lower extremities are excluded due to their reduced relevance for robotics at the present stage. Injury mechanisms of upper extremities are still investigated in current research and since some according literature was already given in Chapter 11, this is also omitted for brevity. As Chapter 6 reflects the state of the art in sharp and acute soft-tissue analysis exhaustively, this part will only be completed by presenting the automotive approach of investigating lacerations.

This chapter is organized as follows. Section 4.1 describes how injury is commonly classified. Then, Sec. 4.2, Sec. 4.3, Sec. 4.4, and Sec. 4.5 present injury indices for the human head, neck, chest, and eye. Apart from giving some background information on the underlying biomechanics used in this monograph, this overview on severity indices is also intended as an outlook on which other indicators will be analyzed in the near future.

4.1 Classifying Injury Severity

A common approach for obtaining severity indices (injury measures) intends to reduce the required measurement into applying a defined input and quantifying its

[1] Federal Motor Vehicle Safety Standard

S. Haddadin: *Towards Safe Robots*,
Springer Tracts in Advanced Robotics 90,
DOI: 10.1007/978-3-642-40308-8_4, © Springer-Verlag Berlin Heidelberg 2014

reaction directly on the struck human body part. After analyzing the outcome, a threshold (or a full scaling if possible) is to be defined that guarantees non-severe consequences on the human body. Before introducing the definition of common severity indices, an intuitive and internationally established generic definition of injury level is described first.

4.1.1 The Abbreviated Injury Scale

A definition of injury level developed by the AAAM[2] and the AMA[3] is the Abbreviated Injury Scale [3]. It subdivides the observed level of injury into seven categories from *none* to *fatal* and provides an intuitive classification, see Tab.4.1. This classification gives no hint as to *how* to measure possible injury. This is provided by so called severity indices. Table 4.1 gives example injuries for the head, thorax, cervical spine, and extremities as described in [8]. An important fact to notice is that the scaling between the levels of AIS is nonlinear. This implies that injury of level AIS 3 is far from being half as life threatening as AIS 6.

Table 4.1 Definition of the Abbreviated Injury Scale

AIS	Severity	Type of injury	Head	Cervical spine	Thorax	Extremities	Lethality rate [%]
0	None	None	-	-	-	-	0,00
1	Minor	Superficial Injury	cranial contusion	distorsion	contusion	skin abrasion	0,00
2	Moderate	Recoverable	mild concussion	dorsal process fracture	simple rib fracture	lower arm fracture	0,07
3	Serious	Possibly recoverable	basal skull fracture	vertebral body fracture	multiple rib fracture	compound fracture shinbone	2,91
4	Severe	Not fully recoverable without care	mild cerebral hemorrhage	incomplete paraplegia	lung rupture	upper leg amputation	6,88
5	Critical	Not fully recoverable with care	extensive cerebral hemorrhage	paraplegia below 4th vertebra	heart perforation		32,32
6	Fatal	Unsurvivable	entire destruction of skull	paraplegia above 4th vertebra	full thorax crushing		100,00

As multiple injuries may be fatal even though each isolated one is non-lethal, the Maximum AIS (MAIS) and Injury Severity Score (ISS) were introduced. MAIS is the maximum occurring AIS score and ISS is defined as

$$ISS = A^2 + B^2 + C^2, \tag{4.1}$$

where A, B, C are the AIS scores of the three most injured body regions. Its maximum value is $\text{ISS}_{\text{max}} = 75$ and if any of the sub-injuries is AIS 6, ISS is automatically set to 75. Important to notice is that a polytrauma is associated with ISS ≥ 16. Furthermore, please also note that the AIS is generally not suitable for rating the potential or duration of injury with respect to convalescence.

Next, the EuroNCAP as the European example of a classification system in automobile crash-testing is described.

---

[2] Association for the Advancement of Automotive Medicine.

[3] American Medical Association.

4.1.2 EuroNCAP

The ADAC crash-tests described later in Chapter 5 are carried out according to the EuroNCAP[4] which is based on the Abbreviated Injury Scale. The EuroNCAP, inspired by the American NCAP, is a manufacturer independent crash-test program uniting the European ministries of transport, automobile clubs and underwriting associations with respect to their testing procedures and evaluations [9]. The outcome of the tests, specified in the program, is a scoring of the measured results via a sliding scale system. Upper and lower limits for the injury potentials are mostly defined such that they correlate to a certain probability of AIS ≥ 3. Between these two values the corresponding score (injury potential) is calculated by linear interpolation. A standardized color code indicates injury potential and is given in Tab.4.2.

Table 4.2 Injury Severity and corresponding color code

Colorcode	Color	Injury potential
Red		Very high
Brown		High
Orange		Medium
Yellow		Low
Green		Very low

In the following, a survey is given on the most important blunt injury criteria of different human body parts.

4.2 Injury Criteria for the Head

4.2.1 Possible Head Injuries and Their Mechanisms

As described in [45], possible injuries of the head can be classified according to Fig. 4.1. Various injury mechanisms may cause these injuries, of which an overview is given in [45]. Generally, the according mechanisms are divided into static ($\Delta t_i > 200$ ms) and dynamic. Depending on whether the injury is caused during contact or non-contact, the impact force with its respective deformation or the inertia with correlating acceleration define the injury. For contact forces one generally distinguishes between bursting (for indirect contact) and bending fracture (for direct contact). For non-contact caused acceleration due to inertial effects the resulting injury mechanisms are focal and diffuse brain injury, respectively.

[4] **E**uropean **N**ational **C**ar **A**ssessment **P**rotocol.

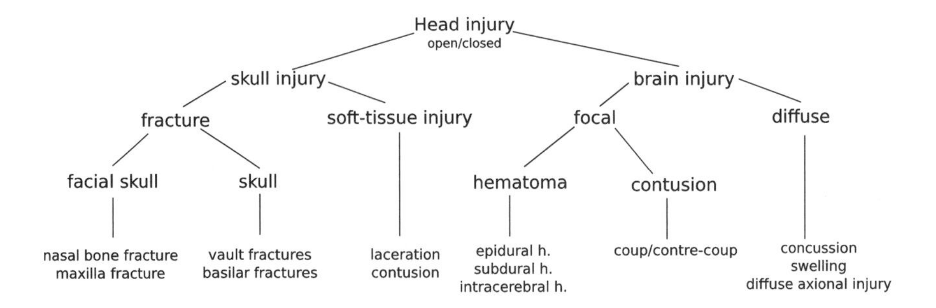

Fig. 4.1 Possible injuries to the human head [45]

According to [27] most research carried out in connetion with automobile crash-testing distinguishes also the two types of head loadings:

1. *Direct Interaction:* An impact or blow involving a collision of the head with another solid object at appreciable velocity. This situation is generally characterized by large linear accelerations and small angular accelerations during the impact phase.
2. *Indirect Interaction:* An impulse loading including a sudden head motion without direct contact. The load is generally transmitted through the head-neck junction upon sudden changes in the motion of the torso and is associated with large angular accelerations of the head.

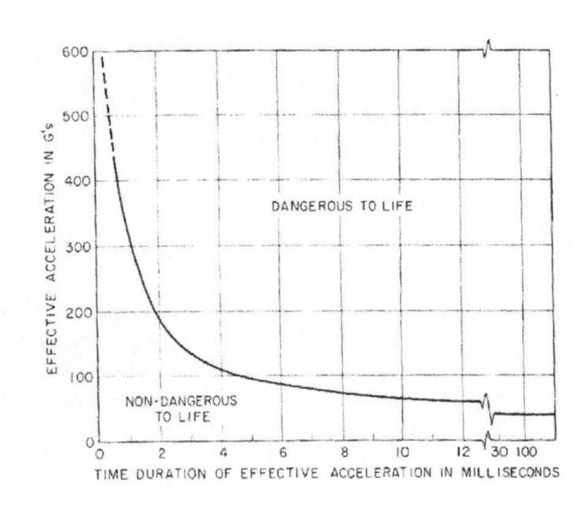

Fig. 4.2 Impact tolerance for the human brain in forehead impacts against plain yielding surfaces, [56]

Table 4.3 The Wayne State Tolerance Curve data basis

Time duration	Object	Configuration	Measured quantity	Criterion
$2-5$ ms	cadaver	drop test: head on steel plate	occiput acceleration	cranial fracture
$5-40$ ms	cadaver, animal	impact test: pressure on open brain	head acceleration	pathological changes
>40 ms	volunteer	sled test: acceleration	sled acceleration	impaired consciousness, concussion

4.2.2 *The Wayne State Tolerance Curve*

For the head many criteria for type 1 interactions are available. Their major theoretical basis is the so called Wayne State Tolerance Curve (WSTC) (also known as cerebral concussion tolerance curve), a fundamental experimental injury tolerance curve [24] forming the underlying biomechanical data of many head criteria, see Fig. 4.2. It consists of data obtained from cadaver, animal, and volunteer tests, see Tab. 4.3. In this early study it was assumed that cranial fractures indicate brain injury. In this sense, it is important to notice that in [16] it is stated that mild to moderate concussion accompanies a linear skull fracture. However, it has been noted that concussions often occur without fracture. Therefore, it is generally assumed that skull fractures occur along with more severe concussions than concussions alone. Thus, it is assumed to form an upper limit for concussion, which should not be exceeded.

There are several aspects to be regarded when interpreting the results of the curve. Especially the long-duration end of the curve, which asymptotic value is 42 g, was mainly obtained from whole body volunteer deceleration tests in the pioneering work of Stapp [54, 55]. As later on other volunteers had survived frontal crash simulations exceeding 45 g in [41], a value of 42 g was considered as considerably too low. Therefore, they recommended to raise the asymptote to 80 g.

Further fundamental biomechanical work that aims at an understanding of the impact dynamics can be found in [49, 48, 50], where the mechanics of head impacts are derived and a theory for the so called Countre-Coup injury is formulated.

Next, the relevant quantities to predict injury due to rotational head motion are discussed.

4.2.3 *Rotational Head Acceleration Limits*

Rotational acceleration thresholds can e.g. be found in [39], which were obtained from Rhesus monkeys and scaled via similarity transformation to the human. The authors found the following tolerance law.

$$\dot{\omega}_{\text{max}} = \frac{k_{\text{rhesus}}}{m_{\text{brain}}^{\frac{2}{3}}} \tag{4.2}$$

The constant factor k_{rhesus} is derived from the Rhesus monkey data.

Table 4.4 Tolerance thresholds for totational acceleration and velocity of the brain [46]

tolerance threshold	type of brain injury	reference
50 % probability: $\dot{\omega}_H = 1800$ rad/s^2 for $t < 20$ ms $\omega_H = 30$ rad/s for $t \geq 20$ ms	cerebral concussion	[39]
$\dot{\omega}_H < 4500$ rad/s^2 and/or $\omega_H < 70$ rad/s	rupture of bridging vein	[26]
$2000 < \dot{\omega}_H < 3000$ rad/s^2	brain surface shearing	[4]
$\omega_H < 30$ rad/s: safe: $\dot{\omega}_H < 4500$ rad/s^2 AIS 5: $\dot{\omega}_H > 4500$ rad/s^2 $\omega > 30$ rad/s: AIS 2: $\dot{\omega}_H = 1700$ rad/s^2 AIS 3: $\dot{\omega}_H = 3000$ rad/s^2 AIS 4: $\dot{\omega}_H = 3900$ rad/s^2 AIS 5: $\dot{\omega}_H = 4500$ rad/s^2	(general)	[38]

In a second approach, published in [40], it was postulated that the investigated "natural" head movements are surely harmless and this should be used as a criterion. The result was a tolerance curve relating rotational acceleration and residence time. An extensive listing of tolerance values for rotational acceleration and velocity of the brain is given in [46], see Tab. 4.4.

Next, the head acceleration-based HIC, 3ms-Criterion, and Generalized Acceleration Model for Brain Injury (GAMBIT) are described. Due their correlation to acceleration these indicators are not able to predict the injury risk of sustaining fracture mechanisms of the facial and cranial bones.

4.2.4 Head Injury Criterion

The most frequently used head severity index is the Head Injury Criterion [56], defined as

$$\text{HIC}_{36} = \max_{\Delta t} \left\{ \Delta t \left(\frac{1}{\Delta t} \int_{t_1}^{t_2} ||\ddot{\mathbf{x}}_H||_2 \text{d}t \right)^{\left(\frac{5}{2}\right)} \right\} \leq 650 \tag{4.3}$$
$$\Delta t = t_2 - t_1 \leq \Delta t_{\text{max}} = 36 \text{ ms}.$$

$||\ddot{\mathbf{x}}_H||$ is the resulting acceleration of the human head[5] and has to be measured in $g = 9.81$ m/s^2. The optimization is done by varying t_1 and t_2, i.e. the start and stop time are both parameters of the optimization process. Intuitively speaking, the HIC weights the resulting head acceleration and impact duration, which makes allowance of the fact that the head can be exposed to quite high accelerations and is still intact as long as the impact duration is kept low. In addition to the HIC_{36} the identically defined HIC_{15} with $\Delta t_{\max} = 15$ms exists. In typical automotive safety applications it is set to ≤ 36ms, [28]. Comparing both likelihood distributions yields that corresponding injury probabilities for HIC_{15} are more restrictive than for the HIC_{36}, see Sec.4.2.6. Further details on the derivation of the HIC can e.g. be found in [56].

A criterion for side impacts proposed by the Economic Council for Europe (ECE) is the Head Protection Criterion (HPC). Its formula is analogue to the HIC, however, it is only evaluated for the duration of contact.

4.2.5 3 *ms-Criterion*

The 3 ms-Criterion, which is also based on the WSTC, requires the maximum 3ms-average of the resulting acceleration to be less than 72 g in the EuroNCAP. Any shorter impact duration only has little effect on the brain. In [14] a threshold of 80 g was proposed.

4.2.6 *Converting Severity Indices to the Abbreviated Injury Scale*

Severity indices do not provide a direct interpretation of injury. Furthermore, they are defined with respect to different physical domains. Thus they are not directly comparable with each other, nor can they be combined. For this purpose various mappings were developed to translate a severity index to the Abbreviated Injury Scale. The National Highway Traffic Safety Administration (NHTSA) specified the expanded Prasad/Mertz curves [36] for converting HIC_{15} values to the probability $p(AIS \geq i)$ of the corresponding AIS level i which are shown in Fig.4.3 (left). In [20] a conversion from HIC_{36} to $p(AIS \geq 2,3,4)_{\mathrm{HIC36}}$ is defined. Since the EuroNCAP underlays its injury risk level definition mainly on the $p(AIS \geq 3)$-level, the corresponding functions for both HICs are illustrated in Fig.4.3 (right):

$$p(AIS \geq 3)_{\mathrm{HIC15}} = \frac{1}{1+e^{3.39+\frac{200}{\mathrm{HIC}_{15}}-0.00372\mathrm{HIC}_{15}}} \tag{4.4}$$

$$p(AIS \geq 3)_{\mathrm{HIC36}} = \Phi\left(\frac{\ln(\mathrm{HIC}_{36})-\mu}{\sigma}\right), \tag{4.5}$$

[5] $||\ddot{\mathbf{x}}||_2$ =Euclidean norm.

with $\Phi(.)$ denoting the cumulative normal distribution with mean $\mu = 7.45231$ and standard deviation $\sigma = 0.73998$. For $p(AIS \geq 2)_{\text{HIC36}}$ and $p(AIS \geq 4)_{\text{HIC36}}$ the numerical values are $\mu = 6.96352, \sigma = 0.84664$ and $\mu = 7.45231, \sigma = 0.73998$. For the very short impacts discussed in this monograph the evaluation of HIC_{15} and HIC_{36} lead to the same numerical value. The original publication of these mappings can be found in [15]. The author analyzed the drop test data documented in [43]. The HIC_{15} indicates a higher risk level than the HIC_{36} for the same numerical value and is therefore more restrictive.

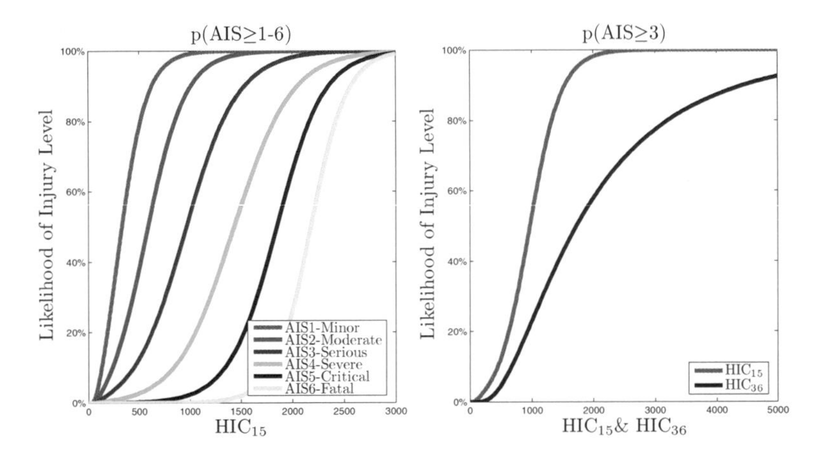

Fig. 4.3 Mapping HIC_{15} to the Abbreviated Injury Scale (left) and comparing $p(AIS \geq 3)_{\text{HIC15}}$ with $p(AIS \geq 3)_{\text{HIC36}}$ (right)

4.2.7 GAMBIT

The GAMBIT was introduced in [34]. It aims at combining translational and rotational head response into one criterion and is defined as follows

$$\text{GAMBIT}(t) = \left[\left(\frac{\ddot{x}_H(t)}{\ddot{x}_{H,c}} \right)^n + \left(\frac{\dot{\omega}_H(t)}{\dot{\omega}_{H,c}} \right)^m \right]^{\frac{1}{s}}, \tag{4.6}$$

with $\ddot{x}_H$ generally measured in [g] and the rotational acceleration $\dot{\omega}_H$ in [rad/s^2]. In [19] a fully parameterized solution is given:

$$\text{GAMBIT}(t) = \left[\left(\frac{\ddot{x}_H(t)}{250} \right)^{2.5} + \left(\frac{\dot{\omega}_H(t)}{25000} \right)^{2.5} \right]^{\frac{1}{2.5}} \tag{4.7}$$

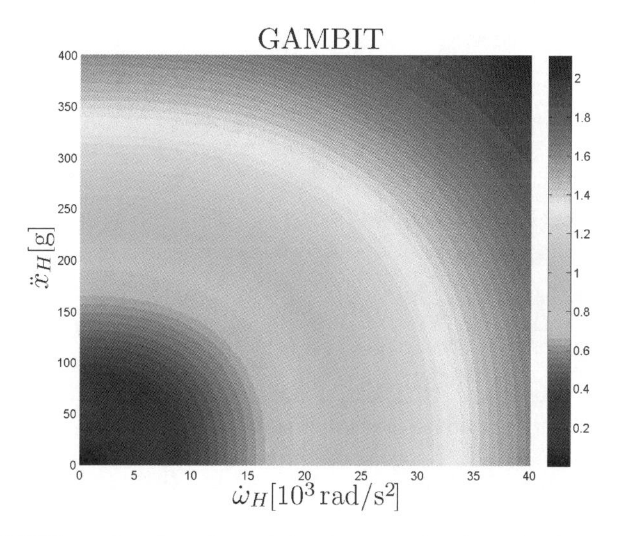

Fig. 4.4 GAMBIT as a function of translational and rotational acceleration

Figure 4.4 shows the iso-lines of the GAMBIT. A considerable simplification of (4.7) was also derived and led to

$$\text{GAMBIT} = \frac{\bar{\ddot{x}}_H}{250} + \frac{\overline{\omega}_H}{10000}, \tag{4.8}$$

with $\bar{\ddot{x}}_H$ and $\overline{\omega}_H$ being the mean translation and rotational acceleration, respectively. The value GAMBIT$=1$ represents the overall tolerance value. However, due to the lack of validation it is hardly ever used and is not included into any regulations so far.

The next two indices were developed for evaluating short impact durations and the third one, the Revised Brain Model predicts injury severity for longer durations of loading.

4.2.8 Vienna Institute Index

The Vienna Institute Index is based on a simple mass-spring-damper model of the human head. The damping is chosen to be $D_H = 1$ and the eigenfrequency $\omega_{M,n} = 635\frac{\text{rad}}{s}$. The according injury index is defined as a displacement relationship [27]:

$$\mathscr{J} := \frac{x_{H,\max}}{x_{H,\text{tol}}}, \tag{4.9}$$

where $x_{H,\max}$ is the maximum displacement x_H for a given acceleration pulse and $x_{H,\text{tol}} = 2.35$ mm is the maximum tolerable value for this displacement. Its critical tolerance level is defined as $\mathcal{J} = 1$. Values $\mathcal{J} < 1$ cause cerebral concussion without permanent after-effects at worst, while $\mathcal{J} > 1$ is considered to be hazardous to life.

4.2.9 *Effective Displacement Index*

The Effective Displacement Index (EDI), introduced in [7], is similar to the Vienna Institute model but is characterized by different damping and stiffness values (eigenfrequency). They are set to $D_H = \frac{\sqrt{2}}{2} \approx 0.707$ and $\omega_{H,n}$ = 482 rad/s. This index differentiates in particular anterior-posterior and resulting displacement, see Tab. 4.5. In [16] this criterion was evaluated and compared to the Severity Index [13, 56, 11], where it was concluded that both indices produced critical values as predicted by their original authors.

Table 4.5 Parameters and tolerance values of the Effective Displacement Index

$x_{H,\text{tol}}$	Anterior-Posterior	Resulting
human	38.1 mm	45.72 mm
dummy	43.18 mm	5.08 mm

4.2.10 *Revised Brain Model*

The Revised Brain Model uses the same dynamics model as the Vienna Institute Index or EDI. The damping and eigenfrequency are selected to be $D_H = 0.4$ and $\omega_{H,n} = 175$ rad/s, respectively. The proposed tolerance criterion is the maximum deformation $x_{H,\text{tol}} = 31.75$ mm for pulse durations of $\Delta t_i \geq 20$ ms and the head velocity $\dot{x}_{H,\text{tol}} = 3.43$ m/s for pulse durations $\Delta t_i < 20$ ms, respectively.

4.2.11 *Maximum Mean Strain Criterion*

The Maximum Mean Strain Criterion (MSC) was introduced in [27] based on modeling the human head as a 2-mass-spring complex, see Fig.4.5. Formally, the criterion is defined as

$$\varepsilon_H = \frac{1}{l_M} \cdot (x_{H_2} - x_{H_1}) \leq 0.0061, \tag{4.10}$$

i.e. it poses a constraint on the elastic deflection of the lumped head representation. Its underlying model was extended by a damper in series to the stiffness [52] and the further revised to the so called Translational Head Model (THM) [51, 53]. In

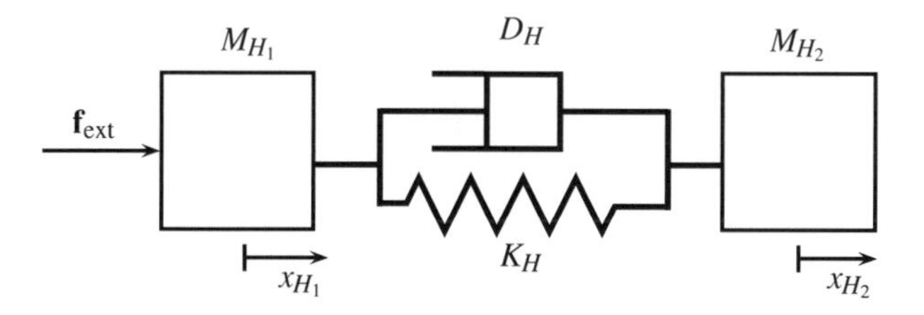

Fig. 4.5 The Mean Strain Criterion model

this work, strain rate was also proposed as a suitable injury criterion. The approach is also applicable for L-R and S-I impacts. However, due to some inconsistent interpretations caused by the chosen formulation the MSC never established itself over the years.

The last distinct severity index for the head described here is the Maximum Power Index (MPI). In contrast to the displacement, velocity, or acceleration-based approaches presented up to now, the MPI concentrates on the change of kinetic energy.

4.2.12 Maximum Power Index

The MPI introduced in [35] is the weighted change of kinetic energy Head Impact Power (HIP) of the human head and the weighting is carried out by two sensitivity matrices $C_x = \text{diag}\{c_{x,i}\}$ with $c_{x,i} > 0$ and $C_\varphi = \text{diag}\{c_{\varphi,i}\}$ with $c_{\varphi,i} > 0$.

$$\begin{aligned} PI &:= C_x M_x \ddot{\mathbf{x}}_H \cdot \dot{\mathbf{x}}_H + C_\varphi M_\varphi \omega_H \cdot \dot{\omega}_H = \\ &= C_x HIP_x + C_\varphi HIP_\varphi \end{aligned} \tag{4.11}$$

M_x, M_φ are diagonal matrices, consisting of the effective mass and moment of inertia of the head, respectively. The MPI is then defined as

$$MPI = \max(PI) \tag{4.12}$$

C_x and C_φ were not yet determined and are therefore set to unity matrices. The *MPI* has not been introduced in any regulations so far. However, this index is validated by analyzing collisions of American football players during a game. Based on this analysis they found a 50 % probability of concussion at $HIP_{\max} = 12.8$ kW. Nonetheless, further analysis is still necessary to concisely correlate HIP with more severe injury mechanisms.

Next, a method from automotive testing for analyzing facial laceration is described.

4.2.13 The Facial Laceration Criterion

Investigating facial laceration in automobile crash-testing originates from designing car glass. For these tests two layers of chamois leather are put over the facial HIII dummy area and after the collision the cut depth in the chamois and the number of cut layers are observed. If the inner layer did not suffer any injury the laceration is classified as minor, while moderate to major laceration injury correlates to large cuts in the inner layer. The original work [17] proposed the so called Chamois Laceration Scale. In [44] a proposal for the tolerance levels for the Facial Laceration Criterion is given that is directly associated to the observed effects on the 2-layer chamois. Table 4.6 shows the correlation between injury level, facial laceration injury criterion, and AIS from [44]. It is significantly simplified in comparison with the original definition given in [17]. A general drawback of the chamois-based methods so far is that they require a skilled subjective interpretation. In [42] the authors proposed the so called Triple Laceration Index (TLI), which is a quantitative assessment of laceration severity. They used two layers of chamois and an underlying layer of rubber. The TLI relates number, length and depth of cuts in the chamois to an according level of laceration severity in the skin.

In the following, the biomechanics of facial and cranial fractures are reviewed.

Table 4.6 Proposed tolerance levels for the Facial Laceration Criterion [44]

Injury level	Facial Laceration Criterion	Equivalent AIS
0	No cuts to outer layer	-
1	No cuts to inner layer	1
2	Moderate to major cuts to inner layer	2/3
3	Moderate to major cuts to inner layer	2/3

4.2.14 Fracture Forces

In [16] it was shown that frontal bone fracture occurs at the same acceleration level as would be predicted by the WSTC. Contact forces were therefore shown to be directly related to fractures of facial and cranial bones. Generally, the human skull consists of cranial and facial bones, which have varying fracture tolerance. Fractures are commonly categorized into linear (well distributed), depressed (fracture area $< 13\ \text{cm}^2$) and depressed with punch through fractures (fracture area $< 5\ \text{cm}^2$). Linear or simple fracture of the skull is rated with AIS $= 2$. Comminuted, depressed fracture of ≤ 2 cm is rated with AIS $= 3$. Complex, exposed or loss of brain tissue fracture corresponds to AIS $= 4$ [6]. As already mentioned fractures are related to impact forces and were investigated quite extensively in the biomechanics literature.

In Tab. 4.7 limits of the facial and cranial bones according to [29, 10, 16, 5, 47] are listed (some measurements are omitted for brevity). The corresponding termi-

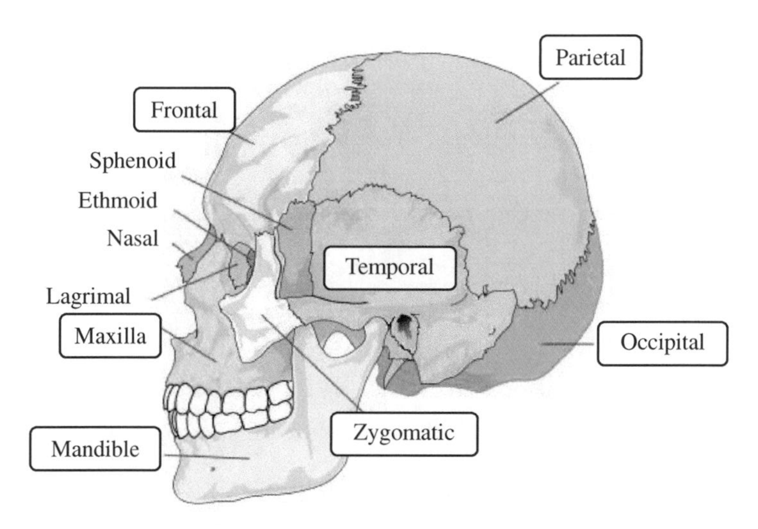

Fig. 4.6 (Simplified) anatomy of the human skull [1]

Table 4.7 Facial Impact Tolerance of cadaver heads

Facial bone	Fracture force	Impactor diameter	Reference
Mandible (A-P)	1.78 kN	0.029 m	[47]
Mandible (lateral)	0.89 kN	0.029 m	[47]
Maxilla	0.66 kN	0.029 m	[47]
Zygoma	0.89 kN	0.029 m	[47]
Cranial bone	Fracture force	Impactor diameter	Reference
Frontal	4.0 kN	0.02 m	[5]
Temporo-Parietal	3.12 kN	0.029 m	[29]
Occipital	6.41 kN	0.017 m	[29]

nology of the head anatomy is illustrated in Fig. 4.6. Generally, the fracture force depends on the contact area used for such tests. Therefore, the impactor size used for the particular experiments are listed as well. [5] showed that the fracture force of the frontal bone is 4.0 kN and [29] that the temporoparietal bone has a tolerance force of 3.12 kN. [47] determined the tolerance force of the mandible (A-P), mandible (lateral) maxilla, and zygoma to be 1.78 kN, 0.89 kN, 0.66 kN, and 0.89 kN, respectively. For the nasal bone [37] measured a tolerance value of 0.34 kN.

An important aspect of safety in automobile crash testing is neck injury, which is described in the following.

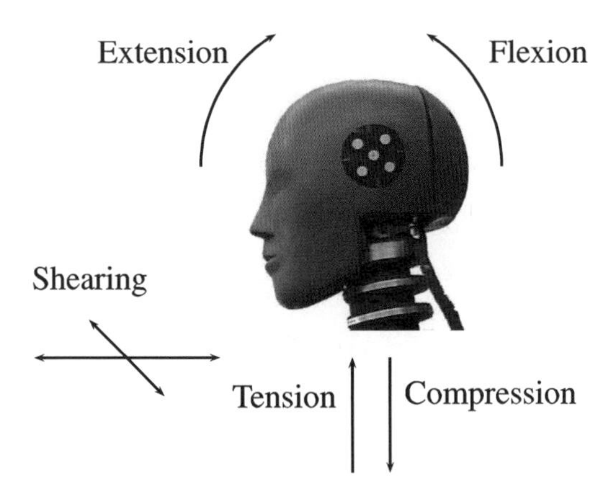

Fig. 4.7 Taxonomy of neck motions

Table 4.8 Higher and lower performance limits specified for the human neck

Load	@0 ms	@25 – 35 ms	@45 ms
Shearing: F_x, F_y	1.9/3.1 kN	1.2/1.5 kN	1.1/1.1 kN
Tension: F_z	2.7/3.3 kN	2.3/2.9 kN	1.1/1.1 kN
Extension: M_y	42/57 Nm	42/57 Nm	42/57 Nm

Table 4.9 Dynamic fracture loads for the thyroid and circoid cartilages

Cartilage	Mean [N]	Range [N]	Nature of Fracture
excised thyroid	180	62 – 377	incipient cracking
excised circoid	248	156 – 302	incipient cracking
simultaneously loaded	490	337 – 810	imminent total collapse
thyroid in situ	400 – 445		marginal fracture
thyroid in situ	400 – 445		marginal fracture

4.3 Injury Criteria for the Neck

In general, inertial injury mechanisms of the human neck are related to forces and bending torques acting on the spinal column. They can be caused by direct impact to the neck or via head inertial loading. In the EuroNCAP corresponding limits for the latter injury class are defined with respect to the positive cumulative exceedance time, see Tab. 4.8. Between these values a linear interpolation is carried out. The corresponding taxonomy of the neck is illustrated in Fig. 4.7, whereas the EuroNCAP specifies limit values only for the motions listed in Tab. 4.8. A good summary of the underlying biomechanical references is given in [29] and further information can also be found in [33, 31, 32].

Apart from these standardized limit values, which do not cover isolated direct neck loading, [29] described tolerance forces for the thyroid and circoid cartilages based on data from [30, 12]. The measured dynamic fracture forces are listed in Tab. 4.9, illustrating the sensitivity of these parts of the body.

4.4 Injury Criteria for the Chest

For the torso the available criteria are divided into four groups, which are understood quite well:

1. Force-based criteria
2. Acceleration-based criteria
3. Compression-based criteria
4. Soft-tissue-based criteria

Before going into the details of the aforementioned classes of chest injury criteria, the standard chest model used throughout this monograph is described.

4.4.1 Lobdell's Chest Model

The standard lumped abstraction of blunt chest impact dynamics is Lobdell's chest model [25]. The authors provided parameters for both human cadavers and HIIIs, respectively. It consists of two lumped masses, two stiffnesses, and two damping elements. Its structure was developed from impact experiments with human cadavers, volunteers, and dummies. M_{C1} is the effective mass of the sternum, a portion of the anterior rib cage, and thoracic contents. M_{C2} is the effective mass of the remaining portion of the thorax. x_{C_1} and x_{C_2} are their position variables. The numerical

Table 4.10 Parameters of the Lobdell dummy chest model

Parameter	Value
M_{C1}	0.45 kg
M_{C2}	27.2 kg
$k_{C,1}$	26.3 kN/m
$k_{C,2}$	78.8 kN/m
$d_{C,1}$	0.525 kNs/m
$d_{C,2}$	1.23 kNs/m
δ_0	0.0318 m

dummy values of these parameters are depicted in Tab. 4.10. The thoracic system can be described by following differential equations:

$$M_{C1}\ddot{x}_{C_1} = f_{\text{ext}} - g(\Delta x_C) - h(\Delta \dot{x}_C) \tag{4.13}$$

$$M_{C2}\ddot{x}_{C_2} = g(\Delta x_C) + h(\Delta \dot{x}_C) \tag{4.14}$$

where $\Delta x_C = x_{C_1} - x_{C_2}$ is the chest deflection. The thoracic spring and damping force can be expressed as

$$g(\Delta x_C) = \begin{cases} k_{C_1}(x_{C_1} - x_{C_2}) & \text{if } 0 \leq (x_{C_1} - x_{C_2}) \leq \delta_0 \\ k_{C_2}(x_{C_1} - x_{C_2}) - f_{\text{ext},0} & \text{if } \quad (x_{C_1} - x_{C_2}) \quad > 0 \end{cases} \tag{4.15}$$

$$h(\Delta \dot{x}_H) = \begin{cases} d_{C_1}(\dot{x}_{C_1} - \dot{x}_{C_2}) \text{ if } (\dot{x}_{C_1} - \dot{x}_{C_2}) \geq 0 \\ d_{C_2}(\dot{x}_{C_1} - \dot{x}_{C_2}) \text{ if } (\dot{x}_{C_1} - \dot{x}_{C_2}) < \delta_0, \end{cases} \tag{4.16}$$

where $f_{\text{ext},0} = (k_{C_2} - k_{C_1})\delta_0$.

4.4.2 Force Criterion

As force-based criteria will be introduced in Chapter 5, only some additional remarks are given at this point.

The human sternum is generally able to withstand high static strains. Contact forces of up tp 3.3 kN pose only minimal risks to the sternum. According to [23] even higher loads are subcritical in most cases.

4.4.3 Acceleration Criterion

The Acceleration Criterion (AC) is also called chest criterion and was applied to whole-body response studies, as well as to the assessment of potential chest injury in frontal impacts. The respective limits are

$$\max(\ddot{\mathbf{x}}_{C_{av}}) \leq 60\text{g} = 588.6\,\frac{\text{m}}{\text{s}^2} \quad \wedge \quad \Delta t_i \leq 3\text{ ms} \tag{4.17}$$

$$\max(\Delta x_C) = \max(x_{H_2} - x_{H_1}) \leq 63\text{ mm}.$$

The NHTSA [2] gives also for the critical chest acceleration and deflection a mapping from $\ddot{\mathbf{x}}_{H_{av}}$ and $x_{M_2} - x_{M_1}$ to probability of injury severity.

$$
\begin{aligned}
p(\text{MAIS2+}) &= \frac{1}{(1+\exp(1.2324-0.0576\ddot{\mathbf{x}}_{C,\max}))}\\
p(\text{MAIS3+}) &= \frac{1}{(1+\exp(3.1493-0.0630\ddot{\mathbf{x}}_{C,\max}))}\\
p(\text{MAIS4+}) &= \frac{1}{(1+\exp(4.3425-0.0630\ddot{\mathbf{x}}_{C,\max}))}\\
p(\text{MAIS5+}) &= \frac{1}{(1+\exp(8.7652-0.0659\ddot{\mathbf{x}}_{C,\max}))}
\end{aligned} \tag{4.18}
$$

$$
\begin{aligned}
p(\text{MAIS2+}) &= \frac{1}{(1+\exp(1.8706-0.04439(\Delta x_C))}\\
p(\text{MAIS3+}) &= \frac{1}{(1+\exp(3.7124-0.0475(\Delta x_C))}\\
p(\text{MAIS4+}) &= \frac{1}{(1+\exp(5.0952-0.0475(\Delta x_C))}\\
p(\text{MAIS5+}) &= \frac{1}{(1+\exp(8.8274-0.0459(\Delta x_C))},
\end{aligned} \tag{4.19}
$$

where $p(\text{MAIS}i+)$ is the probability of the i-th or higher MAIS level to occur.

4.4.4 Compression Criterion

From evaluated cadaver experiments it was derived that acceleration and force criteria alone are intrinsically not able to predict the risk of internal injuries of the thorax. Generally, these tend to be a greater threat to human survival than skeletal injury. Kroell analyzed a large data base of blunt thoracic impact experiments and realized that the Compression Criterion (CC)

$$
CC = ||\Delta \mathbf{x}_C||_2 \leq 22\text{mm} \tag{4.20}
$$

is a superior indicator of chest injury severity. Especially sternal impact was shown to cause compression of the chest until rib fractures occur [21, 23].

For the CC an empirical relationship to the AIS index (the AIS is assumed to be a continuous function) was found.

$$
\text{AIS}(CC,t) = -3.78 + 0.198CC, \qquad \text{for } CC > 21.11\ \% \tag{4.21}
$$

In (4.21) the CC is assumed to be normalized with respect to the initial thorax thickness l_c.

4.4.5 Viscous Criterion

The Viscous Criterion (VC), which is also known as soft-tissue criterion [23, 22] is defined as

$$VC = c_c ||\Delta \dot{\mathbf{x}}_C||_2 \frac{||\Delta \mathbf{x}_C||_2}{l_c} \leq 0.5 \, \frac{\mathrm{m}}{s}. \tag{4.22}$$

In contrast to the CC, it is the product of compression velocity and the normalized thoracic deflection. The scaling factor c_c and the deformation constant (actually the initial torso thickness) l_c depend on the used dummy and are summarized in [57].

In the next section some recent findings from biomechanics of eye injury are shortly described. These are a basis for future investigations.

4.5 Eye Injury

In [18] blunt eye injury was analyzed with respect to its occurrence, cause, and injury mechanisms. According to [18] low severity injury of the eye are e.g.

AIS1:

- Corneal abrasions
- Hyphema: blood in anterior chamber

AIS2:

- Retinal detachment
- Corneal/scleral laceration
- Globe rupture
- Eye enucleation

The authors state that 50.0 % of eye injuries in the United States are caused by blunt objects and occur in home environment with 40 %. This result is based on an eye injury database from projectile tests, which was acquired experimentally and from existing literature. Overall, the authors analyzed data from 8 different studies, consisting of 251 individual tests. They performed a statistical analysis of projectile characteristics related to eye injury risk and developed parametric risk functions for corneal abrasion, hyphema, lens dislocation, and globe rupture. The authors concluded that normalized energy (energy density) is a good indicator of the different injury mechanisms. They verified their results with cadaver testing, developed an FEM model of eye impacting, and contributed to the development of the FOCUS headform, a fully instrumented headform for assessing eye and facial injury risk.

References

[1] Wikipdedia, the free encyclopedia, `http://en.wikipedia.org`

[2] `www.nhtsa.dot.gov/cars/rules/rulings/aairbagsnprm/pea/index.html`

[3] AAAM: The Abbreviated Injury Scale (1990), Revision Update. Des Plaines/IL (1998)
[4] Advani, S., Ommaya, A., Yang, W.: Head injury mechanisms. Human Body Dynamics. Oxford University Press (1982)
[5] Allsop, D., Warner, C., Wille, M., Schneider, D., Nahum, A.: Facial impact response - a comparison of the Hybrid III dummy and human cadaver. SAE Paper No.881719, Proc. 32th Stapp Car Crash Conf., pp. 781–797 (1988)
[6] Brinkmann, B., Madea, B. (eds.): Handbuch gerichtliche Medizin. Springer (2004) (German)
[7] Brinn, J., Staffeld, S.: Evaluation of impact test accelerations: A damage index for the head and torso. In: 14th Stapp Car Crash and Field Demonstration Conference Proceedings (STAPP1970), Paper 700902, vol. 17-18, pp. 188–202 (1970)
[8] Burg, H., Moser, A.: Handbuch Verkehrsunfallrekonstruktion: Unfallaufnahme, Fahrdynamik, Simulation. Vieweg+Teubner Verlag/GWV Fachverlage GmbH, Wiesbaden (2009) (German)
[9] EuroNCAP: European Protocol New Assessment Programme - assessment protocol and biomechanical limits (2003)
[10] European Commission Framework: Improved frontal impact protection through a world frontal impact dummy. Project No. GRD1 1999-10559 (2003)
[11] Gadd, C.: Criteria for injury potential. Publication 977,NAS-NRC pp. 141–145 (1962)
[12] Gadd, C., Culver, C., Nahum, A.: A study on responses and tolerances of the neck. SAE Paper No.710856, Proc. 15th Stapp Car Crash Conf. (1971)
[13] Gadd, C.W.: Use of a weighted - impulse criterion for estimating injury hazard. SAE Paper No.660793, Proc. 10th Stapp Car Crash Conf. (1966)
[14] Got, C., Patel, A., Fayon, A., Tarriere, C., Walfisch, G.: Results of experimental head impacts on cadavers: the various data obtained and their relation to some measured physical paramters. SAE Paper No.780887, Proc. 22th Stapp Car Crash Conf. (1978)
[15] Hertz, E.: A note on the Head Injury Criteria (HIC) as a predictor of the risk of skull fracture. In: 37th Annual Proceedings of the Association for the Advancement of Automotive Medicine, pp. 73–80 (1993)
[16] Hodgson, V., Thomas, L.: Comparison of head acceleration injury indices in cadaver skull fracture. SAE Paper No710854, Proc. 15th Stapp Car Crash Conf., pp. 299–307 (1971)
[17] Jettner, E., Hiltner, E.: Facial laceration measurements. SAE Paper No.860198, Society of Automotive Engineers International Congress and Exposition (1986)
[18] Kennedy, E., Ng, T., McNally, C., Stitzel, J., Duma, S.: Presentation: Evaluating the risk of eye injury using experimental and computational research methods
[19] Kramer, F.: Passive Sicherheit von Kraftfahrzeugen. Vieweg Verlag, Braunschweig, Germany (2006) (German)
[20] Kuppa, S.: Injury criteria for side impact dummies. NHTSA (2004)
[21] Lau, I., Viano, D.: Role of impact velocity and chest compression in thoraic injury. Avia. Space Environ. Med. 56, 16–21 (1983)
[22] Lau, I., Viano, D.: Thoracic impact: A viscous tolerance criterion. In: Tenth Experimental Safety Vehicle Conference, pp. 16–21 (1985)
[23] Lau, I., Viano, D.: The viscous criterion - basis and applications of an injury severity index for soft tissues. Proceedings of 30th Stapp Car Crash Conference, SAE Technical Paper No. 861882, pp. 123–142 (1986)
[24] Lissner, H., Lebow, M., Evans, F.: Experimental studies on the relation between acceleration and intracranial pressure changes in man. Surgery, Gynecology, and Obstetrics 111, 320–338 (1960)

[25] Lobdell, T., Kroell, C., Scheider, D., Hering, W.: Impact response of the human thorax. In: Symposium on Human Impact Response, pp. 201–245 (1972)
[26] Löwenhielm, P.: Mathematical simulation of gliding contusions. Journal of Biomechanics 8, 351–356 (1975)
[27] McElhaney, J., Stalnaker, R., Roberts, V.: Biomechanical aspects of head injury. Human Impact Response - Measurement and Simulation (1972)
[28] McHenry, B.: Head Injury Criterion and the ATB
[29] Melvin, J.: Human tolerance to impact conditions as related to motor vehicle design. SAE J885 APR80 (1980)
[30] Melvin, J., Snyder, R., Travis, L., Olson, N.: Response of human larynx to blunt loading. SAE Paper No.730967, Proc. 17th Stapp Car Crash Conf., pp. 101–114 (1973)
[31] Mertz, H.: Anthropomorphic test devices. Springer, New York (1993)
[32] Mertz, H., Patrick, L.: Investigation of the kinematics and kinetics of whiplash. SAE Paper No.670919, Proc. 11th Stapp Car Crash Conf., pp. 267–317 (1967)
[33] Mertz, H., Patrick, L.: Strength and response of the human neck. In: Proceedings of the 15th Stapp Car Crash Conference, pp. 207–255 (1971)
[34] Newman, J.: A generalized acceleration model for brain injury threshold (GAMBIT). In: International Research Council on Biomechanics of Injury (IRCOBI 1986), Bron, France, pp. 121–131 (1986)
[35] Newman, J., Shewchenko, N., Welbourne, E.: A proposed new biomechanical head injury assessment function - the maximum power index. Stapp Car Crash Journal, SAE paper 2000-01-SC16 44, 215–247 (2000)
[36] NHTSA: Actions to reduce the adverse effects of air bags. FMVSS No. 208 (1997)
[37] Nyquist, G.W., Cavanaugh, J.M., Goldberg, S.J., King, A.I.: Facial impact tolerance and response. SAE Paper No.861896, Proc. 30th Stapp Car Crash Conference, pp. 733–754 (1986)
[38] Ommaya, A.: Biomechanics of head injury. Bioechanics of Traums. Appleton-Century-Crofts (1984)
[39] Ommaya, A., Yarnell, P., Hirsch, A., Harris, E.: Scaling of experimental data on cerebral concussion in subhuman primates to concussion threshold for man. SAE Paper No670906, Proc. 11th Stapp Car Crash Conf., pp. 73–80 (1967)
[40] Parker, A.: Angular acceleration of the head. Humatic Reports PTM 163 (1965)
[41] Patrick, L., Lissner, H., Gurdijan, E.: Survival by design-head protection. SAE Paper No.963-12-0036, Proc. 7th Stapp Car Crash Conference, pp. 483–499 (1965)
[42] Pickard, J., Brereton, P., Hewson, A.: Objective method of assessing laceration damage to simulated facial tissues - the triple laceration index. SAE Paper No.1973-12-0010, Proc. 17th Conference of the American Association of Automotive Medicine (1965)
[43] Prasad, P., Mertz, H.: The position of the US delegation to the ISO Working Group 6 on the use of HIC in automotive environment. SAE Paper 851246 (1985)
[44] OPERAS: Occupation Protection & Egress in Rail Systems OPERAS web site (2001), `http://www.eurailsafe.net/`
[45] Schmitt, K.U.: Trauma biomechanics: accidental in traffic and sports. Springer, Heidelberg (2004)
[46] Schmitt, K.U., Niederer, P., Walz, F.: Trauma biomechanics: introduction to accidental injury. Springer, Heidelberg (2007)
[47] Schneider, D., Nahum, A.: Impact studies of facial bones and skull. SAE Paper No.720965, Proc. 16th Stapp Car Crash Conference, pp. 186–204 (1972)
[48] Sellier, K.: Zur Physik des Schädeltraumas. Int. Journal of Legal Medicine 51(3), 550–554 (1961) (German)

[49] Sellier, K., Müller, R.: Die mechanischen Vorgänge bei Stoßwirkung auf den Schädel. Klinische Wochenschrift 38(5), 233–236 (1960) (German)
[50] Sellier, K., Unterharnscheidt, F.: Mechanik und Pathomorphologie der Hirnschäden nach stumpfer Gewalteinwirkung auf den Schädel. Hefte Unfallheilkunde 76(1-140), 233–236 (1963) (German)
[51] Stalnaker, R., Low, T., Lin, A.: Translational energy criteria and its correlation with head injury in the sub-human primate. In: International Research Council on Biomechanics of Injury (IRCOBI 1987), Birmingham, England (1987)
[52] Stalnaker, R., McElhany, J., Roberts, V.: The application of the new mean strain criterion (NMSC). In: International Research Council on Biomechanics of Injury (IRCOBI 1956), Göteborg, Sweden, pp. 191–209 (1985)
[53] Stalnaker, R., Rojanavich, V.: A practical application of the translational energy potentials. In: International Research Council on Biomechanics of Injury (IRCOBI 1990), Lyon, France (1990)
[54] Stapp, J.: Tolerance to abrupt deceleration. In: Proceedings of Impact Injury and Crash Protection, pp. 308–349 (1955)
[55] Stapp, J.: Human Tolerance to Severe, Abrupt Deceleration. Gravitational Stress in Aerospace Medicine, Little, Brown, Boston (1961)
[56] Versace, J.: A review of the severity index. SAE Paper No.710881, Proc. 15th Stapp Car Crash Conf., pp. pp. 771–796 (1971)
[57] Data Processing Vehicle Safety Workgroup: Crahs analysis criteria version 1.6.1 (2004)

Chapter 5
Crash-Testing in Robotics

Ensuring safety leads to various aspects ranging from preventing electrical threats to coping with human mistakes. Up to now, this monograph focused on developing different methods for collision avoidance, detection, and reaction, i.e. to equip the robot with reactive motion control capabilities to appropriately react to environmental changes and unforeseen collisions. In this chapter however, the focus is on various aspects of physical human-robot contact and their related injury potential. In Figure 5.1 a first overview on relevant contact scenarios which potentially lead to human injury is given. Generally, one can differentiate between *free impacts*, *clamping in the robot structure*, *constrained impacts*, *partially constrained impacts*, and resulting *secondary impacts*. In this distinction it is not differentiated between

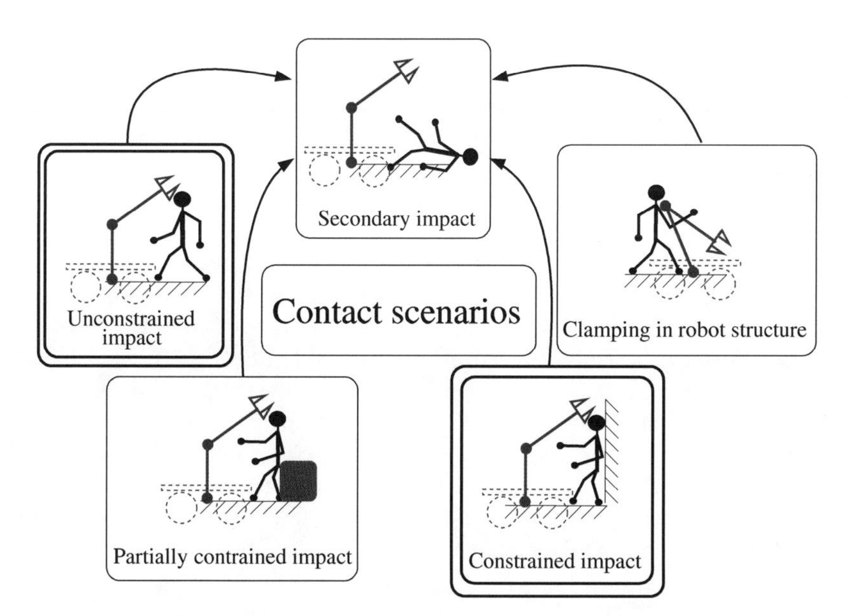

Fig. 5.1 Classification of undesired contact scenarios between human and robot

S. Haddadin: *Towards Safe Robots*,
Springer Tracts in Advanced Robotics 90,
DOI: 10.1007/978-3-642-40308-8_5, © Springer-Verlag Berlin Heidelberg 2014

blunt or sharp contact since the contact *situation* stays untouched in this context. The *unconstrained impact* is characterized by only the robot and human being directly involved into the collision. *Clamping in the robot structure* is e.g. a situation in which a human arm is being crushed between two link segments of an articulated manipulator. The case of a *partially constrained impact* is characterized by only a part of the human being clamped which is not directly in contact with the robot (in contrast to *constrained impacts*). This causes e.g. shearing and potentially large torques on the human body at the shearing point. Apart from the direct effects of collisions *secondary impacts* may cause further injuries, potentially leading to even larger injuries than by the direct impact itself (please note that in the pictogram only one example of this type is given). A combination or sequential order of the contact types is possible as well. Imagine a human that is standing in some distance in front of a barrier (e.g. a table) being hit by the robot in free space, and then being partially clamped against the object. During each of the depicted collisions various injury sources are present, such as fast blunt impacts, dynamic and quasi-static clamping, or cuts by sharp tools.

Up to now, few attempts have been made to investigate real world threats via collision tests and to use the outcome for considerably improving safety during physical Human-Robot Interaction. Although several countermeasures, criteria and control schemes for safe physical Human-Robot Interaction were proposed in the literature as e.g. [28, 7, 18, 21, 8, 9, 32], the main objective of actually quantifying and evaluating them on a biomechanical basis was marginally addressed. In this chapter, an overview is given of the systematic evaluation of safety in Human-Robot Interaction during blunt human-robot impacts, covering various aspects of the most significant injury mechanisms. To actually quantify the potential injury risk emanating from such manipulators, impact tests with various robots were carried out using standard automobile crash test facilities.

In this chapter, it is concentrated on unexpected impacts of a smooth surface related to the three body regions head, neck, and chest. Injury mechanisms caused by sharp tools or similar injury sources were not taken into consideration, since these mechanisms cannot be measured with standard crash-test dummies[1]. To evaluate the resulting injury severity the European testing protocol EuroNCAP was applied. The results of several injury criteria for head, neck, and chest were measured by the German Automobile Club (ADAC). The most prominent index for the head is the Head Injury Criterion [47], which was introduced to robotics in [50, 5] and used as a basis for new actuation concepts. As mentioned in Sec. 2.2.4, work that has been carried out up to now in the field of physical Human-Robot Interaction was mainly based on simulations. These contributions indicated high potential of injury to humans by means of the HIC, already at a robot speed of 1 m/s. This also matched the "common sense" expectation that a robot moving at maximal speed (e.g. due to malfunction) can cause high impact injury. In this regard, this chapter presents very surprising and striking results.

[1] Chapter 6 treats these issues in depth, especially analyzing soft-tissue injury due to cutting and stabbing.

Moreover, one of the main contributions of this chapter is the first experimental evaluation of the HIC in standard crash-test facilities. Additionally to the impact evaluation it will be shown that even with an ideally fast (physical) collision detection one is not able to react sufficiently fast to a stiff collision (e.g. head) in order to decrease the effect of the adverse contact forces for link inertias similar or larger to the ones of the LWR-III.

Based on these tests, several industrial robots of increasing weight were evaluated and the influence of robot mass and velocity investigated. The analyzed non-constrained impacts only partially capture the nature of human-robot safety. A constrained environment and its effect on resulting human injuries are therefore also discussed and evaluated from different viewpoints. Apart from such impact tests and simulations the major problem of a quasi-static constrained impact is analyzed, which poses under certain circumstances a serious threat to the human even for low-inertia robots.

Based on the insights gained from the above analysis, the intention in the last part of the chapter is to provide a crash-test report for blunt impacts for robots in general. Such a procedure is essential for any robot that enters human environments in the future, since its inherent injury potential has to be analyzed and quantified. The same holds for effective human-friendly control and motion schemes which have to be evaluated. For achieving such a representative routine, new findings for the basic understanding of human-robot impacts are contributed with large experimental campaigns. At the same time statements given in the first part of the chapter are verified. These tests provide an extensive set of data for the robotics community. Similarly to reports known from the automobile world[2], a fact based and result oriented view on the results from a large experimental campaign is given.

This chapter is organized as follows. In Section 5.1 a brief overview of injury quantification is given[3], followed by Section 5.2, which describes blunt impact tests with the LWR-III. Then, in Section 5.3 the role of robot mass and velocity is analyzed in detail for unconstrained impacts, followed by constrained impacts in Sec. 5.4. Quasistatic clamping close to singularities is discussed in Sec. 5.5. Finally, the results of the aforementioned large experimental campaign are presented in Sec. 5.6.

5.1 Automobile Crash Testing

A large variety of injuries are possible during an accident of a human with a robot, see Fig. 5.2a. In order to evaluate and categorize all these possible injuries, a common definition of injury severity is needed. In the following analysis, an internationally established definition of injury level and its related pendant in automobile crash testing is used. The Abbreviated Injury Scale (AIS) which is defined in [1, 2]

[2] A well known example from Germany is the *ADAC Motorwelt*.

[3] Please note that Chapter 4 gives a rather extensive overview of injury biomechanics.

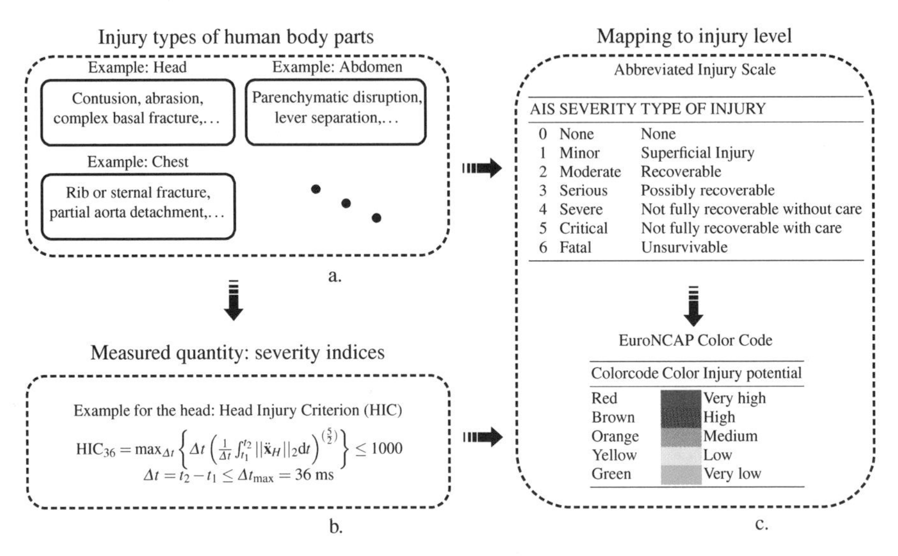

Fig. 5.2 The relationship between possible injuries of different body parts, its quantification and classification. Injury types of the human body parts and their severity can be quantified by severity indices. These in turn are mapped to a generic injury level like the Abbreviated Injury Scale.

subdivides the observed level of injury into seven categories from *none* to *fatal*, see Fig. 5.2c. In automobile crash testing the EuroNCAP[4], based on the Abbreviated Injury Scale and inspired by the U.S. NCAP, is the European automobile crash testing standard. A standardized color code indicates the corresponding injury potential, see Fig. 5.2c (bottom right).

In order to quantitatively evaluate injury, severity indices are used which are widely adopted and accepted measures of injury. Each of them is particularly defined for a certain body region. Defining and validating appropriate injury indices for a certain type of interaction is difficult, since it needs acquisition, biomechanical analysis, and abstraction of data from real human injuries. The biomechanical literature contains a large variety of such indices, but selecting the appropriate ones for robotics is a challenging task, requiring interdisciplinary skills. Already introduced into the robotics literature in [5, 50] was e.g. the HIC [47] which is widely used in automotive crash tests, see Fig. 5.2b. In the present chapter this criterion and other indices will be analyzed in order to assess their use and relevance to robotics. Mappings from a severity index to injury level or probability of injury level exist and are usually expressed by means of AIS/EuroNCAP injury level. For further information on EuroNCAP, HIC, AIS and for the definition of other severity indices (not only for the head but also for the neck, chest, and eye), please refer to Chapter 4.

After introducing relevant aspects from quantification and classification of injury in automobile crash testing, the important class of blunt unconstrained impacts is

[4] **E**uropean **N**ational **C**ar **A**ssessment **P**rotocol

discussed next. First, this will be done based on experimental data acquired with the LWR-III. Apart from evaluating this particular robot general findings are reported as well. They give more general understanding of rigid blunt impacts and some comments on the effect joint stiffness contributes to safety in pHRI will be given.

5.2 Blunt Unconstrained Impacts with the LWR-III

In this section, the experimental setup at the ADAC, consisting of a LWR-III and a standard frontal Hybrid III Crash Test Dummy (HIII), is briefly described.

5.2.1 Experimental Setup

The HIII represents the standard equipment used to measure various front crash injury criteria at a sampling frequency of 20 kHz. The signals are filtered according to the standardized specifications given in [10]. In Figure 5.3 the impact configuration of the LWR-III for head impacts is shown, which was chosen as a tradeoff between high maximal impact velocity and large reflected inertia (≈ 4 kg). The commanded impact velocity was $||\dot{\mathbf{x}}||_{\mathrm{TCP}} \in \{0.2, 0.7, 1.0, 1.5, 2.0\}$ m/s, ranging almost up to full Cartesian speed of the robot. For this experiments the robot is additionally equipped with a high-bandwidth force (1-DoF) and high-bandwidth acceleration sensor

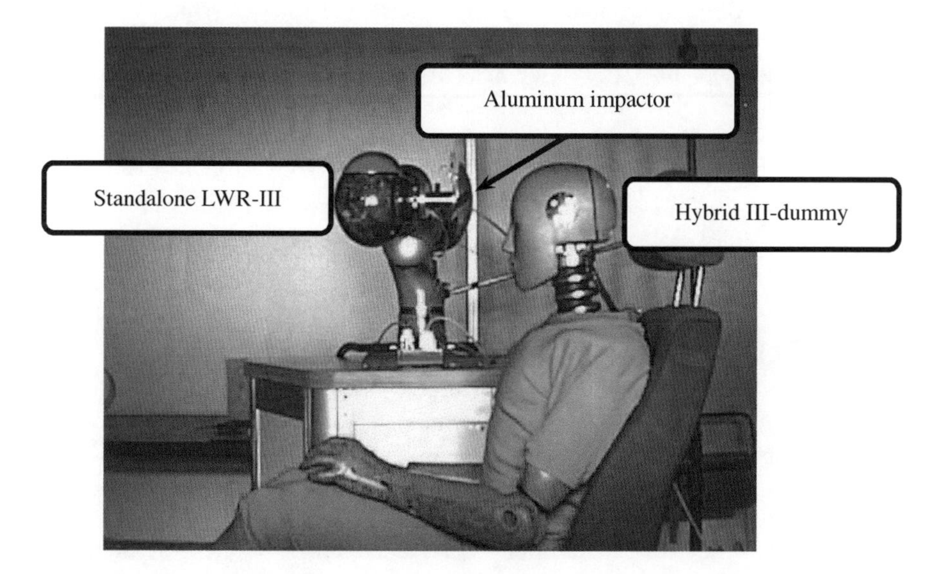

Fig. 5.3 High-speed recording of the impact tests with a Hybrid III-dummy

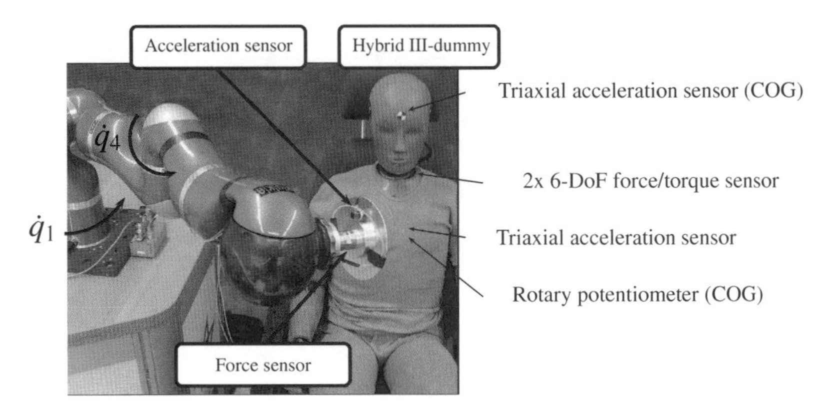

Fig. 5.4 Instrumentation of the LWR-III (additional external force and acceleration sensor) and the HIII

(3-DoF) mounted on a 1 kg impactor which defines the contact geometry. Figure 5.4 indicates the instrumentation of the HIII and the LWR-III.

The desired trajectory was a rest-to-rest motion, which start and the end configuration was given by

$$\mathbf{q}_{\text{start}} = [-45 \quad 90 \quad -90 \quad -45 \quad 0 \quad -90 \quad 147]^\circ$$
$$\mathbf{q}_{\text{end}} = [45 \quad 90 \quad -90 \quad 45 \quad 0 \quad -90 \quad 147]^\circ.$$

In order to maximize the joint mass matrix (reflected inertia was $\approx$ 4kg at the TCP) the trajectory was selected such that the robot hits the dummy in outstretched position. Furthermore, high TCP velocities can be achieved in this impact configuration. In the experiments the robot impact velocities were chosen to be $||\dot{\mathbf{x}}||_{\text{TCP}} \in \{0.2, 0.7, 1.0, 1.5, 2.0\}$ m/s.

A TCP velocity of 2 m/s is already close to the maximal robot speed and, as pointed out later, poses a potential threat to the mechanics of the robot particularly in the case of impact.

5.2.2 *Results for the Head*

In Figure 5.5 the resulting HIC_{36} values are plotted with respect to the impact velocity of the robot. The corresponding injury classification is described in Sec. 4.1.2. In order to classify an impact into the *green* labeled region, the occurring HIC_{36} must not exceed 650, which corresponds to a resulting 5 %-probability of serious injury (AIS $\geq$ 3). This value originates from [39, 41] and differs only slightly from the one obtained by the fitting function (4.5).

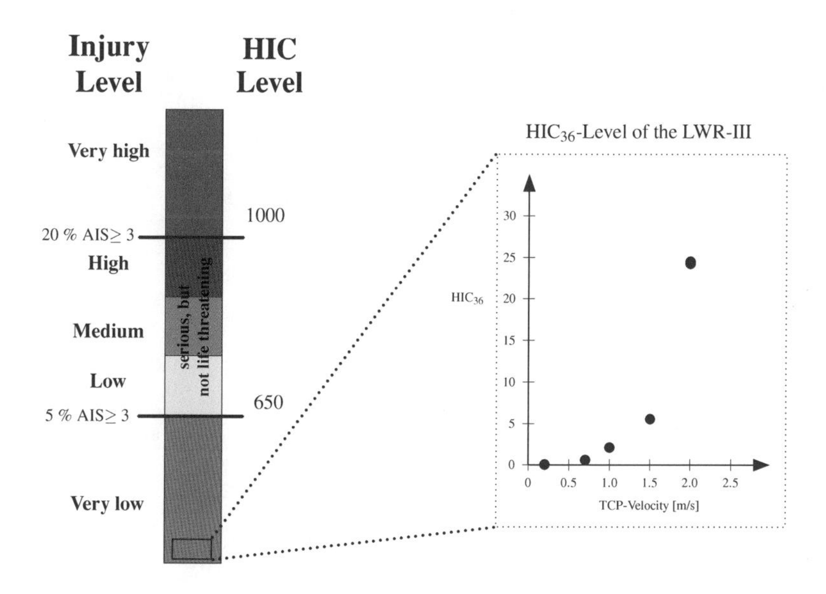

Fig. 5.5 Resulting HIC_{36} values for varying impact velocities, rated according to the EuroNCAP *Assessment Protocol And Biomechanical Limits*

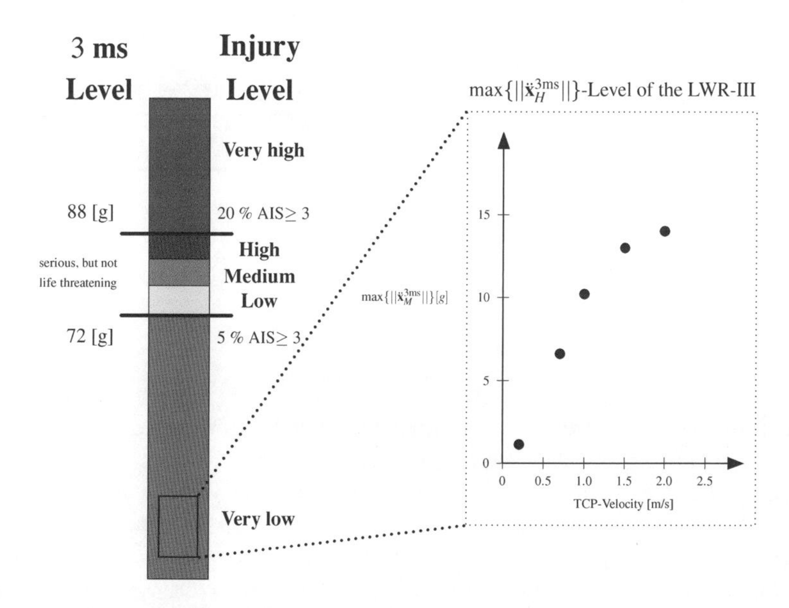

Fig. 5.6 Resulting 3 ms-Criterion values for varying impact velocities, rated according to the EuroNCAP *Assessment Protocol And Biomechanical Limits*

As indicated in Fig. 5.5, the HIC_{36} caused by the LWR-III is below 25 at 2 m/s which corresponds to a *very low* injury level. The resulting probability of injury severity obtained by (4.4) and (4.5) is ≈ 0 % for all categories (more specifically 4.87×10^{-6} and 1.1×10^{-5}). Another aspect that clearly can be deduced from Fig. 5.5 is that the HIC_{36} is rapidly increasing with robot velocity.

Similar to the results of the HIC_{36}, very low potential danger is indicated by the 3 ms-Criterion. Even at a tip velocity of 2 m/s less than 20 % of the lower limit of 72 g are reached, see Fig. 5.6.

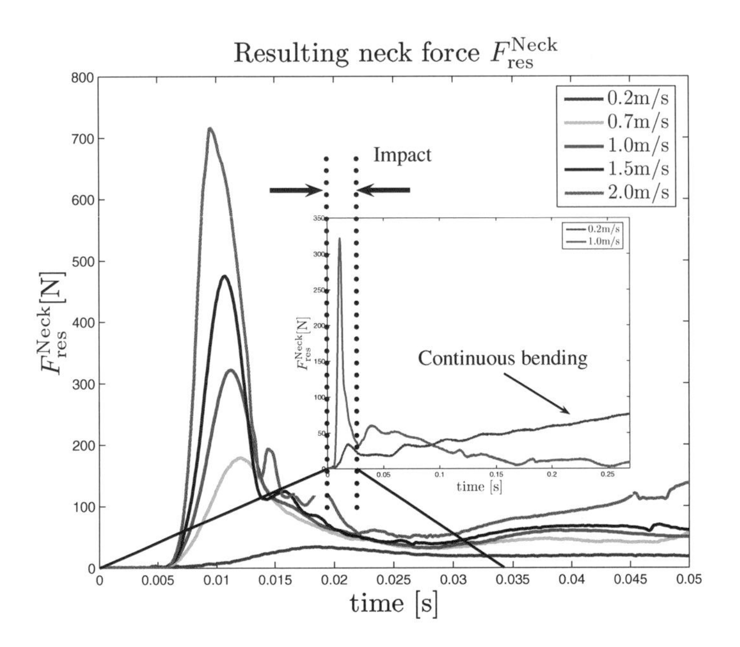

Fig. 5.7 Resulting impact force during head impacts

5.2.3 *Results for the Neck*

The resulting neck force F_{res}^{Neck} for varying robot velocities caused by head impacts is illustrated in Fig. 5.7. The actual impact is characterized by a very short peak with duration and maximum value dependent on the impact velocity. For fast impacts a low-level safety feature of the robot activates and stops it because the specified maximum joint torques are exceeded. Therefore, the *maximum* neck force/torque during the entire collision is determined by this *peak* force/torque occurring within the first $5 - 20$ ms of the impact. On the other hand, if the impact velocity is very low (0.2 m/s), the *impact* force is reduced dramatically and does not trigger the low-level stopping mechanism. Consequently, steadily growing neck bending can take place, increasing neck forces to even larger values than the ones caused by the

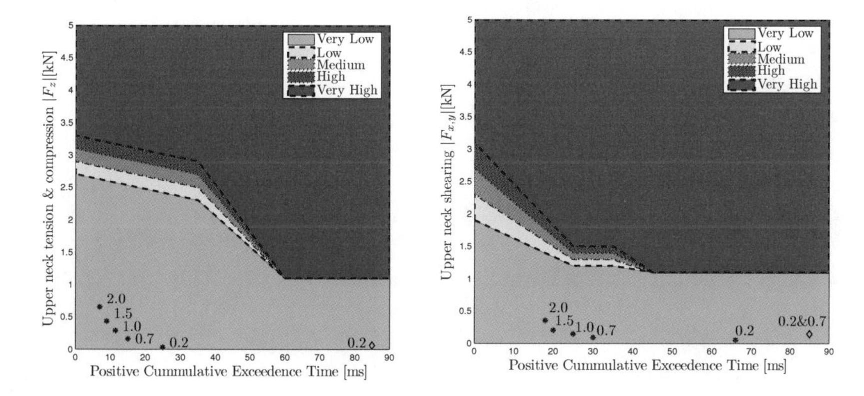

Fig. 5.8 Resulting $F_{x,y}$ and F_z values for varying impact velocities, rated according to the EuroNCAP *Assessment Protocol And Biomechanical Limits*

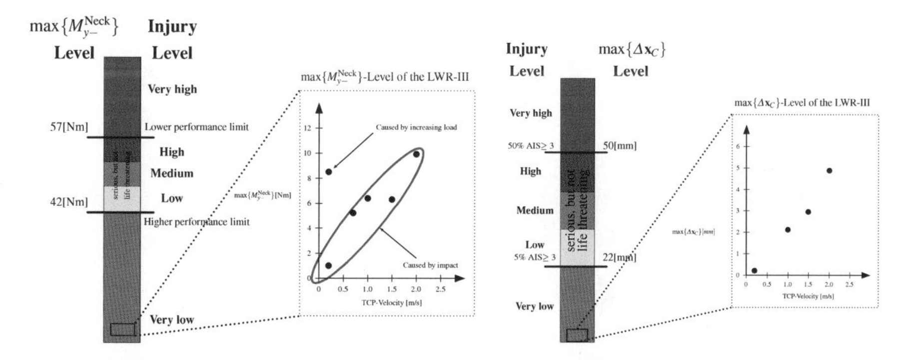

Fig. 5.9 Resulting $\max\{M_{y-}^{\text{Neck}}\}$ values for varying impact velocities, rated according to the EuroNCAP *Assessment Protocol And Biomechanical Limits* (left). Resulting $\max\{\Delta \mathbf{x}_C\}$ values for varying impact velocities, rated according to the EuroNCAP *Assessment Protocol And Biomechanical Limits* (right)

original impact as the robot continues to follow its desired trajectory. This becomes clear if the neck forces for the impact velocities 0.2 m/s and 1.0 m/s are plotted for a longer time period, see Fig. 5.7: After $\approx$ 20 ms both *impact* maxima are over and at 1 m/s the low-level stop of the robot is triggered because the impact forces (up to 2 kN were measured at the aluminum impactor) cause extremely high joint torques. In contrast, at 0.2 m/s the neck force is steadily increasing and might become even larger than impact forces at higher velocities.

In Figure 5.8 the occurring upper neck shearing and tension/compression forces are plotted with respect to the positive cumulative exceedance time. Only tension limits are specified in the EuroNCAP. However, according to [6] tension is more critical than compression and thus applying available limits to both, tension and compression seems to be a reasonable choice.

The tolerance values for neck forces are not constant, but a function of the exceedance time (see Sec.4.3). The particular neck tolerance values used in the EuroNCAP originate directly from biomechanical and forensic literature and are listed in standard textbook literature such as the *The Handbook of Forensic Medicine (German)* [6]. The resulting forces are labeled with the corresponding TCP velocity. In Fig. 5.9, a * indicates the forces caused by the impact and ◇ the ones by continuous bending, if they were larger than the impact forces. In order not to break the dummy neck, the robot stopped at a predefined distance after the collision occurred. This limits the bending forces & torques, which otherwise would further increase. In Figure 5.9 (left) the results of the extension torque are visualized. Similar to the previous head severity indices, the occurring neck forces/torques are totally subcritical, i.e. pose no threat to the human.

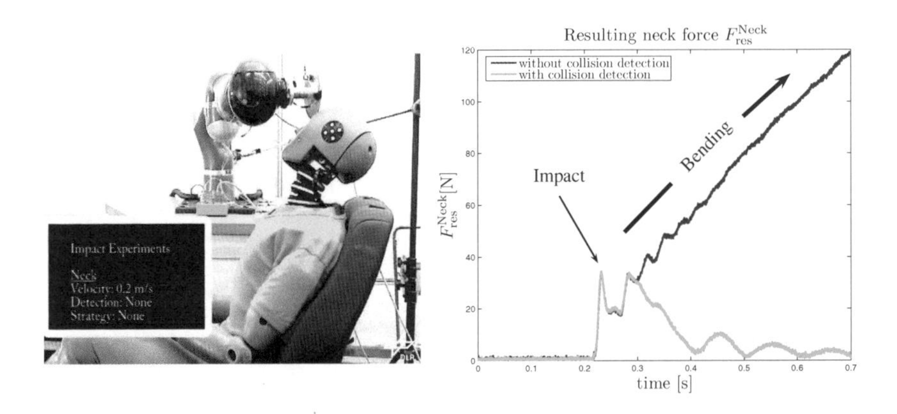

Fig. 5.10 Bending the dummy neck at a robot velocity of 0.2 m/s (left). Resulting dummy neck force with and without collision detection and strategy 2 (right).

The second experiment concerns quasistatic loading for partially constrained impacts. This was analyzed by pushing with the robot against a dummy head with a constrained torso. The experimental setup and the neck force F_{res}^{Neck} caused by head impacts for an impact velocity of 0.2 m/s are illustrated in Fig. 5.10. The actual impact is characterized by a very short peak, which duration and maximum value depend on the impact velocity. After this impact phase, a steadily growing neck bending force arises in absence of a collision detection. The plot with activated collision detection clearly shows the reduction in neck force due to the collision reaction strategy. In case of a constrained human as depicted in Fig. 5.10 (left) one is therefore able to limit the neck forces far below their critical value of 1.1 kN in any direction.

5.2.4 Results for the Chest

According to [40] a 5 %-probability of serious chest injury (AIS$\geq$ 3) corresponds to a compression of 22 mm and 50 % to 50 mm. In Figure 5.9 (right) the resulting compression values are plotted with respect to the impact velocity of the robot. Again, the injury potential is very low, as the values range in the lowest quarter of the *green* area.

The results of the viscous criterion are not presented because the resulting values were located within the range of noise, this criterion is therefore not well suited, nor sensitive enough for the evaluation. This is related to the relatively low velocities, compared to the ones encountered in automotive crashes.

5.2.5 Parenthetic Evaluation and Discussion

During the experiments at the ADAC, the standard measurements for automotive crash tests which can be acquired with a HIII for the head, neck and chest were performed. Injury indices for the head are related to its acceleration, for the neck to forces and torques and for the chest to acceleration and deflection. All calculations of the severity indices were carried out by the ADAC, thus were done according to the EuroNCAP. The main conclusion of the experiments concerning injury severity of humans is that all evaluated severity indices are located in the lowest quarter of the green area in the EuroNCAP color code.

This fact, surprising to the author and other robotics specialists (but not for the ADAC staff), can be explained by the fact that the maximal speed of the LWR-III (as of most industrial robots) is considerably lower than typical car velocities. Automotive crash test velocities usually begin at 10 m/s ($\equiv$ 36 km/h), which is a rather slow car velocity, but is never reached by geared robots. Accordingly, the main source of injury for car accidents is the high velocity; all indices are tailored to reflect this aspect. More specifically, the evaluation of severity indices as the HIC clearly indicates that severe injuries can be excluded during free impacts with a robot moving at speeds up to 2 m/s. The correlation to injury probability of the HIC according to [34] indicates that the probability of suffering from less or equal *minor*[5] injury is $p(AIS \leq 1) = 7.5 \times 10^{-5}$ % for the LWR-III at such velocities. This is a gratifying result and points out that the range of injuries which have to be treated during unconstrained blunt impacts are of very low severity. However, at the same time the need for indicators clearly tailored to low severity injuries seems apparent. To simply use the mapping of the HIC to injury probability [34] appears not differentiated enough since this criterion was clearly developed for much higher injury levels and primarily intended for separating life-threatening from non life-threatening injuries. Due to this re-focus on low injuries during free impacts with

[5] According to the Abbreviated Injury Scale.

robots[6], injury mechanisms need to be analyzed to appropriately represent this class of severity, and corresponding indicators have to be proposed. As outlined in the next section, several dangerous aspects in human-robot crashes can be identified and are worth to be treated in depth.

Apart from the stated results, some further conclusions will now be drawn related to the nature of robot impacts with rigid human body parts such as the head, which to some extent were unexpected. They give some new answers to safety questions posed in the robotics literature. An increase in intrinsic safety was unambiguously related to an introduction of joint compliance in the robotics literature as described in [5, 50]. It was stated that a drastic joint stiffness reduction is desired to realize a decoupling of the motor from the link inertia. In turn this reduces the reflected inertia during human-robot impacts. However, up to now it was unclear what exact joint compliance realizes this decoupling since this is heavily influenced by the contact properties of the human. In this sense, the experiments also give some insight into this question and show that a pure structural compliance (in this case mainly inherent in the Harmonic Drive and the joint torque sensors) as the one of the LWR-III is sufficient to realize this desired behavior.

5.2.5.1 Typical Impact Characteristics

Figure 5.11 (left) shows the recordings of an impact with the dummy head at 2 m/s. It displays the torque τ_4 in the 4-th joint, as well as the acceleration $||\ddot{\mathbf{x}}_{\mathrm{Al}}||$ and force F_{ext} at the tip. The first observation is that the impact peak at the contact between robot and head is very short (only $6-10$ ms), while the propagation of the impulse over the robot inertia and the joint elasticity leads to a considerable delay in the joint torque peak. The consequences shall be discussed in the following.

5.2.5.2 Collision Detection and Joint Stiffness

Before the joint torque starts increasing, the relevant force/acceleration peak period is practically over. Thus, during this particular time interval motor and link inertia are decoupled by the intrinsic joint elasticity, and only the link inertia is involved in the impact. Therefore, decreasing joint stiffness e.g. via antagonistic actuation would not have any effect on a (hard contact) head impact with link inertias similar to, or higher than the ones of the LWR-III. At this point it is implied that the flexible joint assumption holds for similar lightweight designs[7]. For collisions with softer body parts (e.g., the arm as outlined in [15]) the impact duration is higher and decreasing joint stiffness might reduce contact forces. To validate this statement, the resulting contact force was simulated with a dummy head model[8] and

[6] From now on impacts not being faster than 2 m/s are assumed if not stated otherwise.

[7] For a very stiff and heavy industrial robot this is e.g. not the case.

[8] The model is extracted from real impact data.

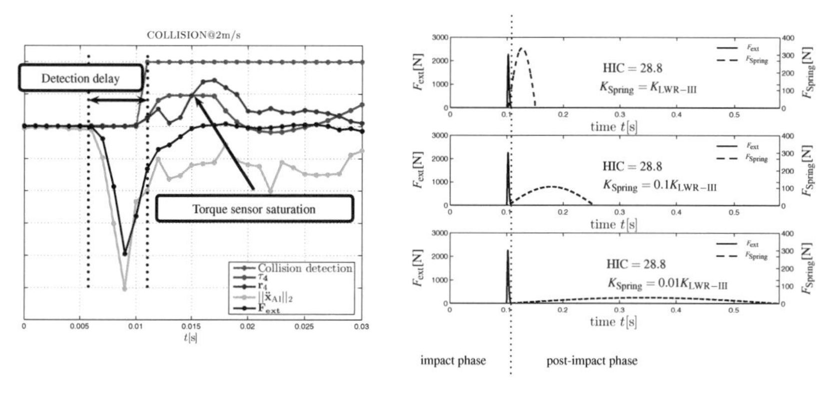

Fig. 5.11 Impact characteristics at 2 m/s. All values are scaled in order to fit into one plot. The plot is intended to show the timing of the signals: While acceleration $||\ddot{\mathbf{x}}_{AI}||$ and impact force $\mathbf{F}_{ext}$ are simultaneous, the joint torque τ_4 and the additional external torque estimation r_4 react delayed to the impact (left). Effect of stiffness reduction on impact force, HIC, and spring force (right). The solid line indicates the contact force and the dashed line the spring force generated by the joint stiffness. The spring force decreases in magnitude and increases in duration when reducing the spring stiffness. The HIC is constant with HIC = 28.8 for all three simulations.

a reduced LWR-III model for three different stiffness values[9]. This shows that the contact force, respectively the HIC is practically invariant with respect to a reduction of joint stiffness to values below the one of the LWR-III, see Fig. 5.11 (right). The spring force starts principally increasing after the maximum contact force was reached, right before the contact to the head is lost. Therefore, neither the reduction of joint stiffness nor of the motor inertia have an influence on the (very short) impact dynamics even for such intrinsic joint stiffness of the LWR-III. Only the link side inertia is influencing the impact force, see Fig. 5.12.

In order to investigate whether a physical collision detection scheme is able to reduce impact characteristics, the collision detection and reaction scheme from Chapter 3 is used in the experiment and indicated in Fig. 5.11 (left). Alternatively, the acceleration signal of the impactor, i.e. an ideally fast detection, was utilized to trigger the reaction schemes. In both cases the resulting values of the injury indices did not differ from the ones obtained without any reaction strategy. This is due to the inability of the motors to extract the kinetic energy fast enough to decrease the impact dynamics.

Three main conclusions concerning severity reduction of impact characteristics can be drawn:

- No physical collision detection and reaction mechanism is fast enough to reduce the impact dynamics of fast and rigid impacts for the considered robot type.

[9] The simulation is one-dimensional, meaning that reflected motor and link inertia as well as reflected joint stiffness are used to simulate this collision.

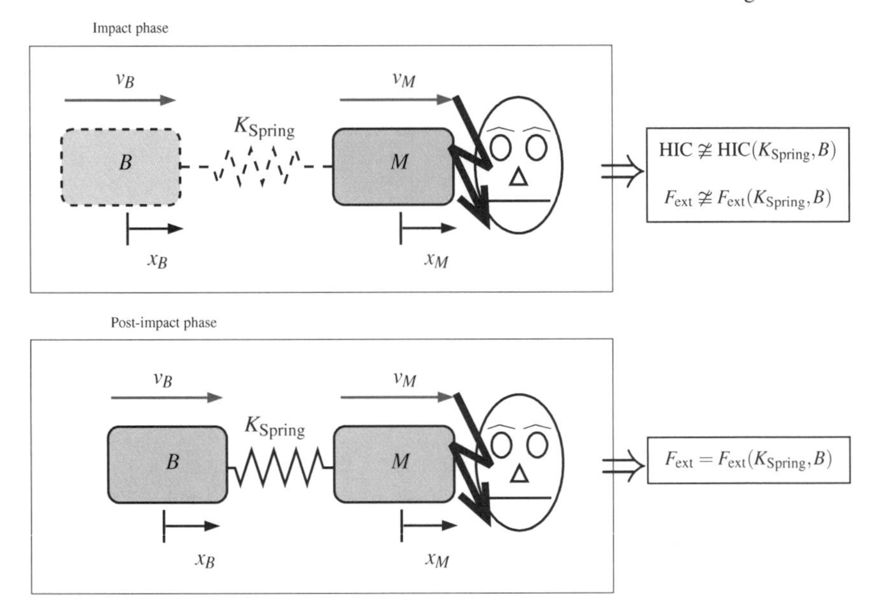

Fig. 5.12 A rigid impact between a compliant joint and the human head is already at moderately high joint stiffness mainly a process between the link inertia and the human head. Please note that it is referred to *Impact phase* and *Post-impact phase* in the sense that the former is relevant for the calculation of the HIC or maximum impact forces and the latter is not. Please compare to Fig. 5.11 (right) as well. On the one hand, due to the short impact duration the link side inertia is basically decoupled already by the intrinsic joint spring even without introducing more joint compliance. On the other hand, the following post-impact phase is highly depending on the joint stiffness.

- For such impacts further joint stiffness reduction does not lower impact forces or severity indices since motor and link inertia are already decoupled.
- Soft covering is an adequate countermeasure to reduce the impact effectively.

Apart from these characteristic properties another important observation, made at impact velocities starting from 1 m/s, is that the specified maximum joint torques of the robot were exceeded for several milliseconds during the impact, see Fig. 5.11 (left)[10]. This shows that the robot is exposed to enormous loads during such contacts and countermeasures are needed for ensuring safety of the robot. Speed limitation to subcritical values is one option, others include reduction in joint stiffness [16, 17] or fast collision reaction strategies. Both measures, though not effective in protecting the human in case of free impacts, can help to protect the robot joints. This is due to the difference of the duration of the impact itself and the joint torque peak, see Fig. 5.11 (right).

[10] A mechanical end stop in the robot limits the deflection range of the torque sensor which then goes into saturation. A low-level emergency stop is initialized as soon as this event is triggered.

At this point another remark concerning the deliberate introduction of mechanical compliance into the joint as e.g. in [33, 5, 46] shall be made:

1. On the one hand it was shown that adding more compliance into the joint does not further reduce the impact characteristics already for the relatively high intrinsic joint stiffness of the LWR-III.
2. On the other hand, it shall be pointed out that introducing an elastic joint element makes it possible to store and release energy during motion[11]. By utilizing the intrinsic joint stiffness it is possible to achieve link velocities above motor levels by choosing an appropriate trajectory. This energy storage and release mechanism gives animals their ability to have outstanding peak performance by means of velocity and was recently used for robots similar to a catapult in [43] and for performance increase in [38, 37, 49, 17].

As will be shown later, impact velocity is the main governing factor during a rigid impact. Thus, a joint design which is intrinsically faster is actually more dangerous by design. One could even argue that a compliant joint is more dangerous than a stiff one in some worst-case conditions (e.g. operated at maximal velocity). Therefore, additional control and planning measures have to be taken in order to keep a compliant joint safe in dynamic operation mode. More details on this issue are discussed in Chapter 10.

5.2.6 Human-Robot Impacts

Due to the results described in the previous sections, and to give the proof for the extremely low injury risk during blunt impacts with the LWR-III, impact tests at increasing robot speed were carried out with a volunteer for the chest, abdomen, shoulder, and the head. Impact speeds ranged up to 2.7 m/s for the first three body parts and up to 1.5 m/s for the head. During the entire experimental series the collision detection was switched off (more accurately: the detection was activated but the robot was programmed to continue its desired trajectory in case of a collision). Only a low-level feature of the robot engaged the brakes in case of exceeding the maximum nominal joint torques of the robot. However, this feature is not able to affect the impact itself due to the delayed increase of the joint torque (see Sec. 5.2.5.2).

As predicted by the dummy tests *no* injury could be observed even at such high speed impacts.

Fig. 5.13 Impact tests with a human chest at 2.7 m/s and head at 1.5 m/s. The impact velocities for the abdomen and the shoulder were 2.7 m/s as well, which is the maximum velocity of the robot. During all these experiments the robot does not react to the activated collision detection. The robot stopped only due to an exceedance of the maximum nominal joint torques. However, as a result from the crash test dummy experiments, the impact forces caused by the very short collision duration cannot be affected by this feature due to its delayed reaction. These test were initially shown in [11] and support, if not even prove the previously given conclusions.

Table 5.1 Human walking/running speeds according to [20]

Running Type	Velocity [m/s]
Slow walking	0.5
Fast walking	2.0
Race walking	4.0
Running (world record 2006)	10.35

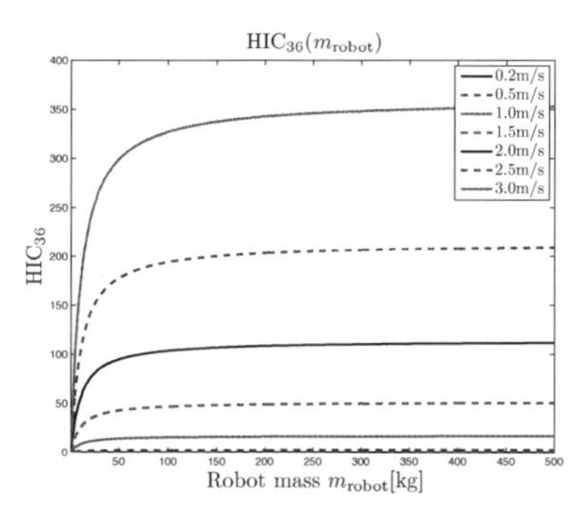

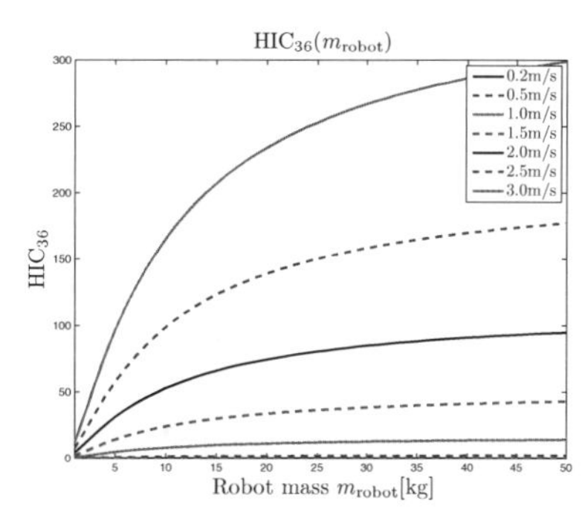

Fig. 5.14 Resulting Head Injury Criterion calculated from simulated 1-DoF impacts between a robot with increasing mass and a dummy head model deduced from real impact data. Clearly, a saturation effect can be observed with increasing robot mass. In other words only the impact velocity is relevant above a certain robot mass. This can be explained by an intuitive analogy: Whether the robot hits the human or the human hits the robot is not relevant. Therefore, being hit by an infinite mass robot at 2 m/s is basically the same as running with 2 m/s $\equiv$ 7.2 km/h (fast walking) against a rigid wall. The intuition already tells from everyday experience that such an impact is certainly hurting but never even close to life threatening.

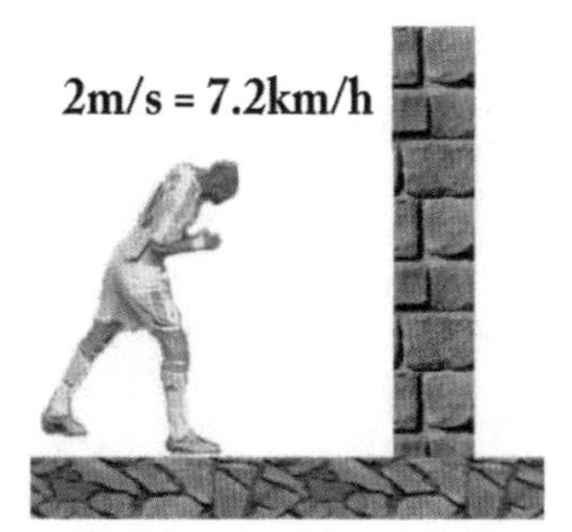

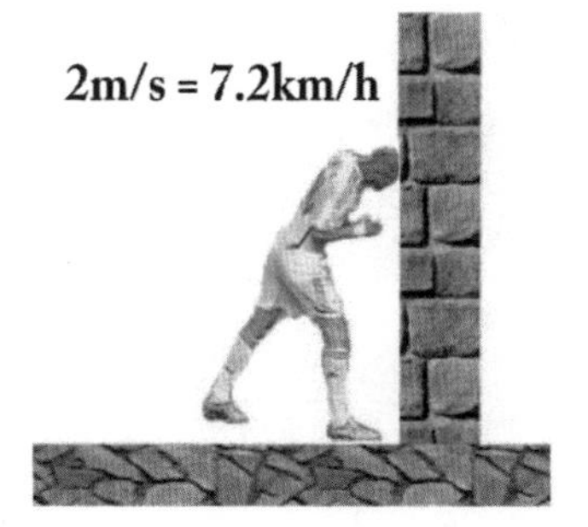

Fig. 5.15 Running against a rigid wall

5.2.7 Influence of Robot Mass and Velocity

Since the LWR-III with its lightweight structure is especially designed for the close cooperation with humans, it is desirable to evaluate the effect of the robot mass on the dynamics of such an impact for a more general class of robots. Apart from the robot's mass, the influence of its velocity is of interest. Figure 5.14 shows the dependency of the HIC on the robot mass up to 500 kg with the graphs being parameterized by impact velocity. Two main statements can be deduced:

- The HIC saturates with increasing robot mass for all impact velocities.
- Impact velocity is the major factor defining the injury severity.

[11] Please note that this is not a discussion about variable joint stiffness but about a low constant joint elasticity.

The first statement was unexpected as it contradicts the intuition of a massive robot being a priori life threatening. However, an intuitive and afterwards obvious interpretation of the saturation effect can be drawn: whether a massive robot collides at 2 m/s with a human head or the human runs with 2 m/s (which is equivalent to 7.2 km/h) against a rigid wall is nearly the same, see Fig. 5.15. This intuitive example already shows that one would not be seriously injured, even though this impact occured at relatively fast walking speed, see Tab. 5.1, where the velocity of human walking up to world-class running according to [20] are listed. Therefore, even the infinite mass robot cannot become dangerous at 2 m/s by means of *impact related criteria* used in the automobile industry (such as the HIC), as long as clamping and impacts with sharp surfaces can be excluded.

To further clarify, assume a simple mass-spring-mass model for the impact between human and robot[12]. M_H and M_R are the reflected inertias of the human and robot. K is the contact stiffness which is in case of a rigid robot mainly the stiffness of the human contact area. $\dot{x}_R^0$ is the relative impact velocity between the robot and human. Solving the corresponding differential equation leads to the contact force

$$F_{\text{ext}} = \begin{cases} \underbrace{\frac{M_R}{M_R + M_H} \dot{x}_R^0 \omega_n \cos(\omega_n t) M_H}_{\ddot{x}_H} & \text{if } |t| < \frac{T}{2} \\ 0 & \text{else,} \end{cases} \tag{5.1}$$

where $\omega_n = \sqrt{\frac{M_R + M_H}{M_R M_H} K}$ and $T = \frac{2\pi}{\omega_n}$. The maximum value of this force is consequently

$$F_{\text{ext}}^{\max} = \frac{M_R}{M_R + M_H} \sqrt{\frac{M_R + M_H}{M_R M_H}} \sqrt{K} \dot{x}_R^0 M_H = \sqrt{\frac{M_R M_H}{M_R + M_H}} \sqrt{K} \dot{x}_R^0. \tag{5.2}$$

If the robot mass is significantly larger than the human head mass[13], i.e. $M_R >> M_H$ this reduces to

$$F_{\text{ext}}^{\max}(M_R >> M_H) = \sqrt{K M_H} \dot{x}_R^0. \tag{5.3}$$

This shows that for a robot with significantly larger reflected inertia than the human head, only the contact stiffness, the impact velocity, and the mass of the human head are relevant but not the robot mass. In other words, the intuitive analogy of "Being hit at a certain velocity by an infinitely large robot is basically the same as if the human is running at this particular velocity against a rigid wall" is confirmed.

In order to help quantify the influence of the reflected inertia of a particular robot during an impact with a mass-spring complex the *inertial saturation coefficient* is introduced.

---

[12] For the HIC a Hunt-Crossley model was assumed but at this point the discussion is kept simple and therefore a linear spring between robot and human head mass is assumed.

[13] Assuming a simplifying decoupling of the head from the torso, which holds for the short duration of the impact. For the post-impact phase, neck stiffness and body inertia have to be considered, which complicates the analysis considerably.

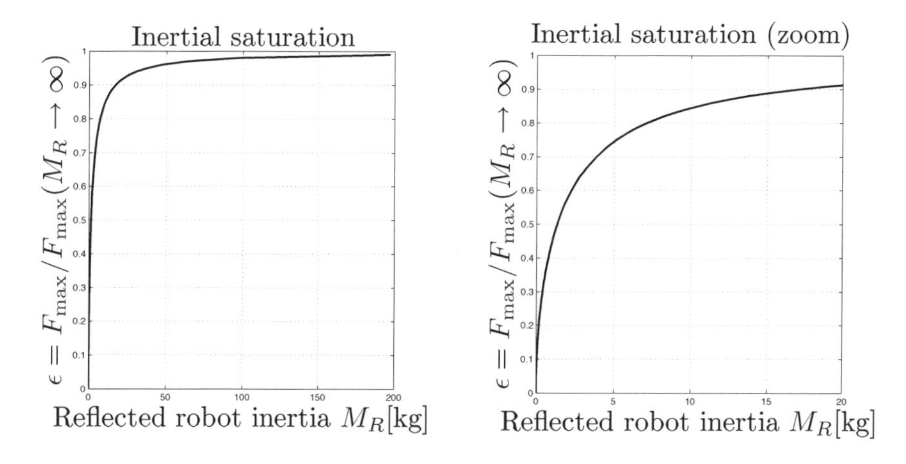

Fig. 5.16 The inertial saturation coefficient ε describes the effect robot mass has on the maximum contact force during an impact between a robot and a human. A reflected inertia of ≈ 17 kg causes already 90 % of the maximum possible contact force.

$$\varepsilon := \frac{F_{\text{ext}}^{\max}}{F_{\text{ext}}^{\max}(M_R >> M_H)} = \sqrt{\frac{M_R}{M_R + M_H}} \leq 1 \tag{5.4}$$

This quantity describes independently from the contact stiffness and impact velocity up to what percentage of the maximum (saturated) contact force is generated by a particular robot, see Fig. 5.16. In other words, it is possible to define the maximum allowable force level (percentaged by means of the saturation force), leading to requirements concerning the maximum reflected inertia of the robot. As will be pointed out in Sec. 5.3.2 it has to be distinguished between different body regions and their characteristic contact parameters and tolerance forces if the actual injury shall be evaluated.

Based on the preceding impact analysis the particular injury heavy-duty robots would cause for rigid blunt impacts is discussed in the next section in more detail .

5.3 Blunt Unconstrained Impacts for General Robots

In this section, the experimental confirmation of the statements given in Sec. 5.2.7 regarding saturation of the HIC with robot mass is presented. The results clearly indicate that the HIC and similar criteria which refer to severe injury have low values. It is crucial to evaluate lower severity injuries and find adequate measures for them. The evaluation of the HIC and related criteria significantly reduces the range of injury severities to be investigated. Now, a closer examination at this lower range injuries has to be taken.

As recorded contact forces during all impact experiments were in the kN range, fractions of facial and cranial bones were identified as a potential injury worth to be investigated due to their correlation to contact force[14].

5.3.1 Evaluated Robots

In order to cover a wide range of robots and to be able to verify the saturation effect explained in Sec. 5.2.7, the 54 kg KUKA KR3-SI (a robot designed for Human-Robot Interaction), the 235 kg KUKA KR6 and the 2350 kg KUKA KR500 (Fig. 5.17) are compared with the LWR-III. The industrial robot tests were carried out with a simplified setup (denoted as Dummy-dummy), mimicking a HIII dummy head[15].

A feature of the KR3-SI, which has to be mentioned, is the safeguarding of the tool by means of an intermediate flange with breakaway function, triggering the emergency stop in case the contact force at the TCP exceeds a certain static force threshold[16]. In combination with the mounted impactor the weight of the flange-impactor complex is 1.4 kg.

5.3.2 Head Injury Criterion and Impact Forces

In Figure 5.32 the resulting HIC values for the different robots are depicted and classified according to the EuroNCAP. The values for the KR3-SI are even lower than for the LWR-III because the intermediate flange decouples the impactor at the moment of impact from the entire robot. Therefore, only the flange-impactor complex is involved in the impact. Furthermore, the saturation effect explained in Sec. 5.2.7 is observed, as the numerical values for the KR6 and KR500 do not significantly differ. The simulation results presented in Fig. 5.14 should be considered as conservative, since the actual saturation value is even noticeably lower than predicted by simulation. This result indicates a *very low* potential injury and the probability of a resulting injury level of AIS ≥ 3 according to [34] is maximally $\approx 0.15\%$, i.e. negligible. The HIC for the KR500 measured at 80 % and 100 % of maximum joint velocity, corresponding to a Cartesian velocity of 2.9 and 3.7 m/s, was 135 and 246. This means that even an impact of such a huge robot as the KR500 cannot pose a significant threat to the human head by means of typical severity indices from automobile crash testing. The injury level for these values are located in the *green* area as well, and the probability of AIS ≥ 3-injuries are 1.2 % and 3.6 % for the faster impacts with the KR500, see Fig. 5.32.

[14] Their fracture tolerance correlates to certain contact forces.

[15] This was due to the high costs of crash tests at certified facilities.

[16] The initiated emergency stop is a Category 0, 1 stop according to DIN EN 60204. Category 0 stop means that the drives are immediately switched off and the brakes engage at the same time. A Category 1 stop lets the robot halt with a hard stop trajectory without using the brakes.

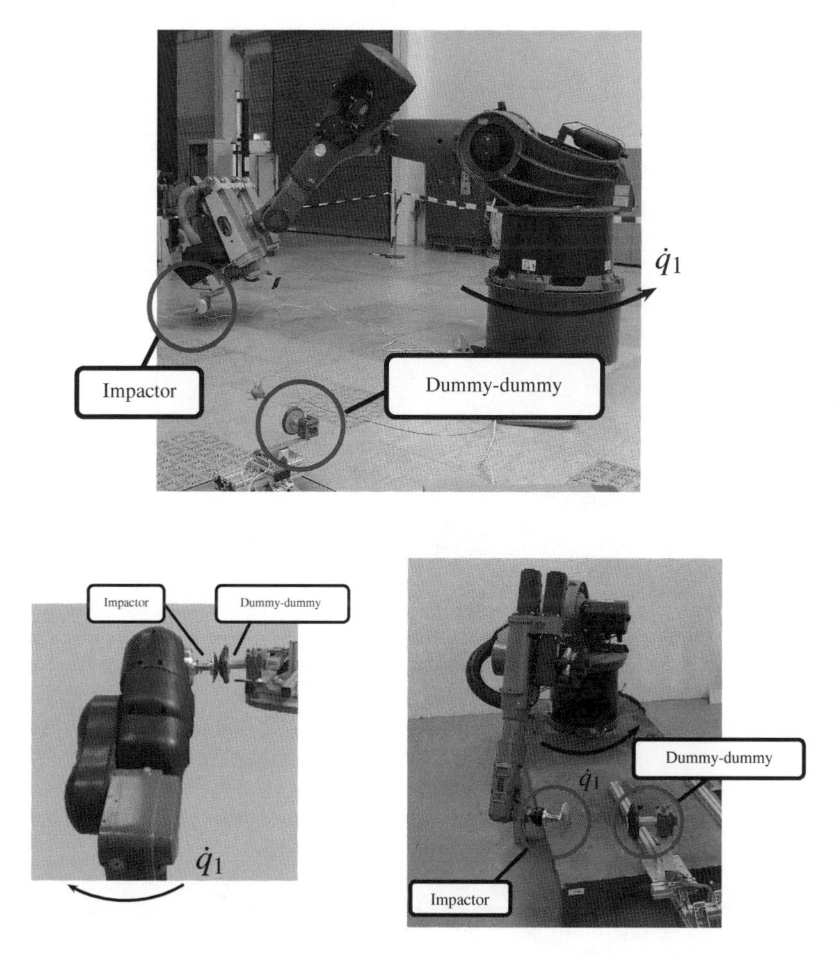

Fig. 5.17 Setup of impact tests with the KUKA KR3-SI (lower left), KUKA KR6 (lower right) and KUKA KR500 (top). Reflected inertias in the direction of impact were $\{12, 67, 1870\}$ kg.

The results indicate that the HIC and similar criteria are apparently not appropriate measures of possible injuries in robotics (by means of relevance for human-robot *interaction*)[17], necessitating the investigation of other injury mechanisms of lower severity like fractions of facial & cranial bones, which could occur during human-

[17] In contrast to the requirements in Human-Robot Interaction it is claimed in Chapter 9 that in *Competitive Robotics* a robot must not be more dangerous than a human [16, 17]. In order to be a peer opponent, as e.g. in the ultimate goal of RoboCup, the robot needs to have similar physical capabilities as a human, leading to extraordinary speed requirements. Since such impacts are coming close to velocities at which automobile crash testing takes place, injury measures as the Head Injury Criterion definitely can be used to evaluate possibly occurring injury there.

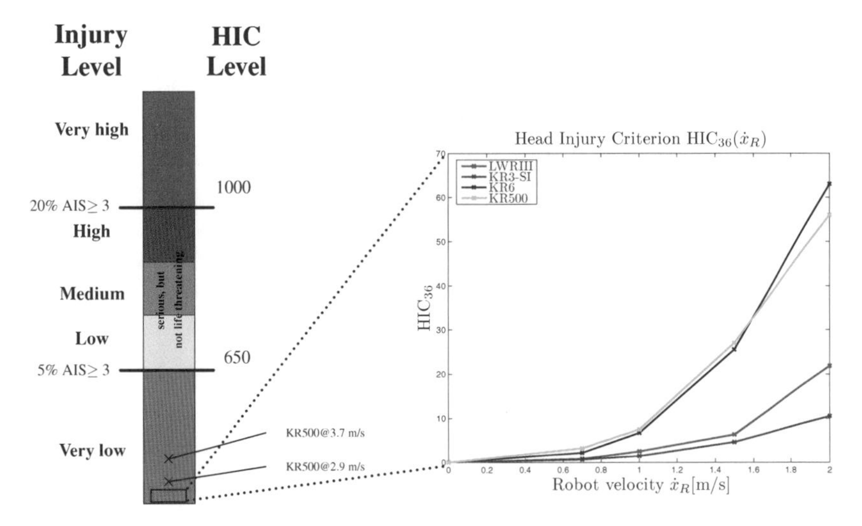

Fig. 5.18 Resulting HIC_{36} values at varying impact velocities for all robots, rated according to the EuroNCAP *Assessment Protocol And Biomechanical Limits*. All robots produced HIC at impact velocities up to 2 m/s values which are ranged in the lowest range of injury level. Furthermore, the previously described saturation effect of the HIC can be observed. In addition, the HIC for the KR500 was measured at 2.9 m/s and 3.7 m/s which is the maximum velocity for this robot. The resulting HIC values are still in the lower half of the *very low* injury level.

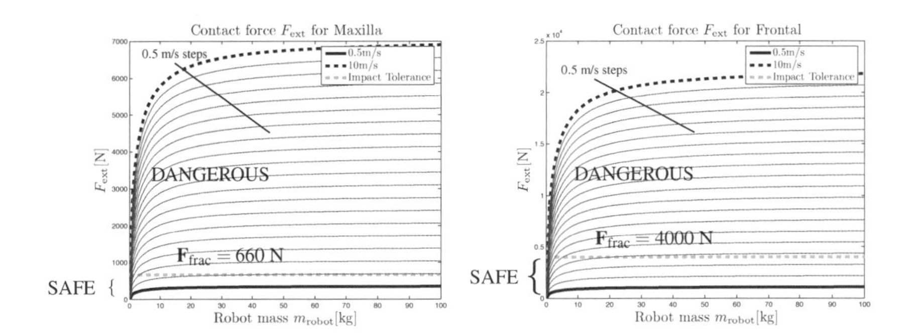

Fig. 5.19 Contact forces for simulated impacts between a robot and the frontal area (right) and the maxilla (left) showing the dependency on the robot mass and velocity. The impact velocity steps are 0.5 m/s. Similar to the HIC, a saturation effect can be observed and it becomes clear that for this conservative estimation already impact forces of 1 m/s potentially break the maxilla.

robot collisions. This is indicated by recorded contact forces of the discussed impact tests which were in the order of the fracture tolerance of these bones [13, 14].

In Figure 5.19 the dependency of the impact force w.r.t. the robot mass and velocity (the robot is assumed to move with constant velocity) for the frontal bone and the maxilla are visualized. Since the goal is to establish safety limits which ensure the prevention of fractures, the simulations were carried out for worst-case conditions[18]. For all bones[19], except the frontal one it appears that starting from the saturation mass value[20], a velocity between 0.5–1.0 m/s is enough to cause fractures. The frontal bone on the other hand is very resistant, generally withstanding impacts approximately up to 2 m/s. Furthermore, it becomes clear that especially for robots with less than 5 kg reflected inertia at the moment of impact the velocity can be significantly higher without exceeding the limit contact force. For weaker bones like the maxilla impact speeds of 2 m/s are already posing a major fracture source even for low-inertia robots.

The experiments described in Tab. 5.2 validate the assumption of a conservative but nevertheless realistic upper bound. According to [23] the correlation between kinetic impact energy and injury severity by means of frontal fractures for cadaver head drop tests on ground were observed.

Table 5.2 Drop tests with cadaver heads

Energy [J]	Resulting injury
$50 - 100$	Drop from $1 - 2$ m height (4.6–9.6 m/s). Resulting in simple linear fracture of AIS $= 2$ or a more severe AIS $= 3$-injury
$100 - 200$	Complicated fracture with AIS ≥ 3 injury severity
≈ 200	Vascular injury, therefore hematoma. Combination of AIS for skull and brain AIS > 3

Below 50 J usually no fractures occur. An impact velocity of 2 m/s would mean a kinetic energy of 10 J at a drop height of 0.2 m. The impact force would be 4.4 kN for the assumed stiffness of the frontal bone in Fig. 5.19 (left), implying a fracture already at 10 J. This can be explained by the conservative estimation of the frontal stiffness which neglects the comparatively slowly increasing force in the beginning of an impact [4, 3]. Therefore, Fig. 5.19 (right) and Fig. 5.19 (left) are overestimating the resulting injury. However, e.g. in [30] it is shown that frontal fracture can already occur at 2–3 kN for smaller contact areas and [44] indicates frontal fractures already at 37 J[21]. Due to the significant biomechanical variation

[18] The contact stiffness is assumed to be the worst-case found in the literature.

[19] Simulations for other facial and cranial bones were carried out as well and show similar behavior.

[20] The robot mass from which on a further increase does not result in significantly higher forces.

[21] An impactor was used, i.e. drop tests with a pre-defined impactor mass were carried out.

found in the literature the most conservative contact stiffness is assumed, leading to an upper bound which is conservative in the range of factor 2. Compared to the *ISO-10218* which is conservative[22] in the range of more than an order of magnitude (for both, the force and velocity), the suggested limits prevent the strong limitation of robot performance demanded by the *ISO-10218*.

In order to estimate the consequences after a fracture occurs one has to take into consideration that the initially applied human model is no longer valid after the fracture. This is because the resistance of the human head is dramatically lowered, possibly causing even more severe injury. A precise statement about these consequences is currently not possible but the experiments according to [23] give first hints. Furthermore, empirical data on cadaver experiments at ≈ 22 km/h (≈ 6 m/s) with an impactor of 23 kg exists [42, 22]. Such impacts lead to maximum AIS $= 3$ injuries for facial impacts, while evaluating the skull, brain, neck, and skin. Note that the authors state that for reality (meaning living humans) AIS $= 4$ is not excluded. Based on these experiments it may be presumed that, due to the increasing injury severity with impact velocity, much less severe injuries occur at the typical robot velocities investigated.

The next section describes clamping simulations based on measurements with several industrial robots to examine at the large injury potential posed by environmental constraints.

5.4 Constrained Blunt Impacts

In the preceding part of this chapter non-constrained blunt impacts were investigated with respect to robot mass and velocity. The effects of these robot parameters in case of clamping are outlined in this section. Robotics literature deals mainly with free impacts [5, 18, 50], only few works as e.g. [26] give a short notion about the injury potential emanating from clamping.

Concerning injuries caused by robots, only little data or literature is available. In [45] the United Auto Workers (UAW) union published a report which provides raw data on various injuries related to robot operations. It indicates that a major fraction of occurring injuries involve somehow clamping of a human body part. Since it is not feasible to adequately treat all different contact types at the same time, this section concentrates again on blunt contact.

A typical situation where a human operator can be clamped is e.g. during maintenance of a robotic work cell. Due to the (partially) confined workspace it is possible to get clamped e.g. between the safety fence or a workbench and the robotic structure[23]. In order to analyze the mechanisms behind such a process it is first explained which types of blunt clamping are relevant to robotics and next the braking distance

[22] The *ISO-10218* imposes a velocity limit of 0.25 m/s, corresponding to a drop height of 2 mm.

[23] Clamping can as well occur within robotic elements, such as two links, but this is not part of the analysis.

of various investigated robots is given. These tests are especially done for estimating the equivalent braking force for a one-dimensional impact simulation[24], which is used to evaluate maximal contact forces and evaluate severity indices. This is necessary to analyze constrained impacts with biomechanical models of the human head and chest. Because unfortunately, real clamping tests with a crash-test dummy (e.g. HIII) and heavy-duty robots are not realizable without destroying the equipment these validated simulations needs to be relied on. In these simulations it is assumed that the robot is able to detect a collision and immediately engages its brakes. It seems clear that (at least) an industrial robot is able to generate forces high enough to kill a clamped human if it is not able to react at all and just continues to follow its desired trajectory.

Furthermore, it is shown that with a robot like the LWR-III, which is especially designed for Human-Robot Interaction, clamping is under normal circumstances not leading to life-threatening injury by means of typical injury measures from the automobile industry, but less severe injuries like fractures of facial and cranial bones can occur (for a conservative analysis)[25].

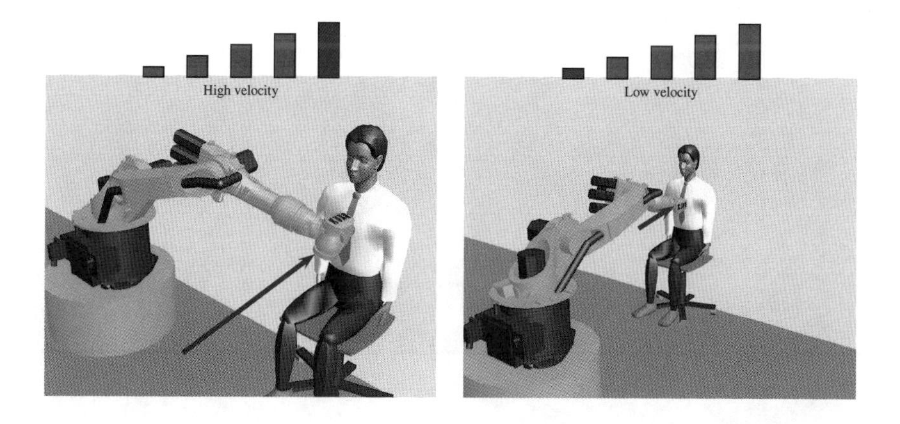

Fig. 5.20 Two different types of clamping: Dynamic clamping at high Cartesian velocities (left) and quasistatic clamping during low velocity movements or near singularities (eventually high joint velocities but slow Cartesian velocity)

5.4.1 Types of Blunt Clamping

Generally, two types of blunt clamping can be differentiated: Dynamic and quasistatic. According to [45] the first one is a major injury source in industrial

[24] A full dynamic model of the industrial robots for simulation is not available.

[25] This does not mean these are the only possible injuries, other ones like Contre-Coup [6] or secondary injuries need further investigation.

applications and will be the focus of this section. The second one occurs if the robot is moving slowly or if the robot is close to a singularity. This is discussed in Sec. 5.5, 5.6.2.2, and 5.6.2.4.

- *Dynamic Clamping:* Dynamic clamping describes the situation where the human is trapped against a rigid object while the robot moves at considerably high Cartesian velocities and hits the human body part as indicated in Fig. 5.20 (left).
- *Quasistatic Clamping:* The injury potential of a quasistatic collision stems mainly from the maximum force the robot is able to exert and the space available to crush the body part[26] as indicated in Fig. 5.20 (right).

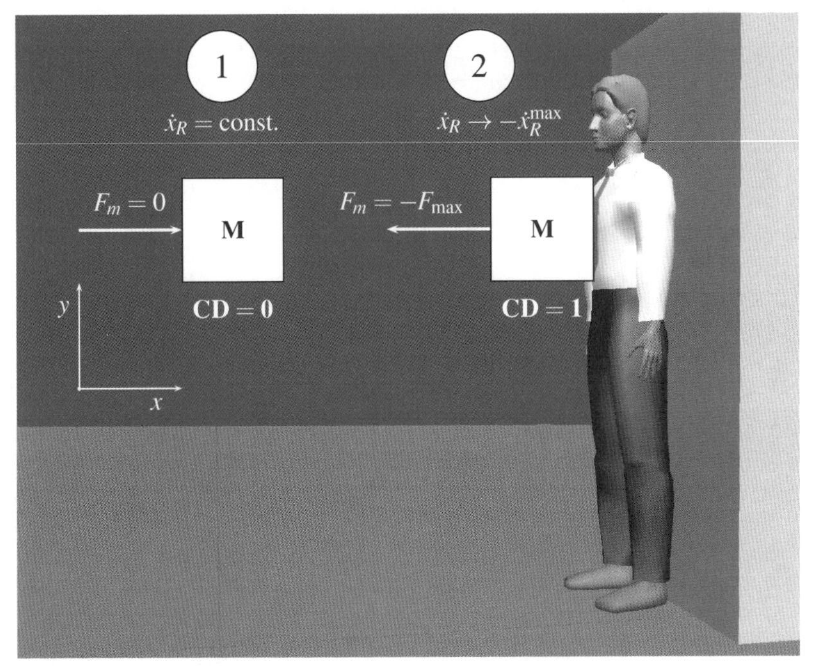

Fig. 5.21 Reduced clamping model for the industrial robots. *CD* denotes the binary collision detection signal. The robot is assumed to approach with constant velocity and as soon as a collision is detected exerts the maximum braking force on the robot inertia reflected at the tip in moving direction until contact with the clamped human is lost.

5.4.2 Braking Tests

The braking distance was measured at various initial velocities, serving two purposes:

---

[26] The space available describes whether enough distance is available with respect to the robot's workspace in order to exceed the particular tolerance values of the body part.

Table 5.3 Inertial key facts of evaluated robots

Robot	Weight [kg]	Nom. Load [kg]	Refl. Inertia [kg]
LWR-III	14	14	4
Kuka KR3-SI	54	3	12
Kuka KR6	235	6	67
Kuka KR500	2350	500	1870

1. Obtain and compare measurements of the braking distances of real robots at typical velocities.
2. Calculate the equivalent braking force for a reduced one-dimensional model of the particular industrial robot.

The one-dimensional model contains the relevant Cartesian direction of the reflected robot mass $M_c \in \mathbb{R}^{6\times 6}$ at the tip [24]

$$M_c = (J_c(\mathbf{q})M(\mathbf{q})^{-1}J_c(\mathbf{q})^T)^{-1}. \tag{5.5}$$

In order to measure the braking distance of the robots (except for the LWR-III) they were abruptly stopped at various (up to full) speeds with and without brakes during their commanded trajectory execution. In this chapter, the braking distance is used, obtained for a particular configuration and velocity, to simulate impacts with clamped humans. In Figure 5.21 the desired model is shown: The robot is represented by its reflected Cartesian inertia, listed in Tab. 5.3, and moves at constant velocity ①. As soon as a collision is detected the robot immediately brakes with maximum available force ②. The braking force acting on the reflected inertia is estimated from the real trajectories (see Appendix A). All models used for the head and chest of the human can be found in [36, 29, 4, 3, 27].

Table 5.4 compares the braking distance and time of all evaluated robots[27]. It shows that increasing the robot mass results in very large braking distances up to 690 mm for the KR500 at robot speeds up to 2 m/s at Category 1 stop. At maximum joint velocity (3.7 m/s Cartesian velocity) the KR500 needs almost 2 m at Category 1 to fully stop, see Fig. A.4. Category 1 stops significantly reduce the braking distance. Furthermore, a comparison concerning idle and stop time ($\Delta t_{\text{stop}} = \Delta t_{\text{idle}} + \Delta t_{\text{brake}}$) and idle and stop distance is given in Tab. 5.4 which already suggests the assumption that collisions could become fatal in case of clamping. Detailed plots of these experiments are given in the Appendix A.

[27] The LWR-III is compared with the KUKA KR3 (54 kg), the KUKA KR6 (235 kg), and the KUKA KR500 (2350 kg).

Table 5.4 Comparison of Cartesian braking distances and time for impact velocities of 0.2–2 m/s for all robots. For the LWR-III an impact reduces the braking distance significantly (shaded grey). Braking characteristics for maximum velocities (shaded red) of KR6 and KR500.

Robot	Δt_{idle}[ms]	Δt_{stop}[ms]	Δx_R[cm]	Δx_{idle}[cm]
LWR-III (link)	11–23	200	0.55–6.8	0.23–4.8
LWR-III (dummy)	11–23	200	0.25–4.2	0.23–4.8
LWR-III (motor)	4	250	not def.	not.def.
KR3-SI (Cat.0)	36–48	200–300	6.5–34	2.6–9.6
KR6 (Cat.1)	36–48	150–200	6–24	2.4–9.5
KR6 (Cat.0)	36	48–132	1–17	0.8–7
KR500 (Cat.1)	60–72	400–650	16–69	4.2–14
KR500 (Cat.0)	12–24	60–336	0.8–42	0.6–7
$\text{KR6}_{\dot{q}_1^{\max}}$ (Cat.1)	36	252	55	13
$\text{KR6}_{\dot{q}_1^{\max}}$ (Cat.0)	36	216	45	13
$\text{KR500}_{\dot{q}_1^{\max}}$ (Cat.1)	85	1000	186	26
$\text{KR500}_{\dot{q}_1^{\max}}$ (Cat.0)	36	564	121	13

5.4.3 Experimental Results with the LWR-III and KR6

Before fully analyzing clamping in simulation, two experiments, which give some important insights, are described in the following.

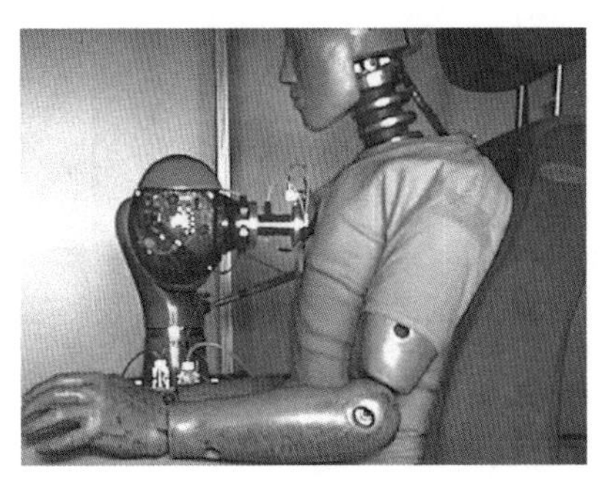

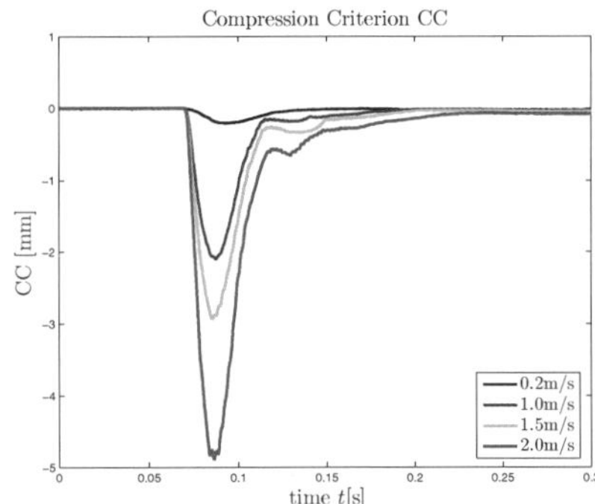

Fig. 5.22 Impact tests with a clamped HIII. The robot hits the dummy in outstretched configuration at various impact velocities (left). Measured Compression Criterion for a clamped HIII with the LWR-III. All values correlate to *very low* possible injury by means of the EuroNCAP (right).

5.4.3.1 LWR-III Chest Impact with HIII

In Figure 5.22 (left) an impact of the LWR-III with an HIII sitting in (and confined by) a car seat is shown. For all impact velocities the maximum nominal joint torques are exceeded and consequently the robot stops. Alternatively, in case the collision detection of Chapter 3.3 is activated the robot reacts compliantly since the reaction scheme is able to limit the joint torques and prevents the previously mentioned low-level stop. This is possible[28] up to impact velocities of almost 2 m/s. From the high-speed videos that were recorded at a frame-rate of 1 kHz it can be observed that the actual impact is completed before the trunk of the dummy starts moving and gets pushed into the seat. Therefore, the compliance of the seat did not influence the impact. In other words, the chest impact dynamics do not differ for the LWR-III, no matter whether the dummy is clamped or not. In Figure 5.22 (right) one can see the resulting Compression Criterion (CC) plots for various impact velocities $||\dot{\mathbf{x}}_R|| \in \{0.2, 1.0, 1.5, 2.0\}$ m/s. The maximal numerical value of 5 mm is far below the threshold value of 22 mm corresponding to *very low* injury by means of EuroNCAP. Therefore, no serious injury of the chest can occur with the LWR-III if the human is clamped because the maximal nominal joint torques are exceeded before the CC values could become critical. This is true, even if the collision detection fails.

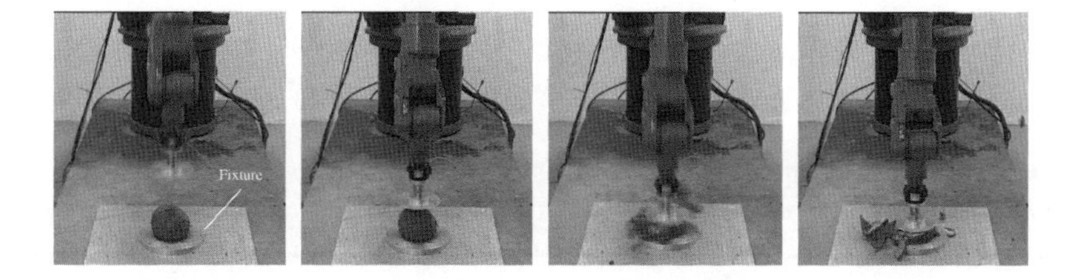

Fig. 5.23 Cracking a coconut with a KR6. An aluminum fixture keeps it centered.

5.4.3.2 Cracking a Coconut

A major drawback of crash-test dummies is that they cannot be used to measure forces acting on the clamped head. In order to illustrate the threat emanating from heavy high-torque robots it is demonstrated what a 6 kg-payload robot like the KR6 is already capable of via an intuitive example: Cracking a clamped coconut, see Fig. 5.23. The robot moves on a predefined trajectory in Cartesian space and impacts the coconut at 0.6 m/s. The coconut is not able to slip away due to an aluminum fixture keeping it centered. The force needed to crack the nut with the blunt impactor

[28] In contrast to the significantly harder impact with the head, where the collision detection and reaction cannot contribute to the reduction of joint torques anymore already at moderate robot speed [12].

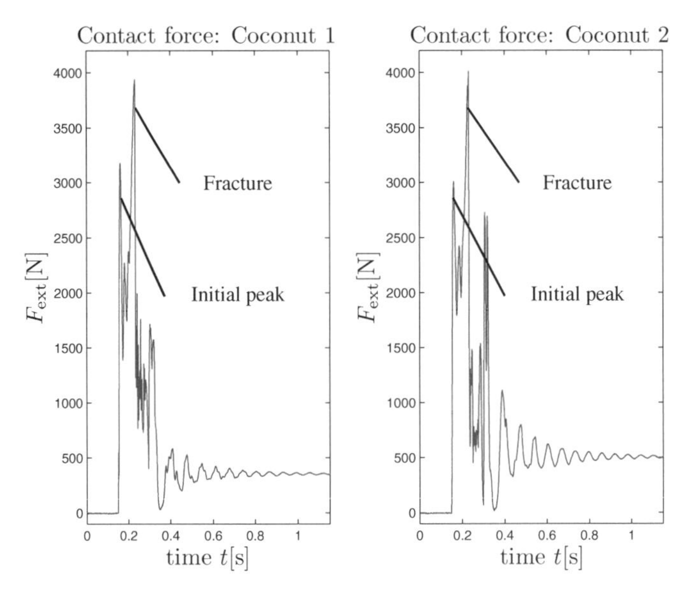

Fig. 5.24 Cracking a coconut with the KR6. Contact force profiles for two different sample coconuts.

is $F_{ext} \approx 4$ kN, as indicated in Fig. 5.24 by the force profiles of such cracks for two different coconuts. It is not entirely clear, whether the initial smaller peak is due to the dynamics of the impact (robot, controller, contact dynamics), slippage, or a first partial crack in the structure. However, slippage appears to be unlikely due to the reproducibility of the experiment, the fixture, and the high stiffness of the robot. An initial crack is also less probable due to the smooth behavior after the initial peak. The measured fracture force corresponds to the typical one of the human frontal bone [13]. The magnitude of the required fracture force shows that this experiment is a sufficient showcase for the clamping of a human head, which would behave similarly[29].

In the next subsection impact simulations are examined that lead to some more general statements about constrained head and chest impacts.

5.4.4 *Simulations*

In this section, the results of the impact simulations with a full model of the LWR-III and one-dimensional representations of the industrial robots are shown. The collision detection and reaction strategy for the LWR-III are as described in Chapter 3.3.

[29] However, according to [23] a human head would usually slip away for quasi-static loading. This was observed as well for the coconut, leading to the usage of the aluminum fixture.

Table 5.5 Conservative impact forces with clamping at 2 m/s obtained for the maxilla and frontal bone

Robot	Contact Force	Maxilla Fracture?
LWR-III	0.6 kN@1 m/s	No
LWR-III	1.2 kN@2 m/s	Yes
KR3	2.2 kN@2 m/s	Yes
KR6 (Cat.0&1)	5.1 kN@2 m/s	Yes
KR500 (Cat.0&1)	23.6 kN@2 m/s	Yes
Robot	**Contact Force**	**Frontal Fracture?**
LWR-III	3.5 kN@2 m/s	No
KR3	6.9 kN@2 m/s	Yes
KR6 (Cat.0&1)	16.3 kN@2 m/s	Yes
KR500 (Cat.0&1)	86.3 kN@2 m/s	Yes

The reflected inertias of the industrial robots and the description of fracture forces and severity indices can be found in Tab. 5.3.

5.4.4.1 Facial Impact Forces with Clamping

In Table 5.5 the clamping forces of the maxilla and frontal bone[30] for impacts at 2 m/s for all robots[31] in their particular impact configuration are listed. The robot reacts to the collision by braking with maximum reverse torque and continuing until contact with the head is lost. The simulations show the vast influence of the relation *robot mass↔braking or motor torque* and already the KR3 produces twice the contact force the LWR-III generates[32]. However, all robots can potentially break the maxilla, including the low inertia LWR-III at 2 m/s. Nonetheless, one should keep in mind that the model and fracture forces assumed in this simulation are kept conservative. The linear model assumption does e.g. not take into account an initial sub-linear characteristic of the real force-deflection relationship of the bone [4, 3]. Furthermore, the fracture forces used in [13] are conservative ones that were found in the literature. For the LWR-III the resulting maximally allowable velocity is ≈ 1 m/s for maxilla impacts if the stop is performed without brakes. With brakes this critical velocity could be significantly higher due to the reduced braking distance, see Appendix A Fig. A.2. For the frontal bone even 2 m/s is still a safe ve-

[30] Other bones were investigated as well, but their analysis would not contribute additional insight.

[31] For this simulation the KR3-SI is assumed to have no intermediate flange with breakaway function, i.e. a KR3 is assumed.

[32] The relation between motor torque and inertia scales disadvantageously when increasing dimensions.

locity in case of the LWR-III. For industrial robots a difference between Cat.0 and 1 stop cannot be observed, showing the inherent danger emanating from such heavy robots (for both evaluated bones). However, not only the force should be considered but the deflection as well. For the KR500 a numerical value of 236 mm is obtained for the maxilla, which is deadly. Additionally, one has to take into consideration that the applied human model is not valid anymore after the fracture occurs. This is because the resistance of the human head is dramatically lowered, possibly causing even more severe injury (higher deflections after the fracture will occur and lead to numerous internal injuries).

Table 5.6 Simulated values for chest severity indices and corresponding AIS values at 2 m/s obtained for the human chest

Robot	CC [mm]	VC [m/s]	F_{ext}^{x} [N]
LWR-III	14.4(0.0)	0.035	741.6(1.3)
KR3 (Cat.0)	31.2(0.0)	0.1	851.9(1.4)
KR6 (Cat.0)	65.5(2.0)	0.25	2836.1(2.7)
KR6 (Cat.1)	66.6(2.1)	0.25	2904.6(2.7)
KR500 (Cat.0)	228.0(6.0)	0.84	14282.0(6.0)
KR500 (Cat.1)	245.0(6.0)	0.89	15491.0(6.0)

5.4.4.2 Chest Impacts with Clamping

In Table 5.6 the CC, Viscous Criterion (VC), and the clamping force F_{ext}^{x} of the chest are listed for all robots at 2 m/s impact velocity. The corresponding EuroNCAP injury level is indicated for CC and VC. For the CC the AIS level, obtained by the mappings introduced in [13], is additionally given in brackets as well. The contact force F_{ext}^{x} is not part of the EuroNCAP evaluation but the corresponding AIS values according to [25] are denoted. The injury level of the CC and F_{ext}^{x} show how increasing robot mass leads to a higher probability of injury level with respect to the EuroNCAP definition and/or AIS. The LWR-III does not pose a threat to the human chest, as indicated in Sec. 5.4.3.1. The KR6 on the other hand can cause *very high* injury level by means of the EuroNCAP classification. The AIS mapping which is less conservative indicates approximately AIS $= 2$, meaning recoverable injury. The KR500 is deadly as intuition already tells. The Viscous Criterion is due to the still low velocities subcritical except for the KR500 because the deflection then dominates the criterion[33]. The same conclusions as for the CC can be drawn from the contact force and its correlating injury level.

Similar to the head it may be sum up that the chest is posed to a continuously increasing threat with growing robot mass if the human is clamped. CC and F_{ext}

[33] This is consistent with the fact that the VC is used for high velocity injuries in automobile crash-testing.

appear to be good indicators of injury for the chest in case of clamping due to their sensitivity in the relevant ranges.

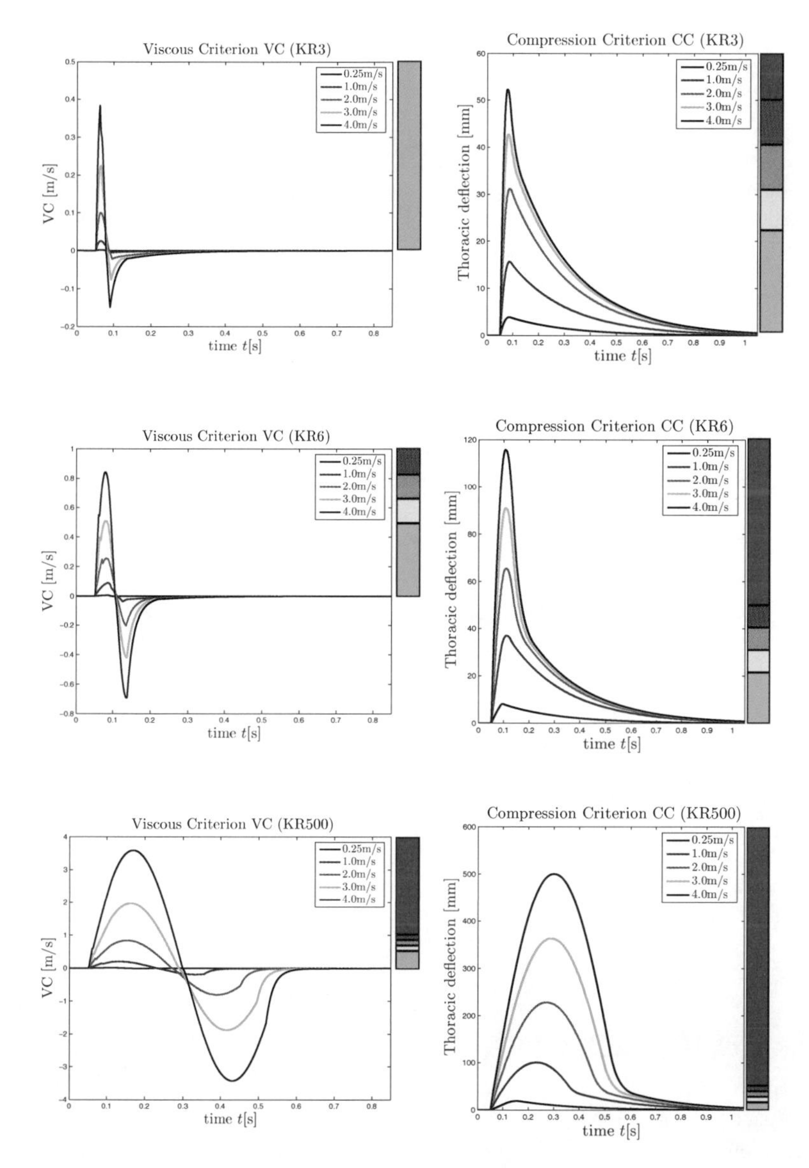

Fig. 5.25 Time courses of severity indices for simulated robot-chest collisions at various impact velocities with a clamped human chest for the KUKA KR3, KR6, and KR500. The left column shows the time evolution of the Viscous Criterion and the right column the one of the Compression Criterion. The colors indicate the injury potential with respect to EuroNCAP.

In Figure 5.25 the full time courses for CC and VC are given. The left column shows the time evolution of the Viscous Criterion parameterized by impact velocities up to 4 m/s and the right column the same for the Compression Criterion. The corresponding injury potential is indicated by the color-coded EuroNCAP bars. The compression criterion is clearly the more sensitive and appropriate criterion for this type of collision. For the KR3 a velocity of 2 m/s exceeds the possibility to keep *very low* injury potential, whereas for the KR6 even less than 1 m/s is enough to exceed this threshold. In case of the KR500 only very low speeds of less than 0.5 m/s are keeping the robot below the *very low* injury threshold.

After this investigation of dynamic blunt impacts with and without clamping, the problem of quasi-static loading is discussed as a case-study on the LWR-III. However, the resulting methodology of investigation can be applied to any robot as well.

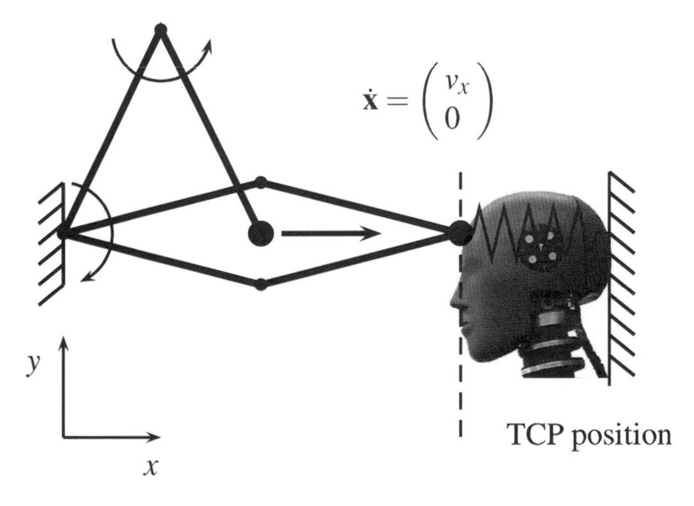

Fig. 5.26 Impact configuration for LWR-III-Dummy crash tests. Clamping the human with the robot in near-singular (almost outstretched) configuration. This is due to reconfiguration, meaning from "elbow up" to "elbow down" or vice versa.

5.5 Constrained Contact with Singularity Forces

At impact configurations with large levers, robots of similar inertias (and maximum joint torques) to the LWR-III do not pose a potential threat by means of the HIC [12]. On the other hand, the almost outstretched arm poses a significant injury threat which shall now be evaluated more in detail, see Fig. 5.26.

The maximum nominal torques for a given robot are represented by a hyper-rectangle. The corners of this hyper-rectangle are then transformed via the pseudo-inverse of the transposed Jacobian to the corners of a hyper-polygon of Cartesian forces. In order to get the maximal applicable force in the relevant worst-case

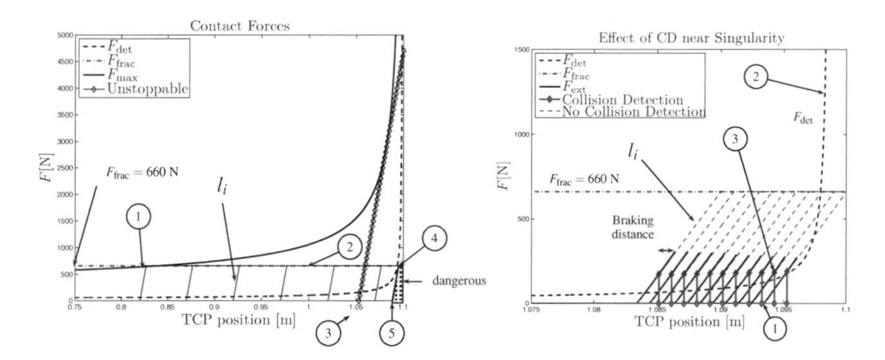

Fig. 5.27 Clamping of the human head with a rigid manipulator. The chosen bone for this analysis is the maxilla, whereas the theoretical analysis for the reconfiguration of the manipulator is shown (left). For better illustration the collision threshold is set to 10 % in this plot. The evaluation of the Collision Cetection (CD) schemes with a full dynamic simulation of the LWR-III confirms their benefit (right). This reconfiguration trajectory (see Fig. 5.26) was carried out at maximum joint velocity of the LWR-III which is 120 ^{o}/s. The deviation of the behavior of the LWR-III from the rigid case is mainly due to the intrinsically flexible joints and the contact modeling. In the lower plot the collision threshold is set to the currently lowest achievable value of 2 %.

direction, the corresponding hyper-rectangle corner has to be evaluated. Again, the Collision Detection (CD) of Chapter 3 is used. Its detection threshold τ_{det} for the external joint torque of the robot is defined as a percentage of the maximum nominal joint torque τ_{max} (e.g., 2 %) which directly leads to the detection threshold of the contact force.

$$\tau_{det} = 0.02\tau_{max} \rightarrow \mathscr{F}_{det} = 0.02\mathscr{F}_{max} = J^{T\#}\tau_{det} \quad (5.6)$$

$J^{T\#}$ is the pseudo-inverse of the transposed manipulator Jacobian[34]. To theoretically analyze the configuration boundaries which can cause fractions of facial and cranial bones the reconfiguration from "elbow up" to "elbow down" is the most dangerous case. The robot can be commanded in such a way that it passes the outstretched position if the clamped head is contacted close to this singularity. Since the human head would get clamped only slowly due to the low Cartesian velocities close to the singularity an acceleration based criterion as the HIC cannot indicate the force that is exerted on the head. Therefore, such criteria drop out entirely for this analysis. Instead, contact forces and related bone fractures are used as injury indicators[35].

[34] Note, that since the torque τ_{det} is produced only by a TCP force, any generalized pseudoinverse will lead to the same value of $\mathbf{F}_{det}$.

[35] This statement cannot be made for high-speed constrained impacts at the current state since the human head is not a rigid body and it cannot be excluded that a significant acceleration occurs during such impact.

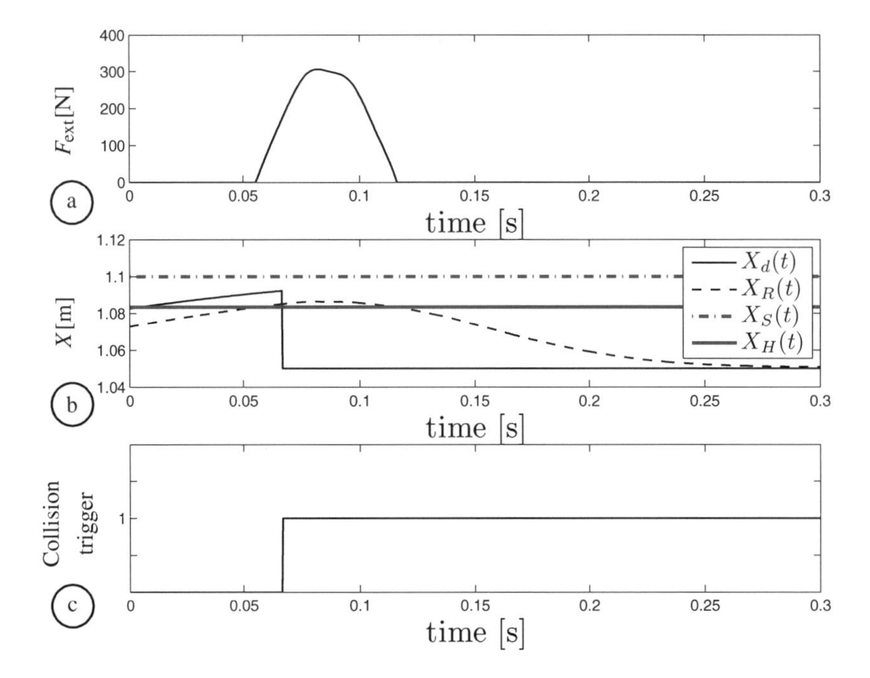

Fig. 5.28 Time courses for the constrained impact with the full dynamic model of the LWR-III and the human head close to the singularity. The robot is reconfiguring from "elbow up" to "ellbow down" configuration at maximum joint velocity in joint 4 and half this velocity in joint 2 and 7. The resulting trajectory is a straight line in x-direction. The human is standing 16 mm before the singularity. The collision reaction consists of setting the desired configuration to the configuration corresponding to 50 mm before the singularity, i.e. a "jump" backwards. The collision threshold is set to 2 %, which is the lowest threshold currently achievable and a realistic maximal joint torque value is assumed. The contact stiffness corresponds to the maxilla which means that it is set to 10^5 N/m. ⓐ shows the time course of the contact force, clearly pointing out that the collision detection prevents an exceedance of the threshold force of 660 N. ⓑ shows the desired trajectory $X_D(t)$, the robot position $X_R(t)$, the position of the singularity $X_S(t)$, and the position of the human $X_H(t)$. ⓒ indicates the collision detection signal which triggers the reset of $X_D(t)$ in ⓑ.

In Figure 5.27 the maximal force which can be exerted on a human maxilla by a rigid, slowly moving robot (no dynamic forces) are analyzed. The stiffness of the maxilla[36] is in the order of 10^5 N/m according to [36, 4, 29]. Thus, the force will linearly increase with position after contact, as represented in Fig. 5.27 for several collision points along the lines l_i. The linear forces are displayed only up to the limit at which the bone will break ②, denoted by $F_{frac} = 660$ N. The curve F_{max} represents the maximal force that can be exerted by the robot, which goes to infinity when approaching the singularity. If this curve is above F_{frac} and if F_{frac} is exceeded

[36] Because the variation of data obtained by human cadaver tests is large and this data cannot be applied to children or elderly people it has to be treated carefully.

before reaching the singularity for a given collision point (this depends on the slope of l_i), the bone will break. For the considered case, this would happen starting with ①, i.e. more than 27 cm before reaching the singularity. Starting with ③, there does not exist even a hypothetical equilibrium point, meaning that the considered stiffness cannot stop the robot from reaching the singularity. Using the collision detection with a threshold of $0.1\tau_{max}$, the maximal forces are lowered, as displayed by the curve F_{det}. In this case, the critical region is substantially reduced to about 2 cm before the singularity ④. Restricting the workspace of the arm such that this configuration is not reached, poses no significant limitation to usual applications. The limit safe configuration[37] is denoted by ⑤. This analysis can be carried out with all facial and cranial bones listed in [13] and yields similar observations for each of them. After this rigid-robot evaluation, the full-dimensional simulation, especially including its intrinsic joint compliance for the LWR-III at maximum joint velocity shall be given and discussed[38].

For the results shown in Fig. 5.27 (right) a feasible collision detection threshold of $0.02\tau_{max}$ was assumed. The implemented collision reaction strategy immediately sets the desired position to a resting position 50 mm before the singularity. In contrast to the rigid robot there exists no significant workspace restriction, since even the last possible impact location ① that could theoretically lead to the fracture force F_{frac} can be handled by the collision detection. The theoretical rigid-case collision threshold ② close to the singularity is slightly below the one obtained from the complete dynamic simulation ③, presumably due to the elasticity in the joints of the real robot. Furthermore, one can see that the Cartesian braking distance decreases the closer to the singularity the contact occurs. This is due to the duality of Cartesian velocity and force. After the collision detection activation and the following braking distance, the robot switches its Cartesian velocity direction and comes to a rest position 50 mm before the singularity. For clarity Fig. 5.28 denotes the time course for such a constrained impact. It shows how the collision detection and reaction can limit the contact force to subcritical values. In this particular simulation, the human is standing 16 mm before the singularity which means that sufficient space remains for achieving the necessary fracture force. A full experimental analysis for head and chest impacts is given in Sec. 5.6.2.2 and 5.6.2.4.

A major goal in safety in pHRI is to develop a general procedure for evaluating blunt impact injury for robotic systems. Hence, a next set of standardized experiments was carried out with the LWR-III and considerably heavier industrial robots, extending the partly standardized investigations presented so far.

---

[37] With an ideal collision detection and an infinitely fast stopping robot.

[38] Maximum joint velocity does not mean that the contact has high speed impact characteristics because close to the singularity the Cartesian velocity is low. Therefore, maximum joint velocities were chosen to show the feasibility of the collision reaction strategy at such high demands.

5.6 Towards a Crash-Testing Protocol

Together with the first part of this chapter, the experiments in this section provide a comprehensive set of the relevant robot-human blunt impact situations and their parameterization in terms of velocity and robot mass, leading to suggestions for a test procedure in Chapter 11. In the following, a large amount of experimental results are described that were gained from blunt impacts with a frontal HIII dummy. Following test scenarios are evaluated:

- Head impacts
  - Dynamic unconstrained head impacts and their influence on the head, neck, and chest with the 235 kg-robot KR6 and the 2350 kg-robot KR500.
  - Quasistatic constrained head impacts with the 15 kg-robot LWR-III.
  - Partial clamping during head impacts and their influence on the head, neck, and chest with the KR500.
- Chest impacts
  - Dynamic unconstrained chest impacts and their influence on the head, neck, and chest with the KR6 and the KR500.
  - Quasistatic constrained chest impacts with the LWR-III

All injury criteria for the head, neck, and chest were evaluated that are measurable with the used dummy type. Furthermore, impact forces are obtained for the evaluation of facial fractures of the mandible and the frontal bone. Generally, the purpose of this last part of the chapter is to

1. Understand the general injury mechanisms and severity behind blunt human-robot impacts.
2. Provide the experimental foundations to propose procedures for a standardized crash-testing protocol, i.e. clarify how the concept of crash-testing is applicable to any kind of robot.
3. Provide safety tolerance values depending on robot velocity to maximize the performance of applications under the safety constraint (e.g. incorporating manual guidance of industrial robots). Production cycle times can be optimized with the knowledge gained from such experiments.

The last aspect is especially relevant for industrial robot manufacturers due to the fact that future applications are focused on physical interaction between humans and industrial robots with additional sensors such as a force/torque sensor in the wrist. As described in Sec. 5.4, constrained impacts with heavy-duty industrial robots are very dangerous indeed. Therefore, the (time) optimal performance is achieved if clamping can be excluded in the particular application. This in turn requires a precise and careful design of the workstation in order to significantly reduce possibilities of the human getting pinched.

After giving an overview of the experimental results, detailed evaluations of interesting aspects related to the experiments are outlined. In order to keep the discussion clearly structured, case discussions are introduced that explicitly focus on particular aspects, which should be treated in more detail.

In the next subsection the overall setup of the impact tests is described.

5.6.1 *Experimental Setup*

In Figure 5.27 the setup for the experiments is shown. Table 5.7 summarizes the instrumentation of the different setups, which is similar to the experiments in Sec. 5.2 and 5.3. In addition to the standard ADAC sensors, the available sensors of the robots are recorded. The reflected inertias during the impacts of the industrial robots can be obtained from Tab. 5.3.

In the next section the experimental data is summarized in a condensed form with the intention to give an overview similar to test reports in automobile crash testing.

5.6.2 *The DLR Crash Report*

5.6.2.1 Dynamic Unconstrained Head Impacts

The first series of tests is dynamic unconstrained head impacts. In Figure 5.30 high-speed recordings of a head impact at full speed are shown to visualize the dynamics of such collisions. The robot is commanded such that it hits the dummy in the face in outstretched configuration while rotating about the first axis. The head is accelerated, followed by the neck being bent while the torso starts moving delayed due to the higher inertia and the elastic coupling to the head. The entire contact phase

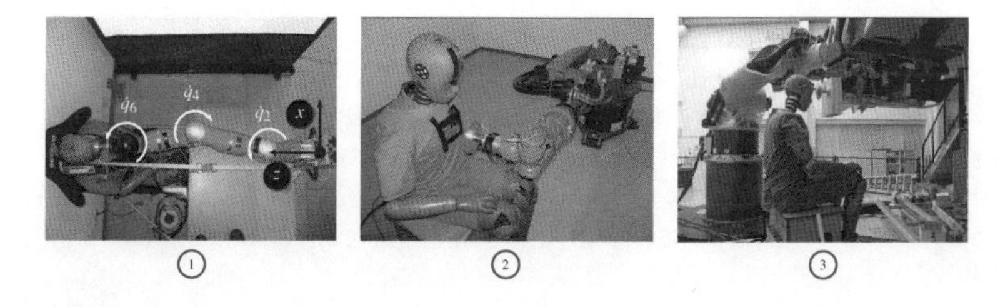

Fig. 5.29 Setup for the impact tests with the LWR-III (①, view from above), the KR6 (②, side view), and the KR500 (③, side view). Since for the LWR-III dynamic impacts were already analyzed in [12], constrained impacts close to a singularity were investigated. Then it is theoretically possible even for a low inertia robot to become severely dangerous. The industrial robots were tested for an outstretched configuration in order to achieve very high Cartesian velocities. The contact force is measured with a high bandwidth crash sensor. The contact geometry is defined by an aluminum impactor with radius $r_I = 120$ mm.

Table 5.7 Measured quantities

	Quantity	Sampling time [ms]
LWR-III	external force $F_{ext} \in \mathbb{R}^1$	0.05
	joint position $\mathbf{q} \in \mathbb{R}^7$	1
	joint torque $\tau_J \in \mathbb{R}^7$	1
KR6	external force $F_{ext} \in \mathbb{R}^1$	0.05
	joint position $\mathbf{q} \in \mathbb{R}^6$	12
KR500	external force $F_{ext} \in \mathbb{R}^1$	0.05
	joint position $\mathbf{q} \in \mathbb{R}^6$	12
HIII	head acceleration $\mathbf{a}_{head} \in \mathbb{R}^3$	0.05
	neck wrench $\mathscr{F} \in \mathbb{R}^6$	0.05
	chest acceleration $\mathbf{a}_{chest} \in \mathbb{R}^3$	0.05
	chest deflection $d_{chest} \in \mathbb{R}^1$	0.05

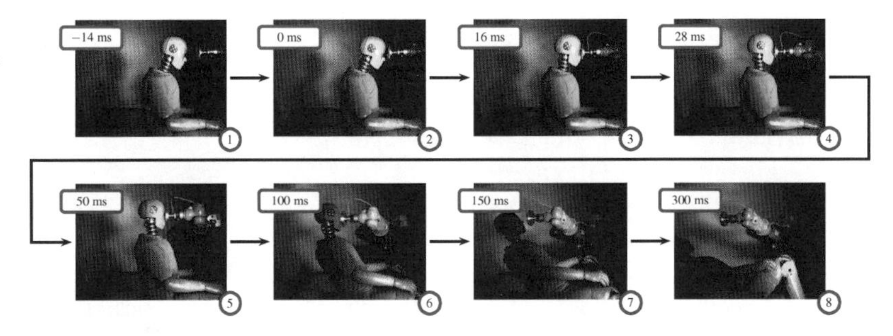

Fig. 5.30 High-speed recordings of an unconstrained head impact test with the KR6 and a Hybrid III dummy at 4.2 m/s

is completed after $\approx$ 100 ms. The following motion of the dummy without having contact with the robot ends in a secondary impact on the floor. In this particular case, the robot stops moving due to an exceedance of the nominal gear torque of the robot, triggered by motor current monitoring. The collision tests were carried out at various Cartesian impact speeds ranging from 0.2 m/s to 4.2 m/s. Contact forces range up to 5 kN. Unfortunately, above $\approx$ 3 m/s the force sensor saturates (indicated by * in the tables).

In Table 5.8 the results for the unconstrained frontal impacts with the KR6 and KR500 are given. The correlation to injury severity by means of the EuroNCAP is indicated by the underlying color. In general, very high robot velocities have to be achieved in order to exceed the threshold from *very low* to *low* injury for all severity indices. Only at maximum velocity the HIC is slightly above the threshold value of 650 but at the same time still significantly below 1000, which denotes the critical value for this indicator. For neck shearing along the x-direction very high values are achieved, which correlate (in the worst-case assumption) to *very high* injury

Table 5.8 Head impact experiments with the KR6 and the KR500

Exp. Nr.	Robot	$\dot{x}_R$ [m/s]	$F_{\mathrm{ext}}^{\mathrm{frontal}}$ [N]	HIC	$a_{\mathrm{max}}^{\mathrm{head}}$ [g]	$a_{\mathrm{3ms}}^{\mathrm{head}}$ [g]	F_x [N]	F_z [N]	$M_{y,OC}^{\mathrm{Flex}}$ [Nm]	$M_{y,OC}^{\mathrm{Ext}}$ [Nm]
11	KR6	0.2	✓118	0.03	1.61	1.53	94.09	−165.73	✓1.93	−2.41
12		0.7	✓783	2.22	13.40	11.00	181.59	−430.26	✓7.77	−7.15
13		1.0	✓1306	6.72	23.52	16.55	320.18	−739.41	✓13.74	−8.61
6		1.3	✓1875	16.65	37.78	20.79	469.45	−861.60	✓25.34	−10.34
7		1.3	✓1766	16.88	37.53	20.82	409.32	−591.72	✓27.42	−1.18
14		1.5	✓2208	25.53	45.64	23.84	476.36	−908.09	✓29.02	−11.50
15		2.0	✓3426	64.36	77.16	25.23	710.72	−1554.63	✓42.96	−13.84
8		2.1	✓3976	96.42	93.25	25.07	691.03	−1483.54	✓45.37	−11.60
9		3.2	×5006*	344.07	167.71	21.71	949.87	−1359.89	✓67.96	−3.45
10		4.2	×5069*	671.98	213.98	38.45	1428.58	−2856.93	✓98.12	−4.09
26	KR500	0.2	✓136	0.04	1.91	1.85	120.69	−86	✓0.54	−2.33
32		0.3	✓168	0.07	2.56	2.45	132.53	−88	✓0.67	−2.80
27		0.5	✓420	0.72	7.21	6.58	112.9	165	✓1.94	−5.03
28		0.7	✓798	3.10	14.87	12.53	215.93	−335	✓6.64	−6.75
29		1.0	✓1200	7.95	23.46	18.10	248.63	−375	✓11.19	−7.31
23		1.2	✓1967	17.82	35.37	24.43	407.28	−899	✓19.47	−11.51
30		1.5	✓2219	28.10	44.96	27.35	493.12	−958	✓22.22	−12.58
24		2.0	×4020	93.77	80.04	38.23	627.23	−1121	✓43.78	−7.99
31		2.0	✓3043	63.33	67.27	33.17	522.13	−758	✓30.48	−29.95
25		3.1	×4965*	248.18	141.48	47.97	967.44	−1575	✓66.22	−22.65
22		4.1	×4963*	560.00	203.66	40.56	1350.79	−2012	✓105.06	−12.35

Table 5.9 Unconstrained hook to the chin with the KR6

Exp. Nr.	Robot	$\dot{x}_R$ [m/s]	$F_{\mathrm{ext}}^{\mathrm{mandible}}$ [N]	HIC	$a_{\mathrm{max}}^{\mathrm{head}}$ [g]	$a_{\mathrm{3ms}}^{\mathrm{head}}$ [g]	F_x [N]	F_z [N]	$M_{y,OC}^{\mathrm{Flex}}$ [Nm]	$M_{y,OC}^{\mathrm{Ext}}$ [Nm]
2	KR6	0.9	✓755	1.48	11.76	9.14	270	417	✓18.26	−9.82
1		1.0	✓965	2.80	16.61	11.53	350	471	✓22.52	−10.63
3		1.8	×1871	12.11	34.21	17.70	525	962	✓30.07	−17.53
4		2.7	×3128	38.92	62.59	25.78	764	1427	✓44.26	−24.60
5		3.6	×4938*	96.64	91.61	44.27	991	2564	✓48.36	−28.36

and for the forces in tension/compression only for the KR6 the threshold from *very low* to *low* injury severity is crossed. For all other EuroNCAP injury indicators the observed potential injury stays within the *green* area.

In Table 5.9 the results for the unconstrained hook to the chin with the KR6 are given. The dummy is hit in cranial direction up to a maximum velocity of 3.6 m/s. All criteria are in the *very low* area except for the maximum resulting head acceleration $a_{\mathrm{max}}^{\mathrm{head}}$ at 3.6 m/s, which is still in the *low* injury severity range. Furthermore, the flexion torques are far below 190 Nm and thus subcritical [12]. Concerning fracture of the mandible it can be stated that the contact force at 1.8 m/s is already slightly above the threshold force and for higher velocities a clear exceedance is observed. However, the human chin behaves differently as it is not rigidly attached to the cranium as for the HIII (its head consists of a rigid aluminum shell). Furthermore, the interface stiffness is presumably far too high as indicated by investigations in [4]. In future work biomechanical faces will be investigated as e.g. applied in [48] and developed in [31].

Table 5.10 Constrained quasistatic head impacts with the LWR-III

Exp. Nr.	Robot	Strategy	d_s [mm]	$F_{\max}^{\text{static}}$ [N]	$F_{\max}^{\text{peak}}$ [N]
$L13$	LWR-III	0	10	0	✓692
$L14$		0	10	674	✓1244
$L15$		1	10	277	✓540
$L16$		2	10	0	✓590
$L19$		0	5	0	✓1593
$L17$		1	5	256	✓505
$L18$		2	5	0	✓617

5.6.2.2 Quasistatic Constrained Head Impacts

In Section 5.2 free blunt impact experiments with the LWR-III were performed. These proved to be non-critical from a safety point of view. As the only possibly dangerous blunt contact situation for the LWR-III the clamping close to a singularity was identified, see Sec. 5.5. Theoretically, a robot is able to exert infinitely large forces at the tip in the singular z-direction, while driving through the singularity. The worst-case seems to be the classical reconfiguration from "elbow up" to "elbow down". In Figure 5.29 the experimental setup for analyzing such a situation is depicted. The LWR-III was mounted horizontally and the position of the dummy was adjusted such that it touched the robot at a certain distance d_s from the singularity. The robot moves from its initial position at maximum joint velocity in joint 4 and half the velocity in joint 2 and 6, see Fig. 5.29. The resulting trajectory is a straight line with constant orientation in z-direction. Therefore, the robot is programmed to pass the singularity in its outstretched configuration.

In addition to commanding the described trajectory and evaluating the resulting injury, the collision detection and reaction for the LWR-III was tested during this worst-case for detection sensitivity. Since in z-direction the sensitivity of this algorithm practically goes to zero close to the singularity, it is necessary to quantify the still achievable benefit for such a situation. The collision detection was evaluated with reaction strategies 0,1,2.

Due to the constrained environment it is not possible to measure any criterion for the head with the HIII (the head is only equipped with an acceleration sensor). Therefore, only the contact force is left for evaluation. High quasistatic forces[39] can be achieved for this impact type as shown in Tab. 5.10. For the experiments $L13$ and $L19$ the robot is able to pass the singularity without exceeding its maximum nominal joint torques[40]. For $d_S = 5$ mm the robot still moves through the singularity

[39] Because the robot approaches a singularity the Cartesian velocity, which defines the impact velocity is low and therefore the contact has no typical impact characteristics anymore. In Figure 5.37 the time courses for a chest impact are shown, pointing out the difference between peak and quasi-static force.

[40] Please note that $L13$ was not using a tension belt, which reduces the effective value of d_S because the seat can give in. For the other experiments a belt was used.

and achieves a maximum quasistatic contact force of 1593 N. However, this is still far below the tolerance force of the frontal bone. The collision detection and reaction can reduce the occurring contact forces by 44 % for d_s =10 mm and by 68 % for $d_s = 5$ mm.

Table 5.11 Unconstrained Chest impact experiments with the KR6 and the KR500

Exp. Nr.	Robot	$\dot{x}_R$ [m/s]	$F_{\text{ext}}^{\text{chest}}$ [N]	CC [mm]	VC [m/s]	$a_{3\text{ms}}^{\text{chest}}$ [g]	HIC	$a_{3\text{ms}}^{\text{head}}$ [g]	F_X [N]	$M_{y,OC}^{\text{Flex}}$ [Nm]	$M_{y,OC}^{\text{Ext}}$ [Nm]
16	KR6	0.2	✓215	2.68	0.00	✓0.41	0.00	0.40	−17.20	✓1.04	−0.14
17		0.7	✓685	7.63	0.00	✓7.95	0.15	1.89	−67.10	✓4.61	−0.59
18		1.0	✓876	10.56	0.01	✓4.80	0.45	2.86	−141.08	✓7.19	−0.92
19		1.5	∼1156	13.97	0.02	✓2.51	1.03	4.02	−145.18	✓9.94	−2.55
20		2.0	∼1528	19.06	0.04	✓3.80	2.16	5.26	−190.13	✓14.49	−5.61
21		4.2	×3277	51.28	0.41	✓8.99	16.89	12.40	−400.71	✓37.89	−18.82
33	KR500	0.2	✓185	3.13	0.00	✓0.38	0.01	0.53	−23.78	✓1.45	−0.12
34		0.7	✓551	4.54	0.00	✓1.94	0.29	2.86	60.56	✓8.45	−1.60
35		1.0	✓847	7.44	0.01	✓4.15	0.77	4.37	56.03	✓12.37	−3.03
36		1.5	∼1400	14.29	0.02	✓5.10	2.7	7.43	−93.30	✓15.65	−3.88
37		2.0	×1939	22.82	0.05	✓5.36	4.09	6.70	−261.30	✓20.43	−6.13
38		4.1	×3962	57.89	0.41	✓36.93	53.26	24.88	−513.53	✓32.25	−23.72

5.6.2.3 Dynamic Unconstrained Chest Impacts

In Table 5.11 the results for the unconstrained frontal chest impacts with the KR6 and KR500 are listed. Apart from the CC and the external chest force, all criteria are subcritical over the entire range of impact velocities. The chest compression reaches for 4.2 m/s with the KR6 and 4.1 m/s with the KR500 potentially lethal values. Forces measured during experiments 19, 20, 36 are within the tolerance band (see Chapter 4) and for 21, 37, 38 the tolerance values are clearly exceeded. The tolerable impact force for the chest[41] is exceeded for 4.2 m/s for the KR6 and already at 2 m/s for the KR500.

5.6.2.4 Quasistatic Constrained Chest Impacts

In Table 5.12 the results of the quasistatic constrained impact of the LWR-III with the HIII chest are shown. The distance to singularity d_S varies from 20 mm to 80 mm, producing a maximum chest deflection (CC) of −11.95 mm at $d_S = 75$ mm. At $d_s = 80$ mm the maximum joint torques of the robot are exceeded, which triggers the low-level safety feature for engaging the brakes of the robot. This shows that (with a granularity of 5 mm) without collision detection and reaction, the worst-case for this robot lies at a distance to singularity of $d_S^{wc} = 75$ mm. The corresponding contact force of 1.04 kN is below the maximum tolerable threshold of the chest and the CC is constantly subcritical in the *very low* region. Similar to the

[41] Please note this is a force humans can tolerate without suffering injury.

Table 5.12 Quasistatic constrained chest impacts with the LWR-III

Exp. Nr.	Robot	Strategy	d_s [mm]	$F_{\max}^{\text{static}}$ [N]	$F_{\max}^{\text{peak}}$ [N]	CC [mm]
$L8$	LWR-III	0	20	0	✓379	−3.42
$L9$		1	20	180	✓300	−2.34
$L10$		2	20	0	✓332	−2.67
$L1$		0	40	–	–	−6.15
$L2$		1	40	130	✓291	−1.86
$L3$		2	40	0	✓300	−2.19
$L4$		0	60	0	✓859	−9.22
$L11$		0	70	0	✓995	−11.38
$L12$		0	75	0	✓1043	−11.95
$L7$		0	80	425	✓824	−7.59
$L5$		1	80	92	✓263	−1.79
$L6$		2	80	0	✓287	−1.99

constrained head impacts, the reactive strategies reduce the contact force and the CC considerably. Even at the configuration closest to the singularity $d_S = 20$ mm (meaning lowest detection sensitivity of all measurements) the robot is able to react effectively.

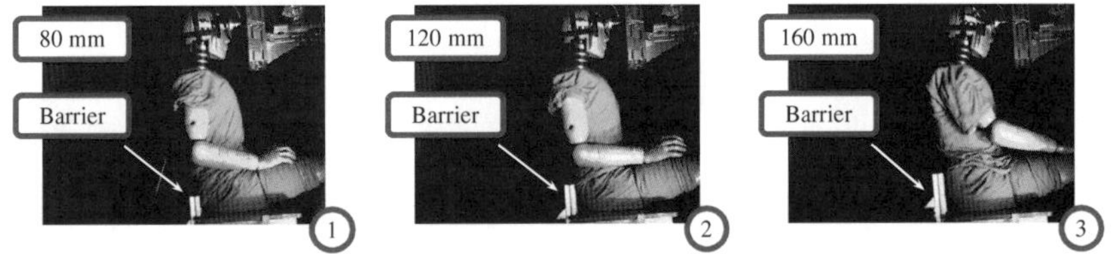

Fig. 5.31 Partially constrained impact. A barrier was mounted on the back of the dummy with 80 mm (1), 120 mm (2), and 160 mm height (2).

5.6.2.5 Partially Constrained Dynamic Head Impacts

Table 5.13 lists the evaluated injury indices for partially constrained impacts. In this experiment, the dummy was sitting in front of a barrier, which height varied from 80 mm to 160 mm, see Fig. 5.31. The impact criteria, such as the HIC, have similar values to the ones obtained for the unconstrained dynamic impacts from Sec. 5.6.2.1. The influence of the barrier mainly results in larger neck forces and torques. Experiment 43 is presumably not comparable as the impact direction contains a significant lateral component.

Table 5.13 Partially constrained head impact experiments with the KR500. The barrier height h_B ranges up to 160 mm.

Exp. Nr.	Robot	h_B [mm]	$\dot{x}_R$ [m/s]	$F_{ext}^{frontal}$ [N]	HIC	a_{max}^{head} [g]	a_{3ms}^{head} [g]	F_x [N]	F_z [N]	$M_{y,OC}^{Flex}$ [Nm]
39	KR500	80	2.0	✓2945	59.48	59.76	36.35	563.42	−479.90	✓37.45
41		120	2.0	✓3059	67.98	67.98	34.83	476.76	−471.44	✓29.02
43		160	2.0	✓2795	40.87	53.67	30.80	482.06	−1137.25	✓16.74
40		80	4.1	×4950*	419.38	170.64	47.67	975.13	−670.60	✓68.07
42		120	4.1	×4978*	408.31	170.45	47.27	864.13	−913.97	✓53.04
44		160	4.1	×5165*	500.37	195.54	45.13	1268.34	−1296.91	✓93.98

After this presentation of the data from the impact tests with standardized crash-test dummies, now various aspects related to these tests are addressed in a case-based discussion.

5.6.3 *Case Discussions*

Cases 1-4 treat unconstrained head and chest impacts, case 5 partially constrained impacts, and case 6 constrained quasistatic impacts. Such detailed discussions are important to extract the relevant information to be taken into account for future standards.

5.6.3.1 Case 1: The Saturation of the Head Injury Criterion

As was extrapolated from robot-dummy impacts with the LWR-III and the experiment with the dummy-dummy, a saturation of the Head Injury Criterion at a certain impact velocity with increasing robot mass is observed. In Figure 5.32 the HIC values for all tests presented in Sec. 5.2 and the ones shown in Sec. 5.6.2 are depicted up to an impact velocity of 2 m/s and classified according to the EuroNCAP [10]. First, it is confirmed that the HIII-head imitation device reproduces similar HIC values to the HIII and thus can be used by other researchers as a tool for simple experimental HIC evaluation. Furthermore, the mentioned saturation effect is confirmed by the fact that the KR6 and the KR500 produce very similar HIC values for equivalent impact speed by means of standardized crash-test measurements. In general, the obtained HIC values for speeds up to 2 m/s are classified as subcritical. By means of the EuroNCAP only *very low* injury can occur. Although this clearly confirms that the human head is not in a critical situation at velocities up to 2 m/s, the question arises whether other body parts, such as the neck, would be posed to a serious threat during the post-impact phase of such a collision. This answer points to the next question: whether the neck stiffness and body inertia are constructed such that the neck is the weak point. This brings to the next case: The description of the head-neck-torso complex dynamics during a rigid blunt impact.

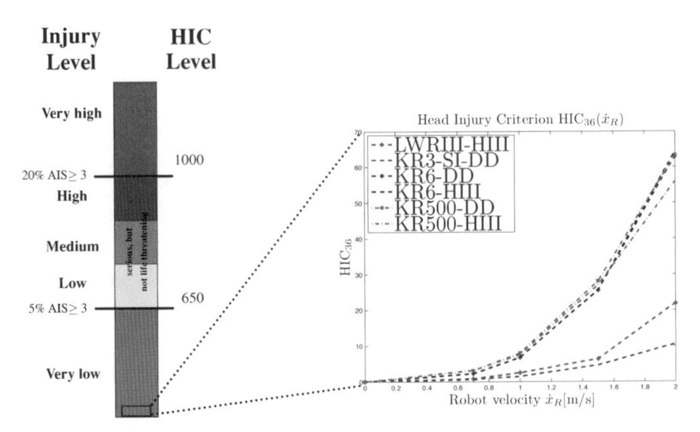

Fig. 5.32 Resulting HIC_{36} values at varying impact velocities for robots of different weights: the 15 kg-robot LWR-III, the 54 kg-robot KR3-SI, the 235 kg-robot KR6, and the 2350 kg-robot KR500. The HIC is rated according to the EuroNCAP *Assessment Protocol And Biomechanical Limits*. All produced HIC values at impact velocities up to 2 m/s range in the green area. This indicates that only *very low* head injury occurs during the impacts. Furthermore, the previously described saturation effect of the HIC can be observed. The HIC is displayed for impact tests with a Hybrid III-dummy (denoted by HIII) and with a simplifying setup (denoted by DD) mimicking the behavior of the Hybrid III-dummy head.

5.6.3.2 Case 2: Timing Properties of the Head-Neck-Torso Complex during Fast Head Impacts

Up to now, the simulated head impact and the corresponding HIC evaluation was treated as an isolated event between the robot and the human head, as was done in [5, 12] as well. The head is generally assumed to act decoupled from the torso during the short acceleration pulse that defines the impact dynamics. In the present case the validity of this assumption is discussed experimentally.

In Figure 5.33 the time courses of the head acceleration, the neck force, and the acceleration of the chest in x-direction are depicted for a head collision at an impact velocity of 4.1 m/s with the KR500. The head acceleration peak occurs timely along with a peak in the neck force (the load cell is mounted between head and neck, which is a stiff construction). Delayed to that, the torso starts accelerating and reaches its maximum value several milliseconds after the head acceleration and neck force passed their peak values. The impact phase is followed by a continuous bending of the neck and a longer acceleration phase of the torso. One can see in x-direction the decoupling assumption really holds to a certain extent. However, the z-acceleration (not displayed here) of the chest was observed to lag only 1 ms behind the maximum impact acceleration. This may be an effect caused by the higher neck stiffness of the HIII compared to a human. Due to this tight neck coupling a clear separation of head and torso during the initial impact does not occur in z-direction for this dummy head.

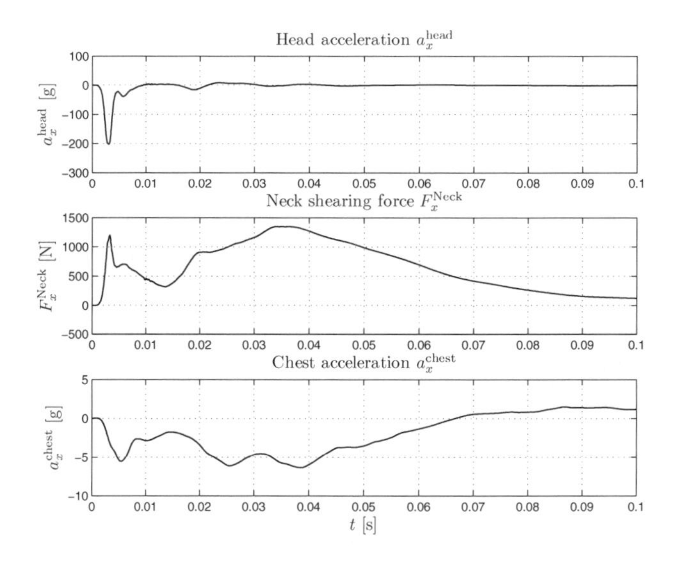

Fig. 5.33 The dynamics behind a frontal head impact with the KR500

In contrast to this observation [35] states that the human head is decoupled during an impact at 3.2 m/s from T1[42]. Furthermore, [19] points out that the neck of the HIII is only to a certain extent able to predict human neck injury due to its much higher stiffness properties. In order to get more realistic dynamics it seems to be desirable to use a dummy, the BIO-RID-II, with a spine that has more biofidelity than the one of the HIII.

After discussing the timing properties of a head impact and the related neck force and chest acceleration, the related question of whether significant neck injury occurs during such a robot-head impact is treated in the following.

5.6.3.3 Case 3: Neck Injury during Head Impacts

In the present case the question is discussed whether the head can be accelerated during an impact powerful enough such that the trunk cannot follow before the neck forces and torques exceed their corresponding tolerance thresholds. The question to be answered is: Can the human suffer severe neck injury despite the HIC being small during an unconstrained impact? With the dummy tests in Sec. 5.2 it was not possible to analyze this because at high velocities the maximum joint torques of the LWR-III are exceeded. This causes the brakes of the robot to engage in order to protect its mechanics. Thus, the robot is not able to further drag the head and potentially injure the neck anymore. During the short duration of the initial impact

[42] The human spine can be divided into the cervical, thoracic, and lumbar spine. T1 is the first thoracic vertebra.

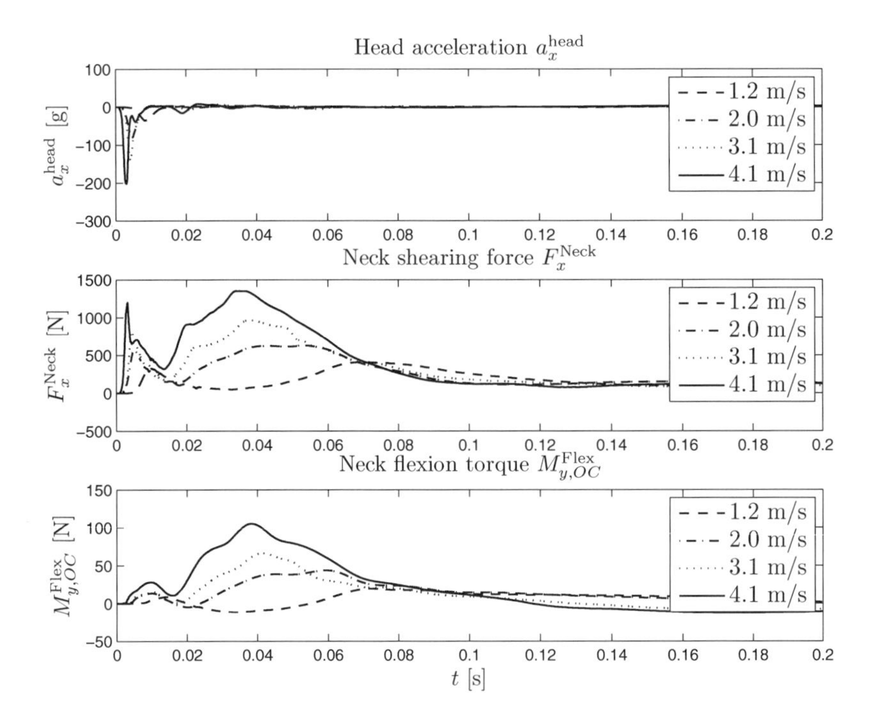

Fig. 5.34 The head acceleration a_x^{head} and neck force F_x^{Neck} in x-direction as well as the neck flexion moment $M_{y,OC}^{\mathrm{Flex}}$ during a frontal head impact with the KR500 at increasing robot speed. The robot behaves due to its large inertia as a velocity source and drags the head further away while the neck is bended and the trunk accelerates due to the transmitted force.

the neck does not reach to critical loads with impact speeds up to 2 m/s. Since the motion of the 2350 kg-robot KR500 is not affected by the collision with the dummy[43] due to its large inertia it can be treated as a velocity source during an impact and is suited to evaluate this question.

In Figure 5.34 the head acceleration, the neck force in x-direction, and the torque about the occipital condyles are depicted for impact velocities up to 4.1 m/s with the KR500. The head acceleration is caused by the short impact, which defines the Head Injury Criterion and the maximum head acceleration. The neck force shows a similar peak in the beginning, followed by a second wider one. Please note that the first and second maximum are similarly large. For the neck torque the first maximum shows only marginal growth with increasing impact velocity while the second peak value increases with impact velocity. The second maximum is in both cases caused by the continuous motion of the robot, which further bends the neck while the trunk begins to accelerate.

[43] Please note that even for the 235 kg-robot KR6 the current monitoring was triggered at high speeds, i.e. also for this robot the maximum joint torques are exceeded. Furthermore, the robot loses significantly momentum during the impact due to its lower inertia compared to the KR500.

In general, neck forces tend to be more dangerous the longer they are applied to the neck (see Chapter 4). Therefore, it is evident that a heavy robot, not affected in its motion during the impact, increases the injury potential significantly. As shown in Sec. 5.6.2.1 the neck forces reach *very high* injury levels only at maximum velocity[44] (above 4 m/s) by means of the EuroNCAP. In case of the neck flexion torques following observations can be made. Although they increase significantly with impact velocity, no more than 100 Nm, which is still under the limit value, are reached. In the limited extend in which a HIII is able to predict neck injury, one is able to conclude that only very high impact velocities could pose a threat to the neck during head impact. Up to 2 m/s, which is believed to be a desirable (high) speed in physical Human-Robot Interaction no significant injury level can be observed by means of the evaluated criteria.

One can therefore conclude that the frontal unconstrained blunt head impact poses no threat below 2 m/s, both in terms of the HIC and indirect effects on the neck. A look at frontal unconstrained chest collisions and their characteristics shall now be taken.

5.6.3.4 Case 4: Chest Injury

In the robotics literature [5, 50, 12] it has been emphasized that the human head has to be treated carefully in a safety analysis due to its fragility. In this sense, the outcome of the tests discussed now is relatively unexpected at a first glance, since it shows that the chest faces to at least the same level of threat as the head, and can reach critical injury levels.

Figure 5.35 depicts the time courses of the Compression Criterion during frontal chest impacts with the KR500 at impact velocities up to 4.1 m/s. The impact duration of more than 150 ms is significantly larger compared to the 5 ms for the head impacts. The main corresponding reasons for this fact are the large inertia of the dummy body and the lower stiffness of the chest compared to the one of the head. In Sec. 5.6.2.1 it was shown that except for the maximum resulting head acceleration, all head criteria during head impacts are in the *very low* injury severity region for an impact velocity of up to 3.2 m/s. Only for the KR6 at maximum velocity of 4.2 m/s an HIC value slightly above the threshold from *very low* to *low* (650) was observed. While facing *low* injury for the head impacts (when not considering the pure maximum acceleration) an aspect that appears surprising is that, according to the chest impact results, the CC indicates *very high* injury severity at maximum velocity for the KR6 and the KR500, see Fig. 5.35. Apparently, the inertia of the

[44] In order to evaluate the injury severity correlating to the measured neck forces on a worst-case basis with respect to the corresponding EuroNCAP rating, the real maximum exceedance interval is determined as an upper bound estimate. Instead of determining the maximum exceedance time the smallest rectangle that fits for the particular index is determined, its width is chosen as the exceedance time and the respective height as the corresponding value of the injury index. This leads to an upper bound and therefore to a more restrictive evaluation of neck forces.

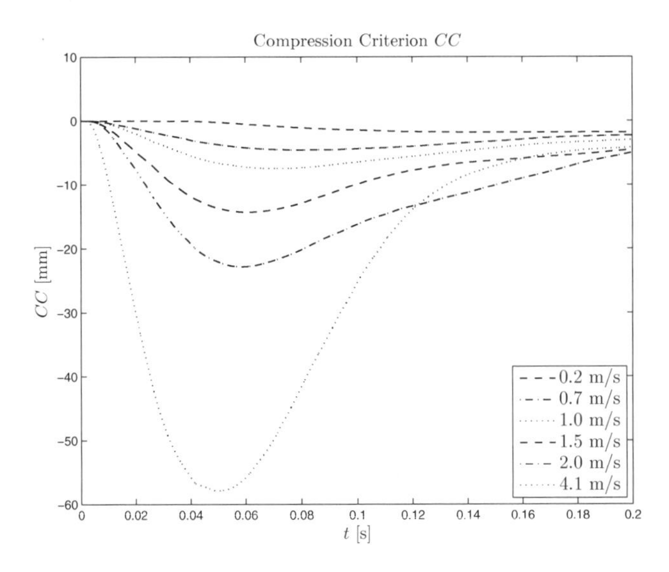

Fig. 5.35 Compression Criterion for chest impacts up to 4.1 m/s with the KR500

dummy trunk delays the motion such that the robot compresses the chest up to potentially lethal dimensions even in the unconstrained case. Furthermore, already at 2.0 m/s the threshold from *very low* to *low* injury is crossed for the KR500, showing that the injury potential starts to become dangerous.

For the unconstrained blunt impact, one can therefore conclude (while excluding the maximum resulting head acceleration from the analysis) that the chest impact is the most critical one for heavy robots.

Now the influence of an increasing barrier is analyzed, i.e. the role of partial constraints.

5.6.3.5 Case 5: The Partially Constrained Impact

Figure 5.36 shows the neck compression force for partially constrained head impacts with varying barrier height h_B. The neck force F_z increases significantly with increasing h_B up to a neck force of -1296 N at $h_B = 160$ mm compared to -670 N at $h_B = 80$ mm. The second peak also shows dependency on the barrier height. Unfortunately, this statement cannot be confirmed as clearly for the neck shearing force and torque. Furthermore, the generally lower impact criteria compared to the unconstrained head impacts are presumably caused by a slightly different location of the dummy during the partially constrained impacts. Nonetheless, although at the current state it is not possible to explicitly determine the lethal threshold, such a height must exist. Further tests are therefore necessary to analyze this effect more in detail and be able to predict the threshold height for a barrier. Furthermore, it is crucial to take a closer look at eventual spine injury during partially constrained impacts.

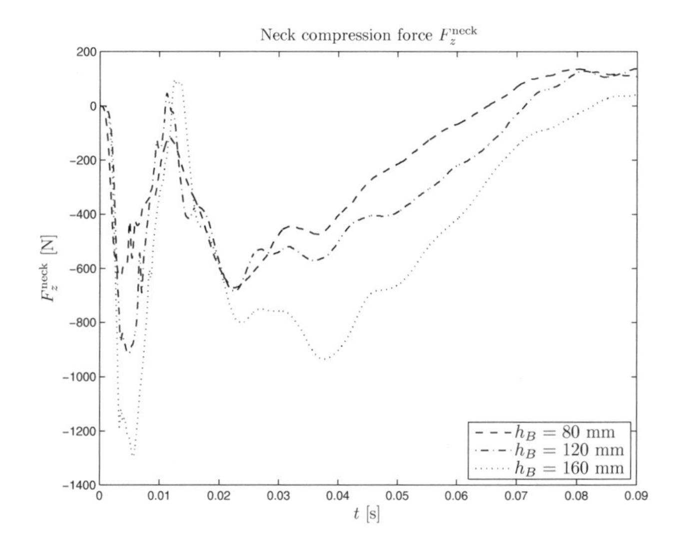

Fig. 5.36 Compression criterion for chest impact up to 4.1 m/s with the KR500

Because the HIII is not able to measure this effect this is left for future research with distinguished equipment.

Another interesting observation made during the partially constrained impact is that a second impact occurs with the barrier obstructing the motion of the trunk. This is not the case for the non-constrained case in which the dummy moves away fast enough to avoid a second impact with the robot.

5.6.3.6 Case 6: The Constrained Quasistatic Impact with the LWR-III

As shown earlier in this chapter, any robot is theoretically able to exceed the fracture tolerance of the facial and cranial bones in case the human head is clamped and the robot drives through a singularity. A prerequisite for this to happen for a particular bone with tolerance force $F_{\text{frac}}^{\text{bone}}$ and stiffness K_{bone}, is for the distance to singularity to be

$$d_s \geq \frac{F_{\text{frac}}^{\text{bone}}}{K_{\text{bone}}}.$$

Although this is theoretically possible the question remains, whether in reality a particular robot would be able to withstand such large forces, or whether unmodeled structural compliances would prevent the occurrence of this worst-case scenario. In Section 5.6.2.2 various constrained head and chest impacts are shown with the LWR-III driving through the singularity in outstretched configuration and leading to the observation that the tolerance values of both, head and chest are not exceeded.

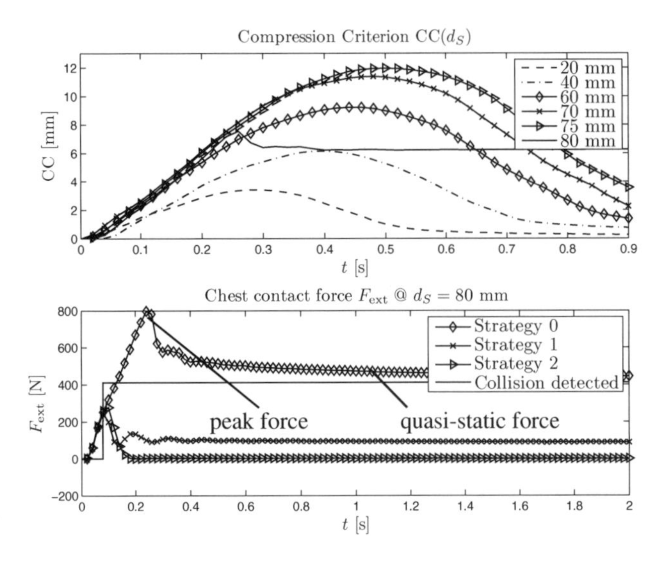

Fig. 5.37 Singularity clamping with the LWR-III and a HIII. The measured CC is displayed for different values of d_S (upper). The chest contact force for $d_S = 80$ mm is depicted for the base case and two collision strategies (lower).

In Figure 5.37 (upper), the CC is plotted for varying values of the distance to singularity d_S. There are two factors affecting the maximally reachable force, depending on the contact point d_s:

1. If the contact point is too close to the singularity (d_s is small), then the maximal force is limited by the compliance of the chest, which deflects and allows the robot to pass trough the singularity.
2. If the contact point is far from the singularity ($d_s > 80$ mm), then the contact force is limited by the maximal joint torques, since the Jacobian is not ill-conditioned anymore. A low-level safety stop is activated when maximal joint torques are exceeded, preventing the further increase of the force.

Under these circumstances, a maximal compression of 11.95 mm was reached for $d_s = 75$ mm. Although an exceedance of the threshold values is not possible for this impact type, the achievable CC value is more than twice as high compared to the unconstrained dynamic impacts presented in Sec. 5.2.4.

Apart from the discussed worst-case behavior, the effect the collision detection and reaction has on such an impact shall now be explained. In Figure 5.37 (lower) the resulting force profiles are plotted and the collision detection signal is indicated. Clearly, the potential threat is cleared quickly after the collision is detected. For every impact configuration the detection is sensitive enough to detect the collision. Both reaction strategies are leading to a significant contact force reduction.

This constrained quasistatic impact can be used as both, a worst-case analysis concerning maximum contact force and as a benchmark problem for a collision detection and reaction scheme, which is only based on proprioceptive sensing as the one treated in the present case.

Up to now various cases were discussed, which treat different aspects relevant for future robotics safety standards defined for physical Human-Robot Interaction. The presented facts may lead to recommendations for standardized crash-testing procedures in robotics. In Chapter 11 a proposal of impact tests is given, which are from the author's perspective necessary for a full safety evaluation of robotic systems.

5.7 Summary

In this chapter, the first systematic evaluation and classification of possible blunt injuries during physical Human-Robot Interaction was presented. It was experimentally shown that potential injury of the head, occurring during a free impact, would saturate at rather low values with increasing robot mass. From a certain mass on potential injury would only depend on the impact velocity. Thus, typical severity indices focusing exclusively on the moment of impact like the Head Injury Criterion or Viscous Criterion do not provide differentiable measures of injury severity in Human-Robot Interaction, as a robot typically cannot exceed these safety critical thresholds. This is mostly due to significantly lower robot velocities compared to car velocities during impact tests of automobile crash-testing. However, the results of this work clearly indicate that the injury to be expected from robot-human impacts is intrinsically very low by means of blunt injury criteria. More specifically, blunt head impacts without clamping at typical robot speed (up to 2 m/s) are indeed, very unlikely to cause severe injury, regardless the weight of the robot. Chest impacts of the same type are even less dangerous, as shown also by real impact data. The presented results are the first systematic experimental evaluation of possible injuries during robot-human impacts using standardized testing facilities. However, other serious head injuries, such as fractures of facial and cranial bones, can (under conservative assumptions) occur already at moderate velocities and may be a more relevant injury mechanism. The appropriate injury indicator for this class of injury is not related to head acceleration but to impact forces.

Drastically different observations can be made in case of clamping, which was evaluated with respect to robot mass and impact velocity. In case of clamping both the head and chest can be severely injured even leading to death for a large robot mass. Nevertheless, the low inertial properties of the LWR-III allow an impact velocity of up to 1 m/s without causing any of the investigated injuries (even under the given conservative assumption).

Apart from the discussed constrained and unconstrained dynamic impacts, it was also shown that even low-inertia robots can become very dangerous in near-singular configurations in case of a constrained collision. This special case is due to the fact that near its singularities a robot is able to generate extremely large quasi-static forces. On the other hand the effectiveness of a collision detection and reaction scheme which can handle this threat was demonstrated.

Based on the insights gained from the first part of the chapter, the evaluation was extended by providing a wide range of impact testing results for future robot crash-test protocols. Again numerous injury indicators were measured with standard automobile crash-testing equipment and rated according to an established crash-testing protocol. Robots of entirely different weight and at various velocities are evaluated for typical and relevant situations. Furthermore, the usage of a HIII gave the ability to analyze full body responses and thus evaluate what happens at remote locations of the body (e.g. the head-neck complex) during an impact with another body part (e.g. the chest). The resulting data basis will help to understand the mechanisms behind injuries in robotics and contribute to a fact based discussion of safety in physical Human-Robot Interaction.

In summary, this chapter presents the first systematic experimental evaluation of injury in robotics. Possibly the largest experimental database for blunt impacts was generated and several new and surprising results were attained. They partially contradict the "common sense" and some previous preliminary simulations from literature. The results of this chapter lead to recommendations for future robotic standards compiled in Chapter 11.

References

[1] AAAM: The Abbreviated Injury Scale (1980)
[2] AAAM: The Abbreviated Injury Scale (1990), Revision Update. Des Plaines/IL (1998)
[3] Allsop, D., Perl T.R.and Warner, C.: Force/deflection and fracture characteristics of the temporo-parietal region of the human head. SAE Transactions, 2009–2018 (1991)
[4] Allsop, D., Warner, C., Wille, M., Schneider, D., Nahum, A.: Facial impact response - a comparison of the Hybrid III dummy and human cadaver. SAE Paper No.881719, Proc. 32th Stapp Car Crash Conf., pp. 781–797 (1988)
[5] Bicchi, A., Tonietti, G.: Fast and soft arm tactics: Dealing with the safety-performance trade-off in robot arms design and control. IEEE Robotics and Automation Mag. 11, 22–33 (2004)
[6] Brinkmann, B., Madea, B. (eds.): Handbuch gerichtliche Medizin. Springer (2004) (German)
[7] De Luca, A., Albu-Schäffer, A., Haddadin, S., Hirzinger, G.: Collision detection and safe reaction with the DLR-III lightweight manipulator arm. In: IEEE/RSJ Int. Conf. on Intelligent Robots and Systems (IROS 2006), Beijing, China, pp. 1623–1630 (2006)
[8] De Luca, A., Mattone, R.: Sensorless robot collision detection and hybrid force/motion control. In: IEEE Int. Conf. on Robotics and Automation (ICRA 2005), Barcelona, Spain, pp. 1011–1016 (2005)
[9] Ebert, D., Henrich, D.: Safe human-robot-cooperation: Image-based collision detection for industrial robots. In: IEEE/RSJ Int. Conf. on Intelligent Robots and Systems (IROS 2002), Lausanne, Switzerland, pp. 239–244 (2002)

[10] EuroNCAP: European protocol new assessment programme - frontal impact testing protocol (2004)
[11] Haddadin, S., Albu-Schäffer, A., Hirzinger, G.: Approaching Asimov's 1st Law. HRI Caught on Film. In: Proceedings of the 2nd ACM/IEEE International Conference on Human-Robot Interaction, Washington DC, USA, pp. 177–184 (2007)
[12] Haddadin, S., Albu-Schäffer, A., Hirzinger, G.: Safety evaluation of physical human-robot interaction via crash-testing. In: Robotics: Science and Systems Conference (RSS 2007), Atlanta, USA, pp. 217–224 (2007)
[13] Haddadin, S., Albu-Schäffer, A., Hirzinger, G.: The role of the robot mass and velocity in physical human-robot interaction - part I: Unconstrained blunt impacts. In: IEEE Int. Conf. on Robotics and Automation (ICRA 2008), Pasadena, USA, pp. 1331–1338 (2008)
[14] Haddadin, S., Albu-Schäffer, A., Hirzinger, G.: The role of the robot mass and velocity in physical human-robot interaction - part II: Constrained blunt impacts. In: IEEE Int. Conf. on Robotics and Automation (ICRA 2008), Pasadena, USA, pp. 1339–1345 (2008)
[15] Haddadin, S., Albu-Schäffer, A., Luca, A.D., Hirzinger, G.: Collision detection & reaction: A contribution to safe physical human-robot interaction. In: IEEE/RSJ Int. Conf. on Intelligent Robots and Systems (IROS 2008), Nice, France, pp. 3356–3363 (2008)
[16] Haddadin, S., Laue, T., Frese, U., Hirzinger, G.: Foul 2050: Thoughts on physical interaction in human-robot soccer. In: IEEE/RSJ Int. Conf. on Intelligent Robots and Systems (IROS 2007), San Diego, USA, pp. 3243–3250 (2007)
[17] Haddadin, S., Laue, T., Frese, U., Wolf, S., Albu-Schäffer, A., Hirzinger, G.: Kick it like a safe robot: Requirements for 2050. Robotics and Autonomous Systems 57, 761–775 (2009)
[18] Heinzmann, J., Zelinsky, A.: Quantitative safety guarantees for physical human-robot interaction. The Int. J. of Robotics Research 22(7-8), 479–504 (2003)
[19] Herbst, B., Forrest, S., Chng, D., Sances, A.: Fidelity of Anthropometric Test Dummy Necks in Rollover Accidents. 16th ESV Conference Paper No. 98-S9-W-20 (1998)
[20] Herman, I.: Physics of the Human Body. Springer (2007)
[21] Ikuta, K., Ishii, H., Nokata, M.: Safety evaluation method of design and control for human-care robots. The Int. J. of Robotics Research 22(5), 281–298 (2003)
[22] Kallieris, D.: Schutzkriterien für den menschlichen Kopf. Forschungsprojekt FP 2.9317/2 (1998) (German)
[23] Kallieris, D.: Personal communication (2007)
[24] Khatib, O.: Inertial properties in robotic manipulation: an object-level framework. The Int. J. of Robotics Research 14(1), 19–36 (1995)
[25] Kroell, C., Scheider, D., Nahum, A.: Impact tolerance and response of the human thorax II. SAE Paper No.741187, Proc. 18th Stapp Car Crash Conference, pp. 383–457 (1974)
[26] Lim, H.O., Tanie, K.: Human safety mechanisms of human-friendly robots: Passive viscoelastic trunk and passively movable Base. The Int. J. of Robotics Research 19(4), 307–335 (2000)
[27] Lobdell, T., Kroell, C., Scheider, D., Hering, W.: Impact response of the human thorax. In: Symposium on Human Impact Response, pp. 201–245 (1972)
[28] Lumelsky, V., Cheung, E.: Real-time collision avoidance in teleoperated whole-sensitive robotarm manipulators. IEEE Transactions on Systems, Man and Cybernetics 23, 194–203 (1993)

[29] McElhaney, J., Stalnaker, R., Roberts, V.: Biomechanical aspects of head injury. Human Impact Response - Measurement and Simulation (1972)
[30] Melvin, J.: Human tolerance to impact conditions as related to motor vehicle design. SAE J885 APR80 (1980)
[31] Melvin, J., Little, W., Smrcka, J., Yonghau, Z., Salloum, M.: A biomechanical face for the Hybrid III dummy. In: Proceedings of the 39th Car Crash Conference (1995)
[32] Morinaga, S., Kosuge, K.: Collision cetection system for manipulator based on adaptive impedance control law. In: IEEE Int. Conf. on Robotics and Automation (ICRA 2002), Washington DC, USA, pp. 1080–1085 (2003)
[33] Morita, T., Iwata, H., Sugano, S.: Development of human symbiotic robot: WENDY. In: IEEE Int. Conf. on Robotics and Automation (ICRA 1999), Detroit, USA, pp. 3183–3188 (1999)
[34] NHTSA: Actions to reduce the adverse effects of air bags. FMVSS No. 208 (1997)
[35] Nightingale, R., McElhaney, J., Camacho, D., Kleinberger, M., Winkelstein, B., Myers, B.: The dynamic responses of the cervical spine: Buckling, end conditions, and tolerance in compressive impacts. Proceedings of the 41st Stapp Car Crash Conference SAE Paper No.973344, pp. 771–796 (1997)
[36] Nyquist, G.W., Cavanaugh, J.M., Goldberg, S.J., King, A.I.: Facial impact tolerance and response. SAE Paper No.861896, Proc. 30th Stapp Car Crash Conference, pp. 733–754 (1986)
[37] Okada, M., Ban, S., Nakamura, Y.: Skill of compliance with controlled charging/discharging of kinetic energy. In: IEEE Int. Conf. on Robotics and Automation (ICRA 2002), Washington, USA, pp. 2455–2460 (2002)
[38] Paluska, D., Herr, H.: The effect of series elasticity on actuator power and work output: Implications for robotic and prosthetic joint design. Robotics and Autonomous Systems 54, 667–673 (2006)
[39] Prasad, P., Mertz, H.: The position of the US delegation to the ISO Working Group 6 on the use of HIC in automotive environment. SAE Paper 851246 (1985)
[40] Prasad, P., Mertz, H.: Hybrid III sternal deflection associated with thoracic injury severities on occupants restrained with force-limiting shoulder belts. SAE Paper 910812 (1991)
[41] Prasad, P., Mertz, H., Nusholtz, G.: Head injury risk assessment for forehead impacts. SAE Paper 960099 (1985)
[42] Rizzetti, A., Kallieris, D., Schiemann, P., Mattern, P.: Response and injury of the head-neck unit during a low velocity head impact. In: International Research Council on Biomechanics of Injury (IRCOBI 1997), pp. 194–207 (1997)
[43] Schempf, H., Kraeuter, C., Blackwell, M.: ROBOLEG: A robotic soccer-ball kicking leg. In: IEEE Int. Conf. on Robotics and Automation (ICRA 1995), Nagoya, Aichi, Japan, vol. 2, pp. 1314–1318 (1995)
[44] Schneider, D., Nahum, A.: Impact studies of facial bones and skull. SAE Paper No.720965, Proc. 16th Stapp Car Crash Conference, pp. 186–204 (1972)
[45] United Auto Workers: Review of robot injuries - one of the best kept secrets. In: National Robot Safety Conference, Ypsilanti, USA (2004)
[46] Vanderborght, B., Verrelst, B., Ham, R.V., Damme, M.V., Lefeber, D., Duran, B., Beyl, P.: Exploiting natural dynamics to reduce energy consumption by controlling the compliance of soft actuators. The Int. J. of Robotics Research 25(4), 343–358 (2006)
[47] Versace, J.: A review of the severity index. SAE Paper No.710881, Proc. 15th Stapp Car Crash Conf., pp. pp. 771–796 (1971)

[48] Viano, D., Bir, C., Walilko, T., Sherman, D.: Ballistic impact to the forehead, zygoma, and mandible: Comparison of human and frangible dummy face biomechanics. The Journal of Trauma 56(6), 1305–1311 (2004)
[49] Wolf, S., Hirzinger, G.: A new variable stiffness design: Matching requirements of the next robot generation. In: IEEE Int. Conf. on Robotics and Automation (ICRA 2008), Pasadena, USA, pp. 1741–1746 (2008)
[50] Zinn, M., Khatib, O., Roth, B.: A new actuation approach for human friendly robot design. The Int. J. of Robotics Research 23, 379–398 (2004)

Chapter 6
Sharp and Acute Contact

Up to now, only blunt impacts were addressed in this monograph (see Fig. 6.2 (left)), leaving open the question of what can happen if a robot with an attached sharp tool can impact with a human, see Fig. 6.2 (right). If robots are supposed to work and help in a useful manner they must be able to handle potentially dangerous tools and equipment. Tasks may range from slicing bread (see Fig. 6.1) or preparing some meal to fulfilling duties of a craftsman. Until a robot actually ful-

Fig. 6.1 What happens if a robot is equipped with a dangerous tool while it is fulfilling a desired task in human proximity? Future domestic and industrial robots will carry out various duties incorporating the usage of sharp tools. As an example the DLR Humanoid *Justin* is slicing bread with a kitchen knife.

S. Haddadin: *Towards Safe Robots*,
Springer Tracts in Advanced Robotics 90,
DOI: 10.1007/978-3-642-40308-8_6, © Springer-Verlag Berlin Heidelberg 2014

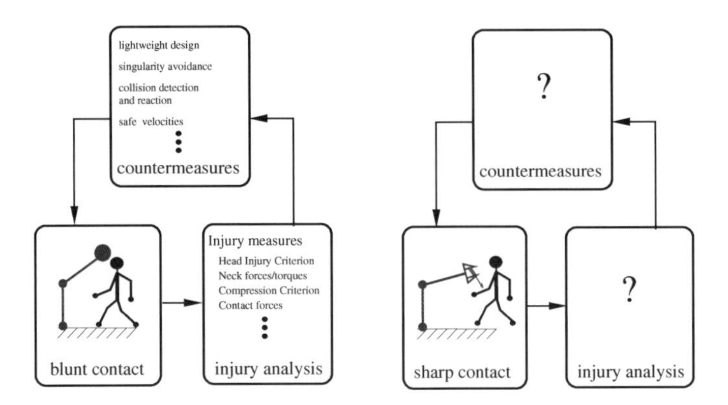

Fig. 6.2 Injury analysis in robotics

fills complex "helper" tasks in domestic environments, using sharp tools, massive safety investigation remains necessary. An important class of injuries to be analyzed in this context are soft-tissue injuries of which typical ones are depicted in Fig. 6.3. They range from less dangerous injuries as contusions or abrasions to painful lacerations and potentially life threatening ones as stab/puncture wounds. Stab/puncture wounds are usually potentially more lethal than laceration. However, for sensitive zones, such as around the area of the underlying arteria carotis, deep cuts can be equally dangerous.

Previous work conducted in [7] and [5] introduced and analyzed skin stress as an injury index for assessing soft-tissue injury. Nevertheless, a real focus shift to the mentioned soft-tissue injuries was not carried out until [9] and [4]. In [9] the need for a full evaluation of soft-tissue injury was given. In this work, the maximum curvature of a robot colliding with a human is approximated with a sphere. This is used to analyze the maximum tensile stress which in turn is the basis to distinguish between safe and unsafe contact.

Generally, soft-tissue injury analysis in robotics has been mainly model based. Knowing how uncertain and contestable simple models (and their parameterization) for such complex biomechanical processes are, the monograph addresses this topic empirically by acquiring real data for injury thresholds. It is believed that these experiments provide reliable facts and can serve as an aid for further evaluation and for designing validation of models.

Several aspects will be addressed in this chapter, leading to four main contributions, namely:

1. To evaluate soft-tissue injuries caused by various possibly dangerous tools. Stab/puncture wounds and incised wounds are considered.
2. To prove the effectiveness of the collision detection and reaction schemes introduced in Chapter 3 for the LWR-III with soft-tissue and volunteer tests. These

countermeasures enable drastic reduction of the injury potential during stabbing and prevent even the slightest cutaneous injury during cutting.

3. To provide empirically relevant limit values for injury prevention for the case of sharp contact.

The chapter is organized as follows. In Sec. 6.1 soft-tissue injuries caused by sharp tools are described and a simulation use-case is discussed. In Section 6.2 various stabbing and cutting experiments are conducted using as test material silicone, pig tissue, and human volunteer tests for situations which prove to be not critical by previous experiments.

6.1 Soft-Tissue Injury Caused by Sharp Tools

In this section, an overview of soft-tissue injury biomechanics is given. Furthermore, measurements concerning the depth of vital organs are introduced.

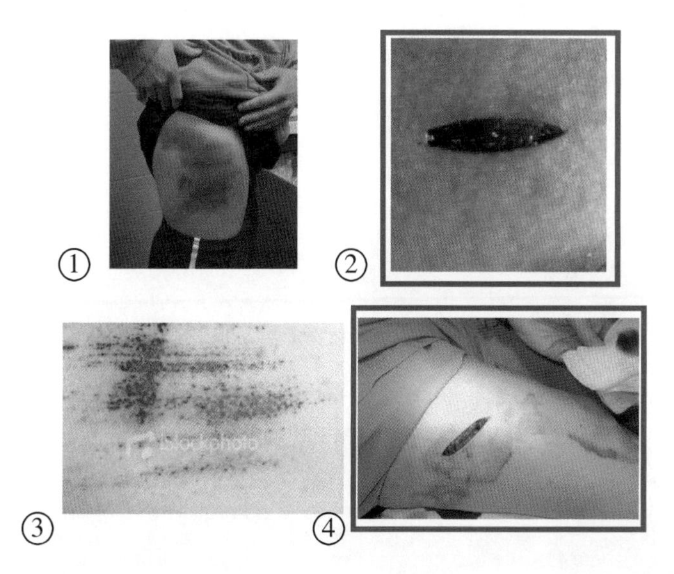

Fig. 6.3 Typical soft-tissue injuries which (at least) should be considered in robotics: ①: Contusion (bruises, crushes, hematoma), ②: Stab/puncture wounds, ③: Abrasion, ④: Laceration (incised wounds/cuts, gashes, contused wounds). Abrasions, lacerations, contusions and stab/puncture wounds can be caused by such different contact mechanisms as blunt or sharp contact surfaces as well as from normal or tangential impact directions. In the context of this monograph ② and ④ are analyzed.

6.1.1 Biomechanics of Soft-Tissue Injury

Sharp contact can cause various characteristic injuries in the context of robotics. The most important ones are abrasions, contusions, lacerations, incised wounds, and puncture wounds.

- Abrasions or excoriations are the ablation of parts or the entire epidermis from the corium.
- Contusions are bleedings into tissue which can be found in the skin, muscles and inner organs.
- A laceration can be described as a tear in the tissue. An incised wound is a transection in skin continuity which is wider than deep.
- A puncture or stab wound is deeper than wide.

In this chapter, the focus lays on stab/puncture wounds and incised wounds/cuts in order to capture the vast threat posed by sharp tools as knifes, scalpels or scissors and leave the low severity injuries caused by a less sharp tool for future research.

The influence of underlying bones is neglected and the evaluation focuses on areas as the abdomen or thigh. This can be considered as a worst-case since the underlying soft-tissue is very sensitive and a bone would (apart from the case of slipping of or impinging) reduce the possible injury severity by means of penetration depth. If e.g. an object hits the human thorax above the heart location and penetrates further it is possible to hit a heart protecting rib. In case the object does not slip or impinge nor is able to exert forces that are able to cause rib fracture, the possible injury is limited to the tissue till the rib and further rib injury. This is much less dangerous than if the robot tip penetrated between two ribs and reached the cardiac tissue. The analysis of these relaxed situations is left for future work.

Stab/puncture wounds were investigated in the forensic literature with different knifes and it was concluded that strain is not an appropriate measure to define a tolerance value for knifes and similar tools because the contact area is too small [1, 3, 10]. Instead, the evaluation of the penetration force F_p is proposed which is in the author's opinion appropriate to be used in the context of robotics as well. Tolerable forces depend on the layers of clothing and range according to [3] between

- mean values of $F_p^1 = 76 \pm 45$ N for uncovered skin
- and $F_p^2 = 173$ N for three layers of typical clothing.

Furthermore, the tolerable force correlates to a skin deflection x_p at which the actual penetration takes place. This deflection is

- $x_p^1 = 1.24 \pm 0.49$ cm for naked skin
- and $x_p^2 = 2.26 \pm 0.61$ cm for multilayered clothes[1].

[1] This evaluation was carried out at low velocities, therefore determining the static stab force. However, in [10] dynamic tests were conducted which produced similar numerical values. In [8] stab tests with three different knifes led on the other hand to significantly lower penetration values.

The relationship is assumed to be linear in first approximation, meaning that the skin can be modeled during stabbing by a stiffness before penetration and a tolerance force, determining the moment of penetration, see Fig. 6.12. Therefore, following contact model results.

$$K_{H,i} = \begin{cases} \frac{F_p^i}{x_p^i} & F_{\text{ext}} < F_p^i \\ 0 & F_{\text{ext}} \geq F_p^i \end{cases} \tag{6.1}$$

What happens after the knife actually penetrates is still not well investigated, and needs further treatment and evaluation. First hints given in [10] show that a second resistance after the initial skin penetration can be observed. As a first indicator the intrusion/penetration depth d_p was considered to be a relevant experimental quantity (depending on the location where the skin is actually penetrated and its underlying tissue) to evaluate the severity of injury.

According to [2] no similar investigation of incised wounds/cuts was carried out up to now. This is presumably due to the lack of forensic necessity. In this sense, the analysis will bring new insights into the understanding of this injury mechanism in a broader perspective and not only for robotics.

Next, results on depth measurements of vital organs are presented, since this seems to be a relevant injury indicator, which is applicable to robotics such that it provides inherent minimum requirements on the robot braking distance.

6.1.2 The Depth of Vital Organs

In order to quantify potentially lethal stabs, ultrasonic measurements with ten human subjects were conducted to estimate the distance from the skin surface to the surface of the human heart. Between the 4th or 5th intercostellar spaces the depth is measurable since the heart borders on the thorax wall. Numerical values of $d_{\text{heart}} = 2.2$ cm to 2.7 cm were measured with a mean of $\bar{d}_{\text{heart}} = 2.4$ cm.

In addition to this initial heart depth analysis, measurements for following vital organs were conducted[2].

- heart
- abdominal aorta
- liver (side)
- liver (subcostal)
- kidney (back)
- soft-tissue throat (right)
- soft-tissue throat (left)
- subclavia
- milt

[2] The measurements had no diagnostic nor therapeutical purpose and did not cause any injury. The subjects were anonymous. The author is grateful to Dr. med. Fahed Haddadin for carrying out the ultrasonic measurements.

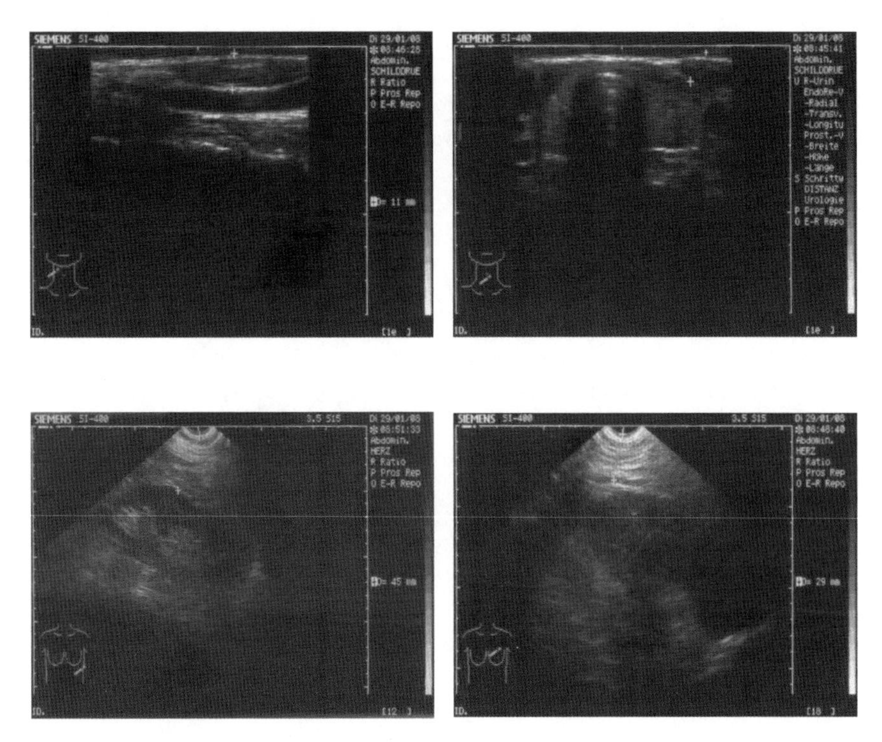

Fig. 6.4 Ultrasonic scans and measurements of the depth of throat/neck soft-tissue (upper) and the heart (lower)

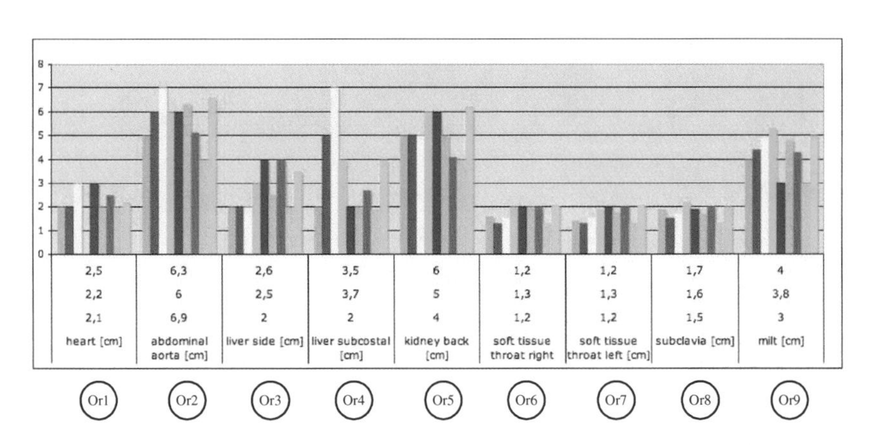

Fig. 6.5 Ultrasonic depth measurements of different vital organs

The used ultrasonic system is a *ESAOTE Megas, yoc 2005*. A 7.5 MHz probe was used for the throat and neck soft-tissue and a 3.5 MHz probe for the other organs. Figure 6.4 shows four sample measurements for the heart (left) and the throat (right). In the soft-tissue of the throat especially the arteria carotis communis and vena jugularis interna are of interest. The lower left figure shows the second area of measurement for the heart depth, where the heart abuts the diaphragm. In Figure 6.5 the results of the depth measurements are visualized for all mentioned organs. The short distances clearly point out how vulnerable human organs are as soon as penetration occurs.

Since it is difficult to estimate the particular injury a human would suffer from sharp contact, it is important to define requirements for robot design and control, which quantify the possible benefits from a collision detection and reaction strategy in an intuitive manner. One important requirement is the maximum braking distance of a robot.

6.1.3 Braking Distance

As shown in the previous subsection, the organ depth d_{organ} can be used as a penetration depth which absolutely needs to be prevented during sharp robot-human contact.

External contact forces caused by the human dynamic response potentially decrease the braking distance especially for low inertia robots. Therefore, the worst-case braking distance is present without taking this into consideration. It consists of three phases:

1. nominal motion before collision detection triggers (system delay, detection sensitivity): $t_0 \rightarrow t_1$
2. nominal motion before stopping reaction strategy activates (system latencies): $t_1 \rightarrow t_2$
3. stopping motion till entire stop of the system (actuator dynamics/saturation): $t_2 \rightarrow t_3$

Therefore the overall braking distance, which should be smaller than d_{organ} is

$$||\mathbf{x}_{\text{stop}}|| = \int_{t_0}^{t_2} \dot{\mathbf{x}}_{\text{nom}}\,\mathrm{d}t + \int_{t_2}^{t_3} \dot{\mathbf{x}}_{\text{brake}}\mathrm{d}t < d_{\text{organ}}. \tag{6.2}$$

This limit is a good indicator to qualify the effectiveness of collision detection and reaction schemes, since it is an absolute limit before life threatening injury occurs if penetration into the body happens. It inherently defines minimum performance characteristics on actuation dynamics by means of maximum joint torque and response time.

To reduce injuries caused by knifes and scalpels the use of collision detection and reaction methods as a countermeasure against soft-tissue injury is first evaluated by simulation.

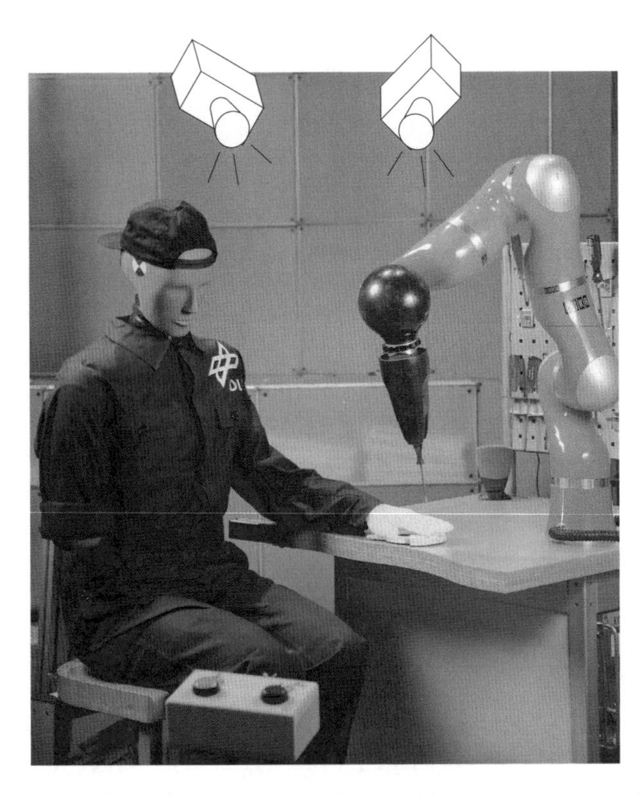

Fig. 6.6 This co-worker setup, presented at the AUTOMATICA 2008 trade fair in Munich was the first contribution to evaluate such co-worker scenarios. The KUKA Lightweight robot, which is equipped with a parallel gripper that is covered by a polyurethane enclosure is supporting the human during an assembly task. The gripper was used to firmly grasp a screwdriver located at a tool wall. The human is modeled by a crash-test dummy, which is equipped with a silicone arm, modeling the soft behavior of the human soft-tissue. The dummy was actuated via pressure valves back and forth and thus one was able to "simulate" the accidental penetration of the robot workspace by the human. The visitors were allowed to trigger the dummy anytime during task execution and for the entire four-day fair the collision detection was preventing a penetration into the silicone duplicate of the human arm. Furthermore, visitors were allowed to get "hands on" experience and a subjective safety feeling was unanimously experienced.

6.1.4 A Simulation Use-Case with the LWR-III

Countermeasures against soft-tissue injury can be manifold but a crucial feature has to be an effective physical collision detection and reaction. If an interaction force is detected, the differentiation whether the robot is currently fulfilling a desired task as preparing food or constitutes a potential threat is still to be done. However, this is a question of higher-level planning and human motion detection involving external sensing as e.g. a vision system, see Fig. 6.6. Therefore, this a separate topic that is not within the scope of this monograph.

In Figure 6.6 a sample scenario is given with the DLR-LWR-III as a co-worker collaborating closely with a human. Distinguishing whether the occurring collision is part of the assembly task or a collision with the human (in this case with the silicone arm of the crash test dummy) was simply solved by switching the collision detection off as soon as clamping of the human can be excluded due to the fact that the distance between the tool and the known environment (table) is lower than a threshold. In this situation, a sufficient world model is necessary which could be the case in an industrial scenario.

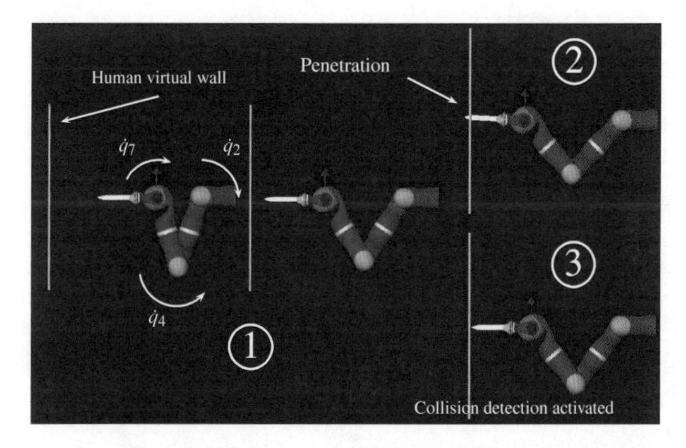

Fig. 6.7 Stabbing simulation with the full dynamic (flexible joint robot) model of the LWR-III equipped with a knife. The human soft-tissue is modeled as a virtual wall with the already mentioned spring constant and is assumed to be clamped, i.e. a worst-case scenario is addressed. The robot is mounted on a fixed base. The maximum joint velocity of the robot is 120 ^{o}/s and the desired motion is a straight line with reconfiguration from "elbow up" to "elbow down". The maximum Cartesian velocity resulting from the maximum joint velocity in the 4th joint, whereas the 2nd and the 7th joint drive at half the velocity, is 0.64 m/s. In this simulation, the Cartesian impact velocity was chosen to be $\dot{\mathbf{x}}_R \in \{0.16, 0.32, 0.64\}$ m/s for fully covered skin. ①: Initial robot configuration. ②: The robot moves towards the human. ②: Without collision detection the robot easily penetrates the human skin. ③: With collision detection the robot is able to detect the collision and stops before the skin is damaged.

In this use-case, the penetration of the human skin with a knife and its prevention will be treated. A simple and reasonable contact model for stabbing is available as mentioned in Sec. 6.1. This simulation shows how easy it is to penetrate the human skin even with a robot moving at moderate speeds. Penetrating the human skin itself appears to be only a marginal injury but at the same time various vital organs as the heart or the liver are located relatively close to the body surface.

In Figure 6.7 the simulated trajectory of the robot is depicted. ① denotes the initial configuration of the robot. ② shows the clear penetration without collision

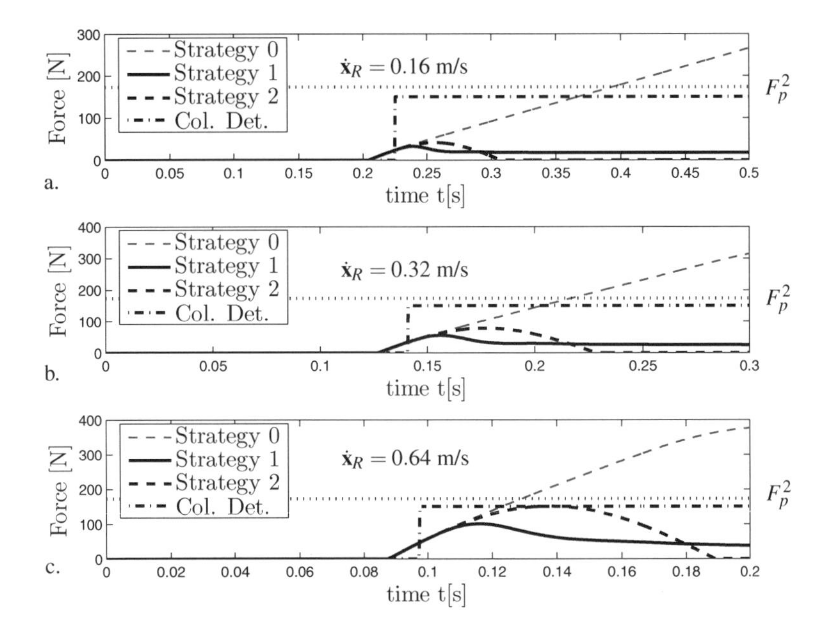

Fig. 6.8 Stabbing simulation with the full dynamic model of the LWR-III equipped with a knife. The **fully covered human** stands 0.3 m before the stretched out singularity of the robot, see Fig. 6.7. F_p^2 denotes the penetration force for fully covered skin. The major difference between Strategy 1 and Strategy 2 is that stopping the robot immediately (Strategy 1) reduces the contact force much faster compared to Strategy 2, whereas the switch in control (Strategy 2) reduces the contact force down to zero in contrast to Strategy 1. Even for 0.64 m/s the collision detection is able to prevent damage from the skin. a.) shows the results for a joint velocity of 30 ^{o}/s, b.) for 60 ^{o}/s, and c.) 120 ^{o}/s in the 4th (elbow) joint. 120°/s is the current maximum joint velocity of the LWR-III.

detection, whereas ③ exemplifies how the human skin can be protected by reacting e.g. with Strategy 1. This particular trajectory is not the worst-case but it corresponds to a typical robot configuration. In Figure 6.8 the results of the simulation are shown. The effectiveness is apparent even for high Cartesian velocities. The skin is not penetrated since the robot is able to react sensitive and fast enough to prevent the human from being hurt. Furthermore, the properties of the collision reaction strategies become apparent: Since Strategy 1 actively stops the robot it reduces the contact force significantly faster than Strategy 2 which reacts delayed. This is due to the passive behavior of the robot in torque controlled mode with gravitation compensation. However, Strategy 2 is able to fully lose external contact in contrast to Strategy 1. A combination of both strategies appears to be the best choice.

After this discussion of a simulation use-case, various experiments are described in the following, giving an insight into the injury mechanisms during contact with various sharp tools.

6.2 Experiments

In this section, various experiments are presented which help analyze the injury severity, possibly occurring if a robot with a sharp tool penetrates a soft material. Especially the dynamics of such an impact is worth to be investigated since during rigid (unconstrained) collisions presented in Chapter 5 the dynamics is so fast that a realistic robot is not able to reduce the impact characteristics by the collision detection and reaction. However, during previous investigations a subjective safe feeling could definitely be experienced by the users. Despite this limitation in reactivity to blunt impacts it was shown as well that the necessity of countermeasures is not absolutely crucial since rigid free impacts with the LWR-III pose only a limited risk at typical robot velocities up to 2 m/s. This is not the case for soft-tissue injuries caused by a stab, since the injury severity due to the penetration can reach lethal dimensions. The particular worst-case depends on the exact location by means of underlying potentially injured organs. Because of the much lower dynamics compared to rigid impacts, the requirements on a reactive robot concerning detection and reaction speed are less stringent and achievable for such situations as exemplified in Sec. 6.1.4. It seems surprising at a first glance that it is not possible to counterbalance rigid blunt robot-human impacts which are non life-threatening by means of control[3], while dangerous or lethal contacts with tools appear manageable to a certain extent. One purpose of the present experiments is to prove this statement.

In the following, the situation in which the robot moves in position control with and without collision detection is considered. The contact force is measured with a JR3 force/torque sensor in the wrist. Please note that this sensor is *only* used for measurement and *not* for collision detection.

6.2.1 *Investigated Tools*

There exists a countless variety of tools one could analyze. Therefore a representative selection[4] was carried out, see Fig. 6.9. The focus was especially on sharp ones so to analyze the problem of stabbing. Furthermore, different blade profiles and lengths were chosen to investigate cutting, which turned out to be a vast injury threat.

6.2.2 *Silicone Block*

As a first experimental contact material a silicone block[5] was used in order to help gain some first hand knowledge for the sensitivity and effectiveness of the collision

[3] Please note that it is referred to impact speeds of up to 2 m/s.

[4] The tools were tested in the same condition they were bought except for the fact they were glued into a rigid mounting to remove eventually beneficial compliances.

[5] The used silicone was *Silastic T2* with a Shore hardness of A40 and manufactured by *Dow Corning*.

detection and reaction for soft contact, see Fig. 6.10 (left). Due to its standardized properties it can be used as a benchmark material, in contrast to some biological tissue. These first tests were conducted at a Cartesian velocity of 0.25 m/s, which is the recommended velocity according to the new *ISO-10218* for collaborative robots [6]. The mounted tool is the kitchen knife. Figure 6.10 (right) shows how effective the collision detection and reaction can help to reduce contact forces and the penetration depth. The desired goal configuration was located at a depth of 8 cm in the silicone block. Without any collision reaction strategy the achieved penetration was 35 mm at a contact force of 220 N with joint six exceeding its maximum joint torque. With activated collision detection and reaction the maximum penetration depth was dramatically reduced to ≤ 6 mm at a contact force of 40 N, i.e. a reduction by a factor of ≈ 5.

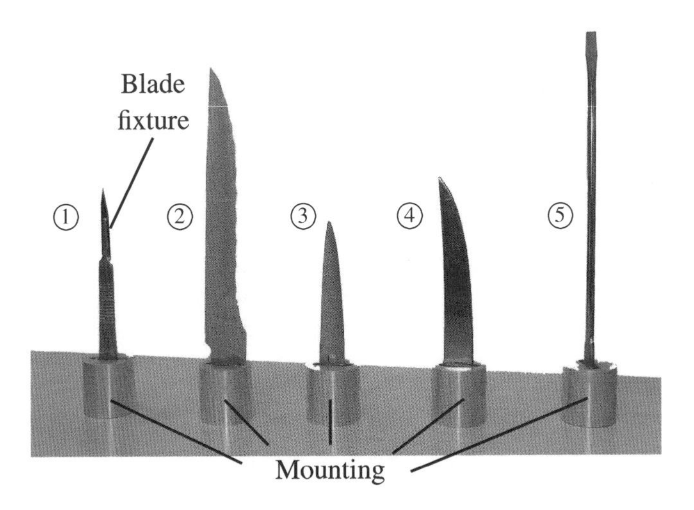

Fig. 6.9 Investigated tools: ① Scalpel, ② kitchen knife, ③ scissors, ④ steak knife, ⑤ screwdriver. These tools were selected as a reasonable choice of potentially very dangerous ones one could think of in robotic applications. They were removed from their original fixtures and glued into new mountings. Therefore, a fixed connection between tool and robot can be guaranteed and no compliance reduces the transferred forces.

6.2.3 Pig Experiments

In order to obtain results with real biological tissue experiments with a pig leg were conducted, see Fig. 6.11. Anatomically, pigs are commonly accepted as being similar to human beings. Both impact experiments in automobile crash-testing and in forensic medicine employ them for first experiments or even for predictions of human tissue results. Differences to humans and changing tissue properties through mortex are apparent but still it seems to be of immanent importance to conduct experiments with natural tissue. These investigations can be fundamental to robotic

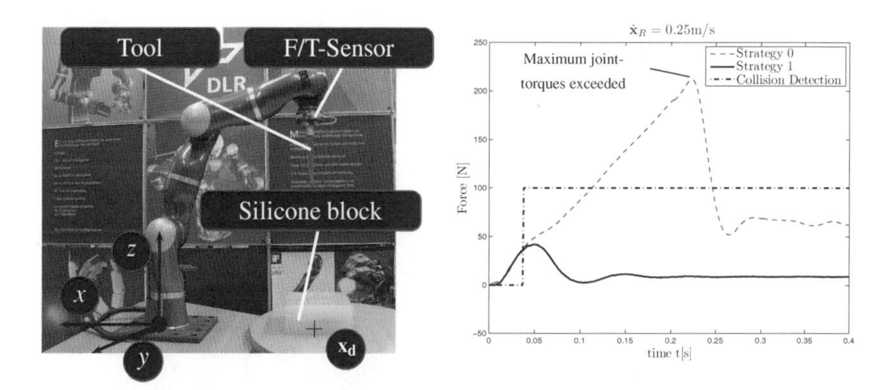

Fig. 6.10 Stabbing tests with a kitchen knife mounted on the robot and a silicone block. The robot moves on a straight line along the z-axis with a target position $\mathbf{x}_d$ about 8 cm inside the silicone. Without collision detection the force reaches a value of 220 N. The force drop is due to the intrinsic joint stiffness of the robot. For activated collision detection and reaction the trajectory stops by setting the desired position to the current position. The contact force can be limited to 40 N.

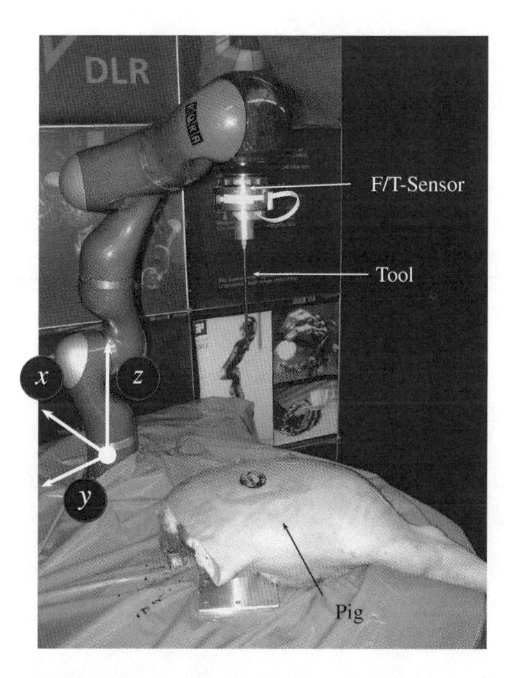

Fig. 6.11 Testing setup for the pig experimental series. The robot is equipped with a JR3 force/torque sensor only for measuring the contact force. The tools are rigidly mounted to the robot such that no significant additional compliance is caused. The stabbing trajectory is a straight line along the z-axis and the desired configuration is slightly above the table. The pig is located on a rigid table, i.e. a clamping scenario is analyzed due to its worst-case properties.

safety since e.g. classical impact experiments with knives in forensic medicine as described in [3, 10] did (of course) not take any robot behavior into account which in turn vastly influences the resulting injury.

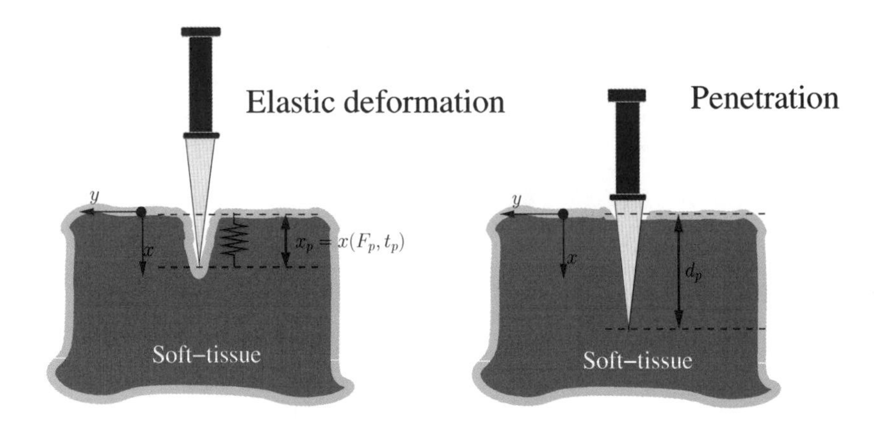

Fig. 6.12 Elastic deformation of the skin before penetration up to x_p at a force level of F_p (left). Penetration depth d_p into the tissue after exceeding the tolerance force F_p (right).

Table 6.1 Results of the stabbing experiments

		$\dot{\mathbf{x}}_R = 0.16$ m/s					$\dot{\mathbf{x}}_R = 0.64$ m/s				
Exp. Nr.	Tool	Strategy	d_p [mm]	t_p [ms]	F_p [N]	x_p [mm]	Exp. Nr.	d_p [mm]	t_p [ms]	F_p [N]	x_p [mm]
A1.1	Steak knife	0	full	100	15	14	A1.2	full	14	11	10
A2.1		1	none/4	–	–	–	A2.2	22	14	11	10
A3.1		2	3 – 5	100	15	14	A3.2	64	14	11	10
B1.1	Scissors	0	full	195	60	25	B1.2	full	47	65	29
B2.1		1	none	–	–	–	B2.2	18	34	45	21
B3.1		2	none	–	–	–	B3.2	42	42	65	25
C1.1	Kitchen knife	0	98	240	76	29	C1.2	135	55	73	32
C2.1		1	none	–	–	–	C2.2	1	48	60	29
C3.1		2	none	–	–	–	C3.2	18	55	76	31
D1.1	Scalpel	0	full	50	5	8	D1.2	full	15	5	10
D2.1		1	17	50	5	8	D2.2	17	15	5	10
D3.1		2	17	50	5	8	D3.2	39	15	5	10

6.2.3.1 Stabbing

Table 6.1 and Fig. 6.13 summarize the outcome of the stabbing tests. The trajectory of the robot was chosen such that it moves on a straight vertical line (compare also Fig. 6.17) contacting the skin in normal direction with the tool axis. The investigated robot velocities were 0.16 m/s and 0.64 m/s. Surprisingly, with the screwdriver mounted, the robot was not able to penetrate the pig skin at all. For this tool

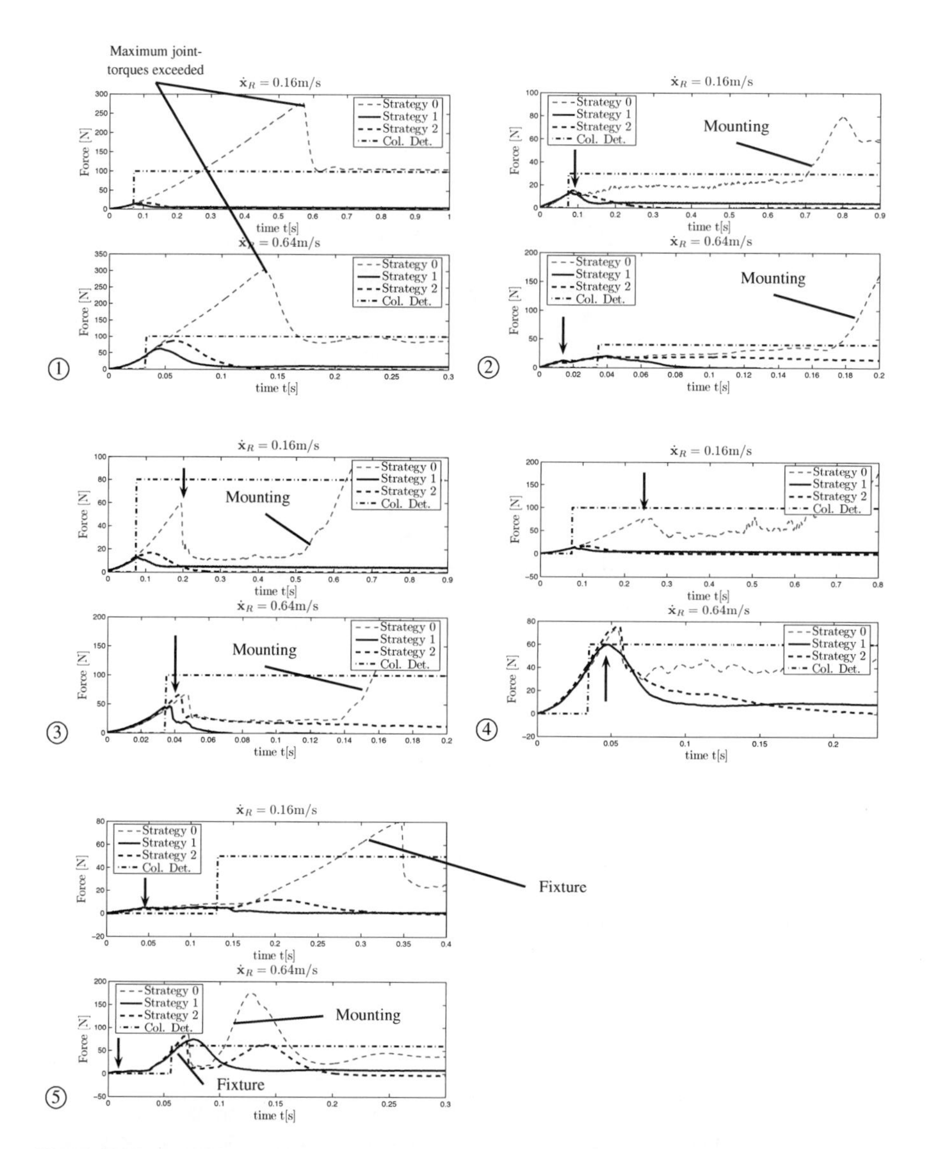

Fig. 6.13 Results of stabbing tests with and without collision detection for the pig tests. ①: screwdriver, ②: steak knife, ③: scissors, ④: kitchen knife, ⑤: scalpel. The arrows denote the moment of penetration.

Table 6.2 Resulting injury for stabbing experiments

	Exp. Nr.(Str.)	Or1	Or2	Or3	Or4	Or5	Or6	Or7	Or8	Or9
steak knife	A1.1(0)	×	×	×	×	×	×	×	×	×
	A1.2(0)	×	×	×	×	×	×	×	×	×
	A2.1(1)	✓	✓	✓	✓	✓	✓	✓	✓	✓
	A2.2(1)	✓	✓	✓	✓	✓	×	×	×	✓
	A3.1(2)	✓	✓	✓	✓	✓	✓	✓	✓	✓
	A3.2(2)	×	×	×	×	×	×	×	×	×
scissors	B1.1(0)	×	×	×	×	×	×	×	×	×
	B1.2(0)	×	×	×	×	×	×	×	×	×
	B2.1(1)	✓	✓	✓	✓	✓	✓	✓	✓	✓
	B2.2(1)	✓	✓	✓	✓	✓	×	×	×	✓
	B3.1(2)	✓	✓	✓	✓	✓	✓	✓	✓	✓
	B3.2(2)	×	✓	×	×	✓	×	×	×	×
kitchen knife	C1.1(0)	×	×	×	×	×	×	×	×	×
	C1.2(0)	×	×	×	×	×	×	×	×	×
	C2.1(1)	✓	✓	✓	✓	✓	✓	✓	✓	✓
	C2.2(1)	✓	✓	✓	✓	✓	✓	✓	✓	✓
	C3.1(2)	✓	✓	✓	✓	✓	✓	✓	✓	✓
	C3.2(2)	✓	✓	✓	✓	✓	×	×	×	✓
scalpel	D1.1(0)	×	×	×	×	×	×	×	×	×
	D1.2(0)	×	×	×	×	×	×	×	×	×
	D2.1(1)	✓	✓	✓	✓	✓	×	×	×	✓
	D2.2(1)	✓	✓	✓	✓	✓	×	×	×	✓
	D3.1(2)	✓	✓	✓	✓	✓	×	×	×	✓
	D3.2(2)	×	✓	×	×	✓	×	×	×	×

the maximum nominal joint torques were always exceeded and a low-level safety mechanism engaged the brakes of the robot as described in Sec. 6.2.2. For the other tools Tab. 6.1 gives the measured values for the penetration depth d_p, the penetration time t_p (which can be interpreted as the *available reaction time* to prevent skin penetration), the penetration force F_p and finally the elastic deflection before penetration x_p, i.e. the deflection of the skin which has to be reached with a particular tool for penetration, see Fig. 6.12.

As shown in Tab. 6.1, without collision detection (Strategy 0) all sharp tools penetrate into the tissue with their entire blade length, pointing out the lethality potential. At the same time it can be deduced that at low speeds a very good chance of detection and reaction exists and especially for the kitchen knife and the scissors a full injury prevention possible. For the steak knife the success depends on the exact location and ranges from no penetration up to a penetration depth of a few millimeters. For the used scalpel there is actually no real chance to detect the penetration of the blade. The collision detection is only triggered by the fixture of the blade, which has a significantly larger cross section, see Fig. 6.9.

For higher velocities, a significant observation confirming the results from the simulation can be made: Switching to Strategy 2 causes a higher penetration depth due to its passive behavior. Because the robot behaves in this control mode as a free floating mass with a certain amount of initial kinetic energy further penetration of the tissue until the robot's energy is fully dissipated takes place. Moreover, only Strategy 1 is able to limit the penetration depth to values which are lethal in absolute worst-case scenarios, i.e. below 2.4 cm. Surprisingly the penetration force appears not to be velocity dependent.

Apart from the characteristic values of Tab. 6.1, the force profiles of the stabbing experiments are depicted in Fig. 6.13. ① shows the obtained graphs for the screw-driver, ② for the steak knife, ③ for the scissors, ④ for the kitchen knife, and ⑤ for the scalpel. The force-time relationship is plotted for all three strategies. Especially following general aspects become clear when evaluating the plots.

- The moment of penetration is characterized by a significant force discontinuity (drop).
- A low resistance can be observed from the moment the tool intruded the subcutaneous tissue.
- Force reduction by Strategy 2 is significantly slower compared to Strategy 1 (compare as well to Sec. 6.1.4).
- After the initial penetration the contact force increases slowly compared to the elastic force of the skin.

The influence of the tool mounting (Fig. 6.9) can be observed for Strategy 0, resulting in a dramatic increase in force and a compression of the entire subject (the tool mounting establishes a blunt contact). The different course taken by the scalpel case ⑤ can be explained in the following. The low penetration threshold is followed by an almost constant section which represents the intrusion of the entire blade. For 0.16 m/s the following increase in force is caused by the fixture of the blade which therefore can be detected. For the graph with an impact velocity of 0.64 m/s the force increase due to the fixture is followed by a second one caused by the mounting as for the previous tools.

Table 6.2 lists the results with respect to each organ and whether it would have been reached or not. Again, the stopping strategy is the most effective strategy which is able to prevent severe penetration.

6.2.3.2 Cutting

The second injury mechanism which is investigated is cutting. The pure cut trajectory with a fixed object can be described by the tool orientation ϕ_1, the desired cut direction ϕ_2 and the cutting velocity, see Fig. 6.14. If ϕ_1 is chosen then the pig position is already determined, since the cut shall be carried out with the full available blade length. ϕ_1 was chosen to be 30 ^{o}. Investigated tools were the steak knife, the scalpel, and the kitchen knife. The question of which cutting angle ϕ_2 is the worst-case was answered experimentally and determined to be 10 ^{o}. Furthermore, it became apparent that cutting velocities must be quite high to cause damage to the skin and the underlying tissue. At a low velocity of $||\dot{\mathbf{x}}_{\mathrm{cut}}|| = 0.25$ m/s more or less no injury was observed and merely a scratch in the skin could be found. However, at $||\dot{\mathbf{x}}_{\mathrm{cut}}|| = 0.8$ m/s this changed dramatically. Figure 6.15 (upper row) shows the large and deep lacerations caused by all tools if no safety feature is activated. Life-threatening depths can be easily achieved. Please note that the subject is fixed, presumably leading to higher injuries compared to a non-fixed subject. Apparently, the blade length can greatly influence the resulting laceration depth. Although a

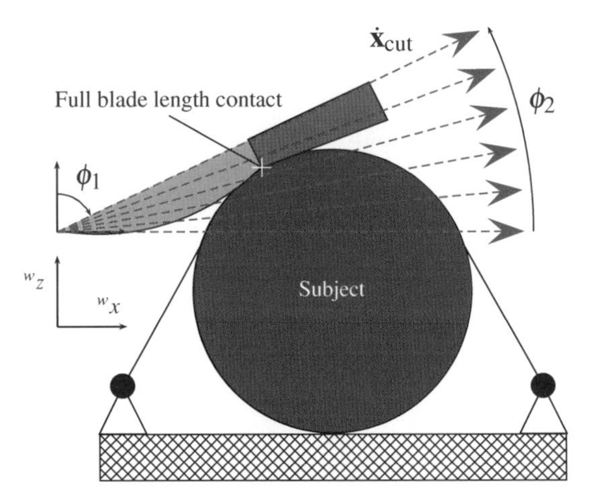

Fig. 6.14 Cutting trajectories for a fixed subject. ϕ_1 is the tool orientation and ϕ_2 the cutting direction. The tool is positioned such that the blade origin contacts the subject. Thus, the full blade length can be used for cutting the tissue.

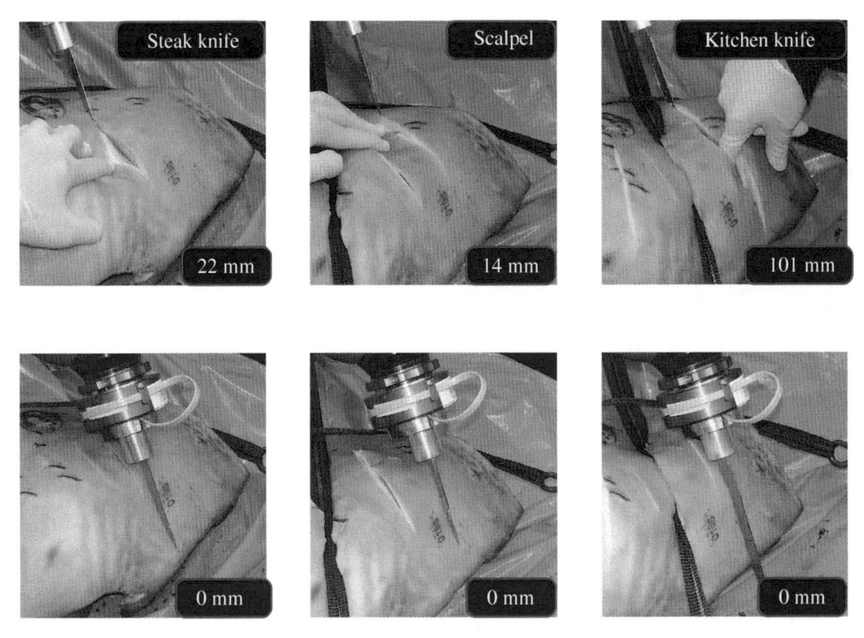

Fig. 6.15 Resulting injury due to cutting. In the upper row the laceration depths caused at 0.8 m/s are indicated. All tools easily penetrated the tissue and cutting depths of up to 101 mm were reached. Such depths are lethal and would pose an enormous threat. In the lower row the effect of reacting to a collision is apparent. The robot stops as soon as a collision was detected (Strategy 1). No cut could be observed. These tests show that the injury level can be reduced from *lethal* to *none*.

scalpel is an extraordinary sharp tool easily penetrating the skin, the small blade length limits the penetration depth to 14 mm. This is almost an order of magnitude smaller than for the large kitchen knife. Thus, for such high velocities long-blade knifes are far more dangerous than e.g. scalpels, which in turn are able to penetrate the skin already at low velocities.

Though the described large and potentially fatal injuries are possible, the risk can be reduced, even as high as 0.8 m/s by collision detection and reaction to almost neglectable levels at which no penetration or cut takes place anymore. Even in case of the scalpel one is able to fully prevent injury of the epidermis, pointing out the surprisingly high sensitivity of the collision detection, see Fig. 6.15 (lower row).

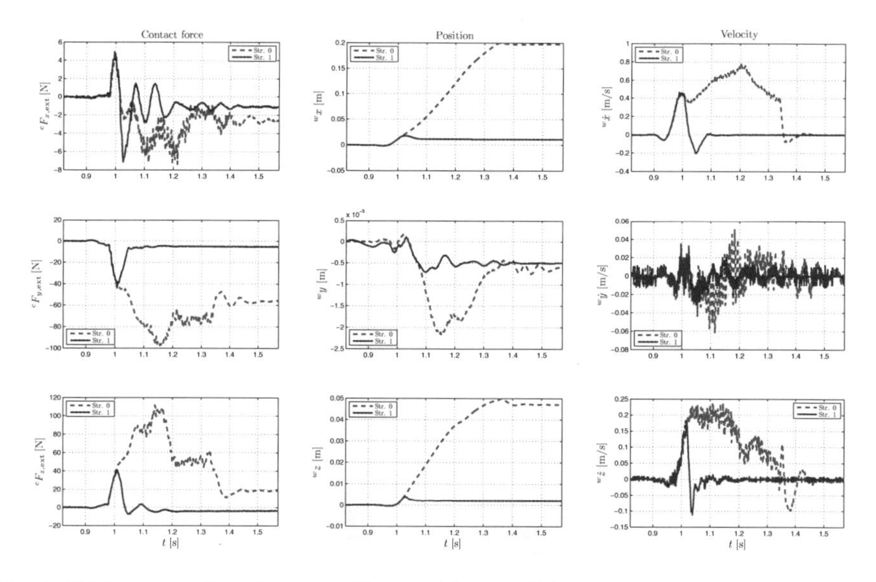

Fig. 6.16 Time evolution of cutting with and without collision detection

Figure 6.16 depicts the force, position, and velocity profiles for the cut motion. The forces are mainly acting in the $({}^{W}x, {}^{W}z)$-plane. The figure shows measurements for Strategy 0 and Strategy 1. Again, the effectiveness of detection is observed. At $t \approx 1.08$ s the beginning of a zagged behavior can be seen, which corresponds to the penetration event. The corresponding contact force is ≈ 80 N. With Strategy 1 activated such large forces are prevented.

In summary, the following main conclusions for cutting can be drawn:

1. Injuries caused by cutting can reach severe or even lethal levels at high velocities. At low velocities the epidermis is barely injured.
2. The achieved level of injury mainly depends on the blade length and the cutting velocity.
3. Collision detection based on joint torque sensing is an effective countermeasure to completely prevent injuries from cutting even at high velocities.

With the knowledge gained from the evaluation of soft-tissue injuries caused by sharp tools, the author was confident to explore the effectiveness of the collision detection in a human experiment.

6.2.4 Human Experiments

Since the presented experiments showed really promising results and proved how reliably one is able to promptly detect and react to collisions, some measurements are shown, where a human holds his arm in *free* space against the moving robot with a mounted knife, see Fig. 6.17. The robot velocity was chosen to be

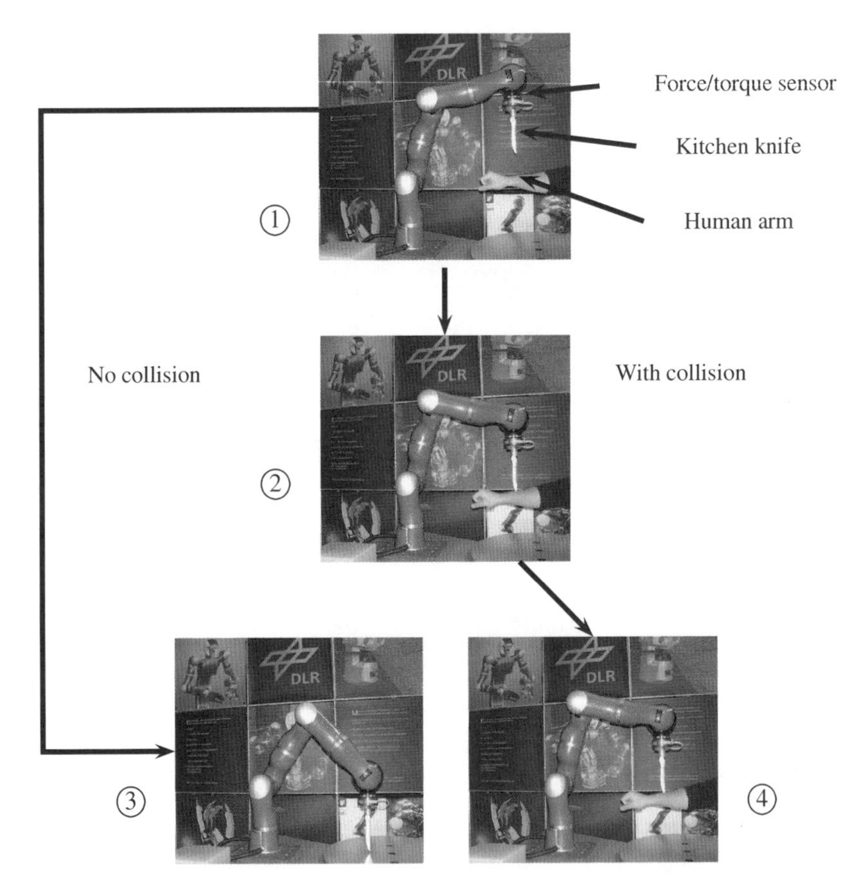

Fig. 6.17 Effectiveness of the collision detection and reaction. The human arm is hit by the robot at $||\dot{\mathbf{x}}_R|| \in \{0.1, 0.25, 0.5, 0.75\}$ m/s. The desired trajectory of the robot is a straight line in vertical direction. ①: Initial robot configuration. ②: The robot moves along its desired trajectory. ③: Desired goal configuration of the robot. ④: The robot detects the collision with the human arm and stops before hurting the human.

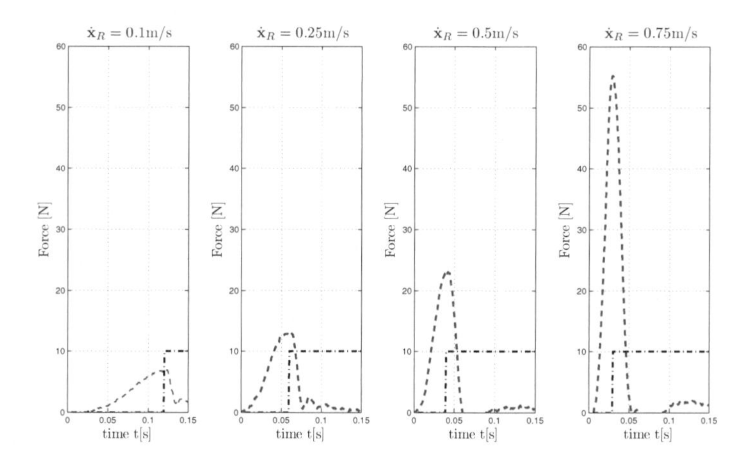

Fig. 6.18 Stabbing tests in free space with human volunteer. The force can be limited to subcritical values. The dashed line is the measured contact force and the dashed-dotted line the collision detection signal.

$||\dot{\mathbf{x}}_R|| \in \{0.1, 0.25, 0.5, 0.75\}$ m/s. In Figure 6.18 the measured force during the collision with the human is plotted. Due to the collision detection the robot is able to prevent the human from being injured at all. The contact force was limited in this experiment to 7 N for 0.1 m/s, to 13 N at 0.25 m/s, to 23 N at 0.5 m/s and to 55 N at 0.75 m/s. Only for 0.75 m/s a minimal scratch in the epidermis could be observed. This experiment strongly supports the results obtained from simulation and experimental evaluations. It points out that, although intuitively it seems unrealistic to prevent injury from humans during sharp contact by means of control, there is a clear chance to greatly reduce danger to the human even up to velocities of 0.75 m/s.

6.3 Summary

In this chapter, the biomechanical basics of soft-tissue injury were introduced. Furthermore, experimental data regarding the depth of vital organs was presented, which is used as a realistic requirement for the braking distance of a robot. Then, a simulation and experimental evaluation of soft-tissue injuries in robotics and a verification of possible countermeasures by means of control were carried out. The treatment of such injuries is a crucial prerequisite to allow robots the handling of sharp tools in the presence of humans. In this chapter, stab/puncture wounds caused by sharp tools were addressed. The fact that a knife can penetrate into deeper human inner regions and therefore threaten sensitive organs mainly motivated this evaluation. Various increasingly sharp tools were tested ranging from a screwdriver to a scalpel and showed the huge benefit of the collision detection and reaction.

References

[1] Brinkmann, B., Madea, B. (eds.): Handbuch gerichtliche Medizin. Springer (2004) (German)

[2] Eisenmenger, W.: Handbuch gerichtliche Medizin - Spitze, scharfe und halbscharfe Gewalt. Springer (2004) (German)

[3] Fazekas, I., Kosa, F., Bajnoczky, I., Jobba, G., Szendrenyi, J.: Mechanische Untersuchung der Kraft durchbohrender Einstiche an der menschlichen Haut und verschiedenen Kleidungsschichten. Zeitschrift für Rechtsmedizin 70, 235–240 (1972) (German)

[4] Haddadin, S., Albu-Schäffer, A., Hirzinger, G.: Safe physical human-robot interaction: Measurements, analysis & new insights. In: International Symposium on Robotics Research (ISRR 2007), Hiroshima, Japan, pp. 395–408 (2007)

[5] Ikuta, K., Ishii, H., Nokata, M.: Safety evaluation method of design and control for human-care robots. The Int. J. of Robotics Research 22(5), 281–298 (2003)

[6] ISO10218: Robots for industrial environments - Safety requirements - Part 1: Robot (2006)

[7] Ulrich, K., Tuttle, T., Donoghue, J., Townsend, W.: Intrinsically safer robots. Tech. rep., 139 Main Street, Kendall Square (1995), `http://www.barrett.com/robot/`

[8] von Prittwitz, Gaffron, A.: Bestimmung der Kraft durchbohrender Einstiche am menschlichen Thorax mit einem "in situ" - Messverfahren. Ph.D. thesis, University of Heidelberg (1974) (German)

[9] Wassink, M., Stramigioli, S.: Towards a novel safety norm for domestic robots. In: IEEE/RSJ Int. Conf. on Intelligent Robots and Systems (IROS 2007), San Diego, USA (2007)

[10] Weber, W., Schweitzer, H.: Stichversuche an Leichen mit unterschiedlicher kinetischer Energie. Beiträge Gerichtliche Medizin 31, 180–184 (1973) (German)

Chapter 7
Reactive Pre-collision Strategies

From a control point of view this monograph dealt to a large extent with physical collisions, their detection and following reaction up to now. Apart from such physical analysis and control, immanent injury can be diminished if the robot is able to reduce its impact speed or change its moving direction prior to the collision. Locally, the robot would circumvent the human or obstacle and avoid the impact completely. Therefore, it is of major importance to provide flexible motion generation methods, which take into account the possibly complex environment structure and at the same time can react quickly to changing conditions.

Motion generation methods can be divided into path planning algorithms and reactive motion generation. On the one hand (probabilistic) complete, highly sophisticated offline path planning methods are used, which provide complete collision free paths for potentially complex scenarios [4] with multi-DoF open or closed chain kinematics. On the other hand, reactive motion generators, which usually show a more responsive behavior, are simpler and have short execution cycles. Usually, these methods associate virtual forces to obstacles that act on virtual dynamics assigned to the robot. Both classes mostly treat the entire motion generation problem from a purely geometric/kinematic point of view. However, with the recent advances in pHRI it becomes more important to be able to plan complex motions for task achievement and cope with the proximity of dynamic obstacles under the absolute premise of safety to the human at the same time. However, under these constraints both existing approaches have significant drawbacks. Complex motion planners cannot match the real-time requirements of the low-level control cycle due to their computational complexity. Reactive methods on the other hand do usually not provide completeness and are (some more, others less) prone to get stuck in local minima. Most importantly however, both approaches do not incorporate physical forces into their according behavior. Therefore, they are not able to treat forces not as a failure but as an additional sensory input that provides valuable information. This dilemma necessitates to treat motion planning, collision avoidance, and collision detection/reaction in a unified approach. Global planning methods have to generate some valid path for the coarse motion of the robot, but it seems that absolute path optimality and strict collision avoidance at the planning stage do not have the top priority in highly dynamic environments, since the overall execution

S. Haddadin: *Towards Safe Robots*,
Springer Tracts in Advanced Robotics 90,
DOI: 10.1007/978-3-642-40308-8_7, © Springer-Verlag Berlin Heidelberg 2014

time, robustness, and flexible reaction are of higher interest. In order to satisfy the requirements posed by quick and safe reaction cycles, real-time methods have to be used for local motions, which can fully exploit the capabilities of the robot and ensure the collision avoidance through local reactions. However, it is no longer satisfactory to only circumvent objects while avoiding contact. On the contrary, contact has to be an integral part of the reactive motion scheme since it could be the vital part of the task. Therefore, contact force information should be integrated into the collision avoidance schemes so that in case unexpected contact occurs, e.g. due to incomplete/inconsistent knowledge of the environment or unpredicted behavior of the human, the robot can retract and circumvent the sources of external forces. Such an approach would require a well balanced interaction of collision avoidance and interaction control. Furthermore, a common severe problem with existing purely reactive strategies is their unpredictable behavior in case of virtual/physical external forces in dynamic settings. This behavior may lead to dangerous situations and was mostly ignored in the robotics literature.

In this chapter, two solutions to the avoidance problem are proposed that are also able to cope with contact forces. First, it is shown how the measurement of distance to the human can be used to reduce and even prevent a collision with the human without deviating from the particular desired geometric path. Then, an approach is discussed that combines trajectory generation and reactive collision avoidance by online motion deformation. The algorithm is also capable of coping with external forces and furthermore is able to serve as a general purpose interpolator with arbitrary desired velocity profile. Even in case of external contacts, a clear behavior of the robot is provided and contact information is used for deforming the trajectory safely in real-time. Important to notice is the fact that both methods treat proximity and contact in a consistent manner and thus do not strictly separate collision avoidance from collision reaction anymore. Generally speaking, the first method is strictly task consistent, while the second scheme is of task relaxing type.

The chapter is organized as follows. Section 7.1 introduces the first avoidance algorithm that is based on the trajectory scaling method from Chapter 3. In Section 7.2 an overview on the concept of the second algorithm with some simple simulations to illustrate it is given. Furthermore, simulation results for the LWR-III are shown and finally, the experimental performance of the proposed method for static and dynamic scenarios is outlined.

7.1 Reaction Strategy with Task Preservation

As described in Chapter 3 trajectory scaling can be used to provide task consistent compliant behavior during contact with the human. There, estimated external torque in combination with a properly designed shaping function is used to scale the dynamics of the trajectory execution. This leads to continuously slowing down, stopping, and reverting of the robot motion along its desired trajectory, depending on the magnitude and direction of the disturbance. The extension of the trajectory scaling methods to this of sensory input is rather straightforward. Apart from utilizing these physical contact signals it is of major importance to use the proximity information of the robot structure with respect to its environment and the human.

The same information on human proximity (or any other object) may be used for collision avoidance in both joint and Cartesian space based on geometric residuals (which can e.g. be transformed via potential fields into virtual forces) in the same sense as contact forces/torques.

7.1.1 Proximity Disturbance Signals

In order to apply similar trajectory scaling techniques to proximity, as already done for force information, the obstacle residual vector $\mathbf{p} \in \mathbb{R}^3$ of a geometric object $\mathbb{GO}$ into the robot hull $\mathbb{S}$ is defined as the residual $\mathbf{p} = \mathbf{x}^{\mathbf{p}}_{\mathbb{GO}} - \mathbf{x}^{\mathbf{p}}_{\mathbb{S}}$. The vector $\mathbf{p}$ denotes the maximum instantaneous penetration of $\mathbb{GO}$ into the hull $\mathbb{S}$, see Fig. 7.1. $\mathbf{x}_{\mathbb{S}}(\mathbb{S})$ is the parametric vector description of the surface and $\mathbf{x}_{\mathbb{S}_I}(\mathbb{S}_I)$ describes the full reverse volume of $\mathbb{S}$, which is the orthogonal projection of $\mathbb{S}$ by $p_{\max} \in \mathbb{R}^+$ along the surface normal. Each $\mathbb{S}$ is associated with so called control points cps distributed along the robot structure. This allows to define a particular distance profile for every cp below which the residual is activated. $\mathbb{S}$ is not necessarily an iso-surface. It can depend on the current robot mode, the instantaneous velocity, or the configuration. For calculating the scaling function that is then used to shape the interpolation time $\Psi(\mathbf{p})$ is defined similar to (3.119):

$$\Psi(\mathbf{p}) = \frac{1}{\alpha}\left(\mathbf{p} \cdot \frac{\Delta\mathbf{x}^i_d}{||\Delta\mathbf{x}^i_d||}\right)_+ \tag{7.1}$$

Then, this signal is directly applied to scale the time increment Δt via (3.120).

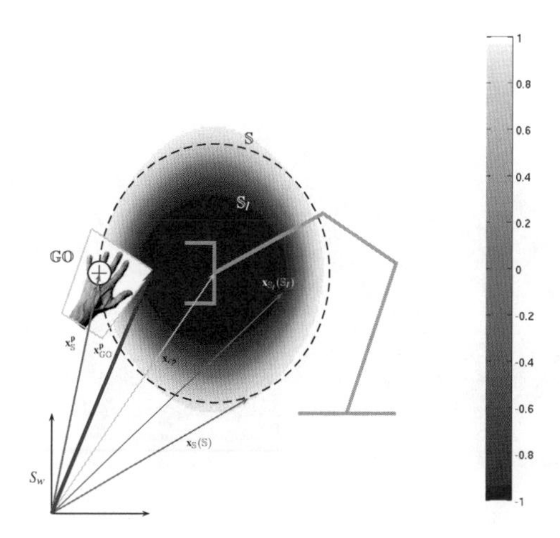

Fig. 7.1 Penetration residual based on proximity

In the example given in Fig 7.1 the endeffector pose is used as the relevant control point $\mathbf{x}_{cp}$. This in turn is directly associated with a hull $\mathbb{S}$. The same methodology can be applied to multiple control points. The residual $\mathbf{p}$ causes the robot to slow down, stop, and revert motion with increasing penetration depth $||\mathbf{p}||$. This behavior is denoted by the indicated scaling value, ranging from 1 (nominal speed) to -1 (full reverse). When the surface $\mathbb{S}_I$ is breached, the normalized residuum is -1. This leads to a full nominal speed reverse of the robot motion until leaving this "emergency" area. In order to use both proximity and contact force (7.1) and (3.119) are combined by

$$\Psi_{\text{res}} = \min\{\Psi(\mathbf{p}), \Psi(\hat{\mathbf{r}})\} \tag{7.2}$$

and use this for scaling interpolation time.

Next, an experiment is discussed that shows the effectiveness of the approach for both virtual and physical residuals acting at the same time.

7.1.2 Experiment

Figure 7.2 shows the experimental setup. The robot performs a Cartesian motion with trajectory scaling being activated for both, contact force and human proximity to the TCP of the robot. The contact force is obtained by (3.98) and the human pose (more specifically the wrist frame) is measured with an Advanced Realtime Tracking (ART) passive marker tracking system [1]. Figure 7.3 depicts the corresponding measurement. On the left side the translation is depicted for the consecutive goal configurations $\mathbf{x}_{d,1} = (-0.45\ 0.5\ 0.4)^T$ and $\mathbf{x}_{d,2} = (-0.6\ -0.2\ 0.2)^T$. During the motion towards the first goal a human pushes against the robot. This causes the

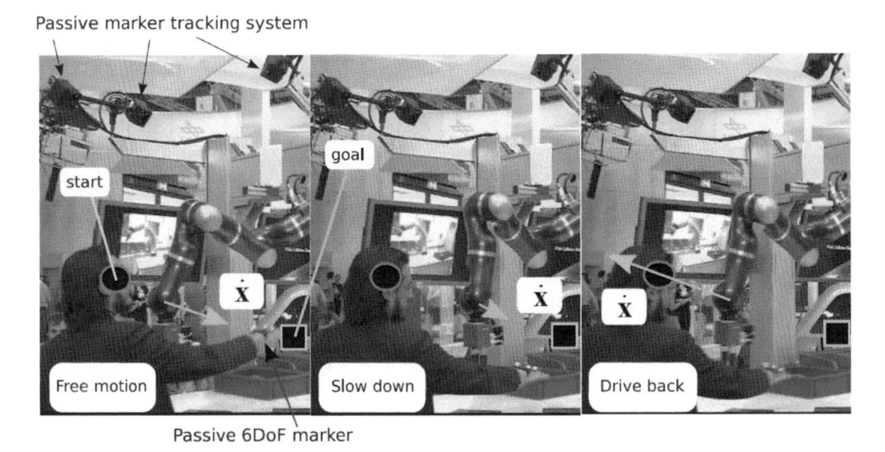

Fig. 7.2 Trajectory scaling for virtual residuals that are calculated from the penetration of the point object "human wrist" and a spherical surface $\mathbb{S}_O$ associated with the robot endeffector

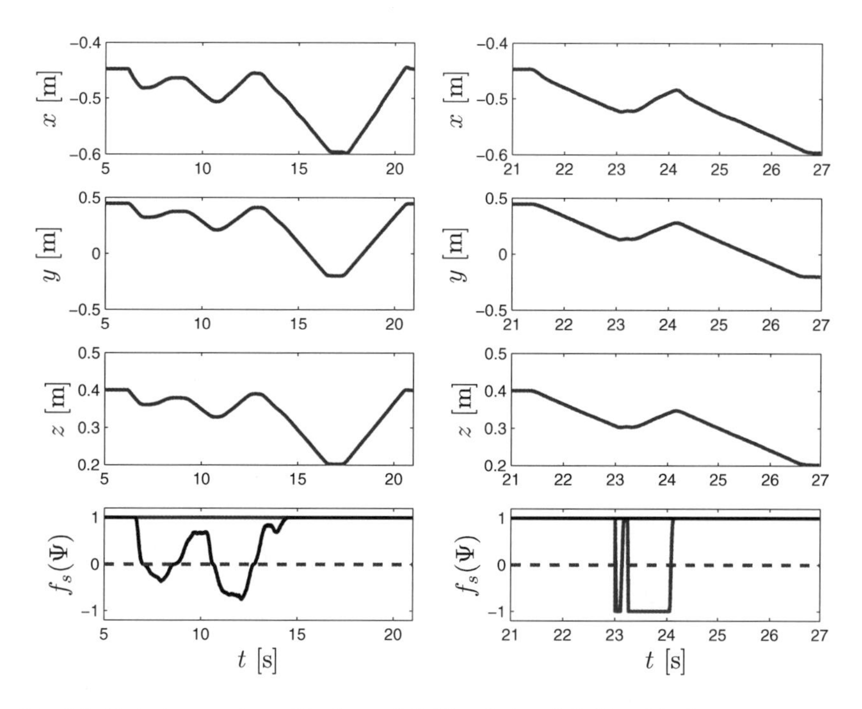

Fig. 7.3 Trajectory scaling with physical (left) and virtual (left) residuals

residual to decrease (slow down), become zero (stop), and become negative (drive backwards)[1]. A similar behavior is observed for proximity measurements. If the residual $||\mathbf{p}|| = 0$, the scaling factor is $f_s = 1$, i.e. normal interpolation time is active (see also Fig. 7.2). If the human breaches the critical proximity, the robot behaves according to the scaling with physical residuals. Important to notice is the coordinated behavior in all axes caused by the scaling of the scalar time variable, which is unique in the interpolation process and thus consistently affects all degrees of freedom.

In the next section a reactive collision avoidance method is developed, which is suitable for real-time collision avoidance and retraction while generating arbitrary desired velocity inputs at the same time.

7.2 Reaction Strategy without Task Preservation

In order to design a method for task relaxing collision avoidance, the desired behavior is first described on an abstract level. Having these requirements makes the derivation of the proposed method more intuitive. Figure 7.4 depicts the overall

[1] This behavior depends on the residual magnitude and was already shown in Chapter 3.

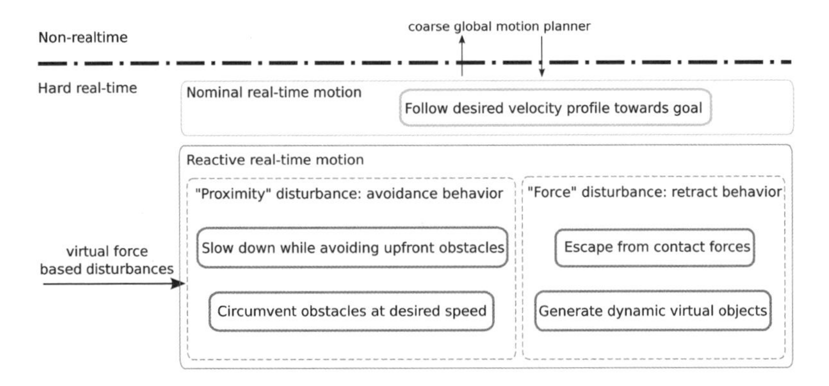

Fig. 7.4 Desired behavior of the proposed task relaxing collision reaction

concept. Please note that it is not intended to solve the global motion planning problem, but to provide an easy to use flexible real-time collision avoidance that is also able to deal with external forces acting on the robot, while at the same time ensures even during the occurrence of disturbances a desired path velocity/acceleration. A coarse motion planner is assumed to provide desired via points that serve the local (hard) real-time motion planner as landmarks to be converged to. The main behaviors that are sought to be provided are as follows.

1. generate motion of arbitrary (useful) path velocity to move from start to goal configuration if no disturbance is present
2. be able to treat both, obstacle proximity and contact force by appropriate reaction behavior
3. be able to integrate force-based disturbance signals for generating various avoidance behaviors (e.g. potential fields or circular fields)
4. escape from contact forces if desired
5. use external forces for the generation of virtual objects in order to prevent future collisions with the particular obstacles

Next, the concrete algorithm design steps are described in more detail.

7.2.1 Algorithm Design

The collision avoidance technique presented in this section is based on the attractor idea of the potential field method. Several further developments/improvements to help overcome some of its major drawbacks are introduced. Figure 7.5 depicts the consecutive desired behaviors (①-⑤), visualizing the design process of the algorithm. In addition, the proposed schemes that were selected to fulfill the requirements are given.

7.2.1.1 Main Concept of the Method

The main goal is to obtain a real-time collision avoidance method that can run in the inner most control loop of the robot (at least 1 kHz). The virtual and physical forces should serve as the input for avoiding collisions or retract from them ①. Therefore, a decoupled impedance equation is chosen for motion calculation and for the sake of smoothness of the generated motion ②. In order to be able to follow arbitrary desired velocity profiles, the predicted path of the resulting attractor dynamics is traversed every time step and the configuration along this predicted trajectory that matches the associated desired velocity value ③ is chosen. This enables to use only the geometric properties of the calculated path, having the favorable characteristics of the attractor, while forcing the motion generator to produce the commanded desired velocities along this path.

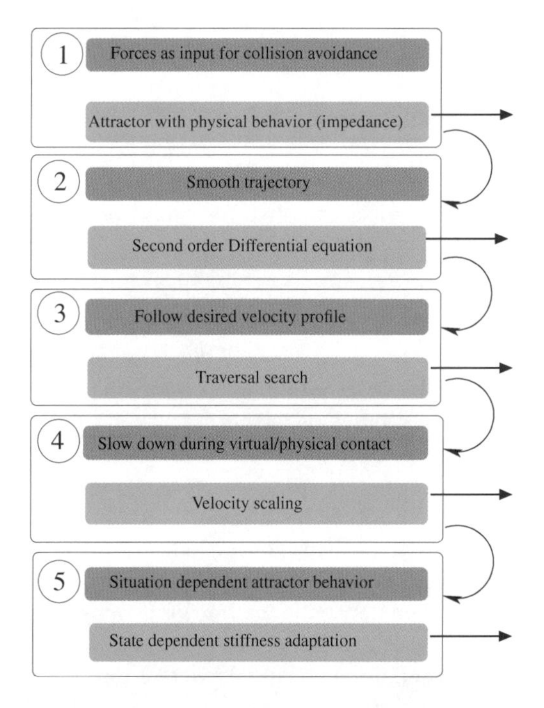

Fig. 7.5 Design steps for the proposed algorithm

Especially during physical contact it is often desirable to slow down the motion, which is ensured by velocity scaling ④. Finally, the coupling to the goal (the attractor stiffness) is altered depending on the current state ⑤. This leads to a temporal detachment from the goal configuration during (virtual) contact and prevents unnecessary fighting between attractive and repulsive forces, which typically leads to bouncing effects.

7.2.1.2 Attractor Design

Potential Field methods as introduced in [6] are well known for their computational efficiency and general applicability. As a result, they have become a standard method in robotics [8]. In the original work a potential field was introduced that consists of a driving attractor for reaching the target configuration, while the robot is being deviated form its desired motion by virtual objects that generate repelling virtual forces. The total force is described by

$$\begin{aligned}\mathscr{F}(\mathbf{x}_d, \mathbf{x}_d^*, \mathbf{x}_o) &= \mathscr{F}_a(\mathbf{x}_d, \mathbf{x}_d^*) + \mathscr{F}_r(\mathbf{x}_d, \mathbf{x}_o) \\ &= \mathscr{F}_a(\mathbf{x}_d) + \sum_k \mathscr{F}_{r_k}(\mathbf{x}_d, \mathbf{x}_{o_k}),\end{aligned} \tag{7.3}$$

with $\mathbf{x}_d, \mathbf{x}_d^*, \mathbf{x}_{o_k} \in \mathbb{R}^n$ being the position of the virtual particle[2], the desired goal configuration and the closest point of the surface $\mathbb{S}_k$ of the k-th repulsive object. $\mathscr{F}$: $(\mathbb{R}^n \times \mathbb{R}^n \times \mathbb{R}^n) \rightarrow \mathbb{R}^n$, $\mathscr{F}_a, \mathscr{F}_r : (\mathbb{R}^n \times \mathbb{R}^n) \rightarrow \mathbb{R}^n$ are the resulting driving, attractive, and repulsive forces associated with the potential $V : (\mathbb{R}^n \times \mathbb{R}^n \times \mathbb{R}^n) \rightarrow \mathbb{R}^n$ via

$$\mathscr{F}^T(\mathbf{x}_d, \mathbf{x}_d^*, \mathbf{x}_o) = -\frac{\partial V(\mathbf{x}_d, \mathbf{x}_d^*, \mathbf{x}_o)}{\partial \mathbf{x}_d}. \tag{7.4}$$

The overall repulsive force usually consists of the sum of the k repulsive components $\mathscr{F}_{r_k} : \mathbb{R}^n \times \mathbb{R}^n \rightarrow \mathbb{R}^n$. The attractive force is expressed by the first order differential equation

$$\mathscr{F}_a(\mathbf{x}_d^*) = K_v(\mathbf{x}_d - \mathbf{x}_d^*) + D_v\dot{\mathbf{x}}_d, \tag{7.5}$$

with $K_v = \text{diag}\{K_{v,i}\} \in \mathbb{R}^{n \times n}, i = 1 \ldots n$ being a diagonal stiffness matrix and $D_v = \text{diag}\{D_{v,i}\} \in \mathbb{R}^{n \times n}, i = 1 \ldots n$ the diagonal damping matrix. In order to bound the resulting velocity, which could in principle become high, [6] proposed to set bounds on the desired velocity, based on the norm of the desired velocity vector. So it is achieved to travel at constant maximum velocity after acceleration and before deceleration phase.

In most cases the repulsive forces are expressed as a function of the distances from the virtual particle to the repulsive elements. These objects are often chosen to be of simple geometric shape as e.g. spheres, cylinders, or planes. For the ease of use, $\mathbf{x}_d, \mathbf{x}_d^*, \mathbf{x}_o$ are omitted from now on in the force functions (using e.g. $\mathscr{F}$ instead of $\mathscr{F}(\mathbf{x}_d, \mathbf{x}_d^*, \mathbf{x}_o)$).

Apart from the slight redefinition of virtual external forces, the classical Potential-Field method is changed by assigning a mass to the virtual particle, producing a trajectory that could in principle take into account the robot's inertial properties into the commanded motion. The resulting particle dynamics are therefore defined by a second order mass-spring-damper system.

$$\mathscr{F}_r = M_v\ddot{\mathbf{x}}_d + K_v(\mathbf{x}_d - \mathbf{x}_d^*) + D_v\dot{\mathbf{x}}_d, \tag{7.6}$$

with $M_v \in \mathbb{R}^{n \times n}$ being the virtual mass matrix.

---

[2] Please note that in the original work, the operational forces were directly projected into motor commands via the Jacobian transpose.

As mentioned earlier, it is important to incorporate real physical forces into the avoidance scheme to provide a more general disturbance response. Therefore, the real external forces $\mathscr{F}_{\text{ext}} \in \mathbb{R}^n$ that act along the robot structure are used as well (in combination with an appropriate positive definite diagonal scaling matrix $G_{\text{ext}} = \text{diag}\{\text{G}^{\text{i}}_{\text{ext}}\}, G^i_{\text{ext}} > 0$). Equation (7.6) becomes

$$\mathscr{F}_{r_{\text{total}}} = \mathscr{F}_r + G_{\text{ext}}\mathscr{F}_{\text{ext}} = M_v\ddot{\mathbf{x}}_d + K_v(\mathbf{x}_d - \mathbf{x}^*_d) + D_v\dot{\mathbf{x}}_d. \tag{7.7}$$

These contact forces are provided by a force sensor in the robot wrist or by the accurate estimation by an observer. In case of joint torques, these can then be transformed into estimations of external forces as described in Chapter 3. Now, they may be used as an input for task space avoidance[3]. By handling the real external forces acting on the robot structure the same way as the virtual repulsive elements, their effects are comparable and can be designed in an unified manner. This makes it possible to introduce more advanced contact responses than if forces are purely used for control purposes in force feedback loops.

This type of second order differential equation is usually unsolvable for dynamic environments, producing highly nonlinear and rapidly changing virtual forces together with basically unpredictable physical forces.

$$\mathscr{F}_{r_{\text{total}}} = \mathscr{F}_r + G_{\text{ext}}\mathscr{F}_{\text{ext}} \approx \mathscr{F}_r + G_{\text{ext}}\hat{\mathscr{F}}_{\text{ext}} = f(\mathbb{S}_R, \dot{\mathbb{S}}_R, \mathbb{S}_i, \dot{\mathbb{S}}_i, t, \dots) + G_{\text{ext}}\hat{\mathscr{F}}_{\text{ext}} \tag{7.8}$$

$\mathbb{S}_R, \dot{\mathbb{S}}_R$ are the relevant surface representations of the robot and their velocity, respectively. $\mathbb{S}_i, \dot{\mathbb{S}}_i$ are the positions and velocities of static and dynamic environment objects.

Due to the mentioned induction of highly nonlinear system behavior, forward simulation of (7.7) needs to be used for achieving a smooth motion with simultaneous collision avoidance, utilizing the input of the repelling virtual and physical force. $t_\varepsilon \in \mathbb{R}^+$ is the time horizon used for calculating the desired motion.

Object motion can be given in terms of observation and prediction, so that $\mathscr{F}_r$ is representing the predicted virtual dynamics during numerical integration of (7.7). External forces act as a constant bias force during each sample .

Double integration of (7.7) every sample time t_n leads to the predicted path $\mathbf{m}_{\varepsilon,n}(t)$, $t \in [t_n, t_\varepsilon]$ of the virtual particle:

$$\mathbf{m}_{\varepsilon,n} := \mathbf{x}_d = \iint_{t_n}^{t_\varepsilon} M_v^{-1}\left[\mathscr{F}_{r_{\text{total}}} - K_v(\mathbf{x}_d - \mathbf{x}^*_d) - D_v\dot{\mathbf{x}}_d\right] \mathrm{d}t + \dot{\mathbf{x}}_d(t_n)\, dt + \mathbf{x}_d(t_n). \tag{7.9}$$

The simplest thing to do is to set $t_\varepsilon = t_n + \Delta t$ with Δt being the discrete interpolation sample time. In other words, each integration step is calculated and the outcome is directly used as the desired trajectory. However, such a simple solution leads for most cases to undesired velocities and accelerations of the generated path.

[3] This estimation degrades when approaching kinematic singularities due to the Jacobian becoming singular.

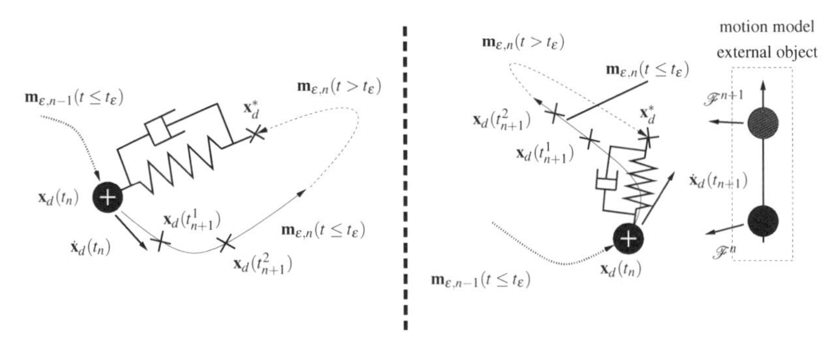

Fig. 7.6 Schematic views of the collision avoidance for two consecutive iteration steps. The left figure denotes free motion during the first step, whereas the right one takes into account a motion model of a suddenly appearing external virtual object for the next iteration.

In order to eliminate this unfavorable property (7.9) is applied with a forward Euler integrator for a limited amount of $s \in \mathbb{N}^+$ steps within a certain predefined time interval $t_\varepsilon = s\Delta t$. The constant s has been chosen such that the real-time condition of the inner most control loop is not violated. This way the system path $\mathbf{m}_{\varepsilon,n}(t' \leq t_\varepsilon)$ is predicted every time step, incorporating the dynamic behavior of the environment and the external forces, which are assumed to be a vector field in this prediction step. The time information associated with it is dismissed and instead a new input variable is used, the desired track speed $\dot{x}'_d \in \mathbb{R}_0^+$. In order to match this desired velocity $\dot{x}'_d$, the configuration $\mathbf{x}_d(t_{n+1})$ along the path $\mathbf{m}_{\varepsilon,n}$ that ensures this velocity is searched for.

This yields $s+1$ sampling points $\mathbf{x}_d(t_{n+1}^0)\dots\mathbf{x}_d(t_{n+1}^s)$ with the starting configuration $\mathbf{x}_d(t_{n+1}^0) = \mathbf{x}_d(t_n)$, $\dot{\mathbf{x}}_d(t_{n+1}^0) = \dot{\mathbf{x}}_d(t_n)$ being also the starting configuration of the robot. The following algorithm interpolates between the bracketing sampling points for the desired track speed $\dot{x}'_d$ and produces the according ordered configuration $\mathbf{x}_{\text{ord}} \in \mathbb{R}^n$.

$i = 0$;
$v_0 = 0$;
while $(v_i < \dot{x}'_d) \wedge (i \leq s)$ **do**
  $i = i + 1$;
  $v_i = v_{i-1} + \frac{\|\mathbf{x}_d(t_{n+1}^{i-1}) - \mathbf{x}_d(t_{n+1}^{i})\|}{t_{n+1}^{i} - t_{n+1}^{i-1}}$;
end
if $i \leq s$ **then**
  $\mathbf{x}_{\text{ord}} = \mathbf{x}_d\left(t_{n+1}^{i-1}\right) + \left(\mathbf{x}_d\left(t_{n+1}^{i}\right) - \mathbf{x}_d\left(t_{n+1}^{i-1}\right)\right)\frac{\dot{x}'_d - v_{i-1}}{v_i - v_{i-1}}$;
end
if $i > s$ **then**
  $\mathbf{x}_{\text{ord}} = \mathbf{x}_d(t_{n+1}^s)$;
end

When $\dot{x}'_d$ cannot be reached, as the integrator steps were not sufficient, the last sample point is chosen $\mathbf{x}_{\text{ord}} = \mathbf{x}_d(t^s_{n+1})$. This usually happens when the virtual particle gets stuck in a local minimum or near the goal position $\mathbf{x}^*_d$ as the goal is asymptotically approached, $\dot{x}'_d$ accidentally commanded to jump, or $\dot{x}'_d$ is inappropriately high. A visual description of the principle is depicted in Figure 7.6.

In summary, keep the smooth properties and the inherent collision avoidance capabilities of the generated local path, but the velocity of the robot can be commanded independently, even arbitrarily. However, the absolute assurance of collision avoidance is given up. On the other hand, as physical forces are already incorporated into the design of the process physical collisions can be easily addressed. Some rough similarity of the proposed algorithm are observed with Model Predictive Control (MPC) [2, 3]. Nonetheless, there are significant differences. In MPC the discrete model of the process to be controlled is used to calculate future system states due to control inputs. This prediction is used to optimally alter the control input for a given cost function.

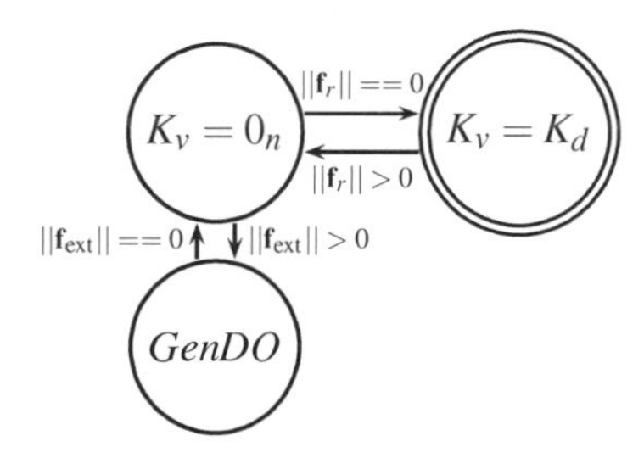

Fig. 7.7 State depending scaling of the attractor stiffness. The attractor switches its stiffness K_v depending on the virtual or physical contact. State *RUN*:$K_v = K_d$ denotes some desired attractive behavior between particle and goal. State *AVOID*:$K_v = 0_n$ denotes zero coupling between particle and goal. *GenDo* denotes the generation of virtual obstacles based on contact force information.

As argued already, continuous fighting between attraction and repulsion/avoidance would occur for any attractor-based method. In order to eliminate this effect, which usually causes bumping behavior during virtual contact, the virtual dynamics of the particle are allowed to instantaneously change its coupling with the goal state K_v. Discrete states are defined which the attractor stiffness can occupy, as *RUN* and *AVOID*, see Fig. 7.7. The state *GenDo* generates virtual obstacles based on the contact force direction and magnitude. Some details of this work are published in [5].

In addition to the designed behavior so far, the attractor is also aimed for being able to alter its velocity magnitude if disturbances occur. As shown in Chapter 3, trajectory scaling could e.g. be applied here. As this basically results in a velocity scaling, the choice of $\mathbf{x}_{\text{ord}}$ may also be directly affected. The easiest way to achieve this is to shape $\dot{x}'_d$ before performing the search. The concrete behavior is basically a free design choice as it may e.g. be desirable to only reduce speed for particular directions or magnitudes of the residual (see Sec. 7.2.2 for implementation details).

Next, the design of the different inputs and parameters of the algorithm are outlined as a possible implementation choice.

7.2.2 Implementation

There are numerous ways to implement the described concept. Therefore, more details on the chosen concrete realization, which performed well during the experiments is given in the following.

7.2.2.1 Repulsive Forces

The particular design of repulsive forces is a choice to be made. In this monograph, a classical choice of range limited, cosine shaped force profiles is used for all simulations and experiments.

$$\mathscr{F}_{r_k}(\mathbf{x}_d,\mathbf{x}_{o_k}) = \begin{cases} \frac{(\mathbf{x}_{o_k}-\mathbf{x}_d)}{d_k} \frac{\cos\left(\frac{d_k}{d_{\max_k}}\pi\right)+1}{2} f_{\max_k} & \text{if } d_k \in [0\ldots d_{\max_k}], \\ 0 & \text{else,} \end{cases} \quad (7.10)$$

with $d_k = \|\mathbf{x}_d - \mathbf{x}_{o_k}\|$ and $d_{\max_k}$ being the maximum distance of influence of a repulsive element. $f_{\max_k}$ is the maximum repelling force of the k-th repulsive element.

7.2.2.2 Velocity Profiles

Since the proposed method allows the use of arbitrary time-based input velocity profiles $\dot{x}_d'(t)$, e.g. classical trapezoidal or sinusoidal motion can be realized with inherent collision avoidance or any other desired profile. However, time-based profiles are only of limited use during virtual or physical collisions, since they are intrinsically violated when deviation from the nominal path takes place. Therefore, a distance-based profile would be a better choice. A combination of both could be selected as well, depending on $\mathscr{F}_{r_{\text{total}}}$. Here, the following desired velocity profile is used.

$$\dot{x}_d'(e_d) = \begin{cases} (v_d-\delta)\frac{1}{2}\left(1-\cos\left(\pi\left(\frac{e_d}{c_1}\right)\right)\right)+\delta & \text{if } e_d < c_1 \\ v_d & \text{if } e_d \geq c_1 \wedge e_d \leq c_2 \\ (v_d-\delta)\frac{1}{2}\left(1+\cos\left(\pi\left(\frac{e_d-c_2}{1-c_2}\right)\right)\right)+\delta & \text{if } e_d > c_2 \wedge e_d < (1-\delta) \\ 0 & \text{else,} \end{cases} \quad (7.11)$$

where v_d denotes the nominal constant track speed, and $c_1, c_2 \in \mathbb{R}^+$ the acceleration and deceleration boundaries, respectively. $\delta \in \mathbb{R}^+ \ll c_1$ is a tolerance value for arrival and $e_d \in [0,1]$ is defined as

$$e_d := \frac{\mathbf{x}_d - \mathbf{x}^*_{d,i}}{\|\mathbf{x}_0 - \mathbf{x}_d\| + \|\mathbf{x}_d - \mathbf{x}^*_{d,i}\|}, \tag{7.12}$$

where $\mathbf{x}^*_{d,i}$ denotes a desired goal configuration. e_d can be interpreted as a normalized "distance to travel". This definition is chosen since it possible to change the goal online without having to re-adapt the boundary values, see Fig. 7.8 (upper left). When changing the goal from $\mathbf{x}^*_{d,1}$ to $\mathbf{x}^*_{d,2}$ during travel, it is not sufficient to only use $e_d := \frac{\|\mathbf{x}_d - \mathbf{x}^*_{d,i}\|}{\|\mathbf{x}_0 - \mathbf{x}_d\|}$. This is due to the fact that $d_2 < d_1 + d_2$ is only valid while the first target is chosen but $d_3 > d_4$ when switching to the second one, potentially leading to $e_d > 1$.

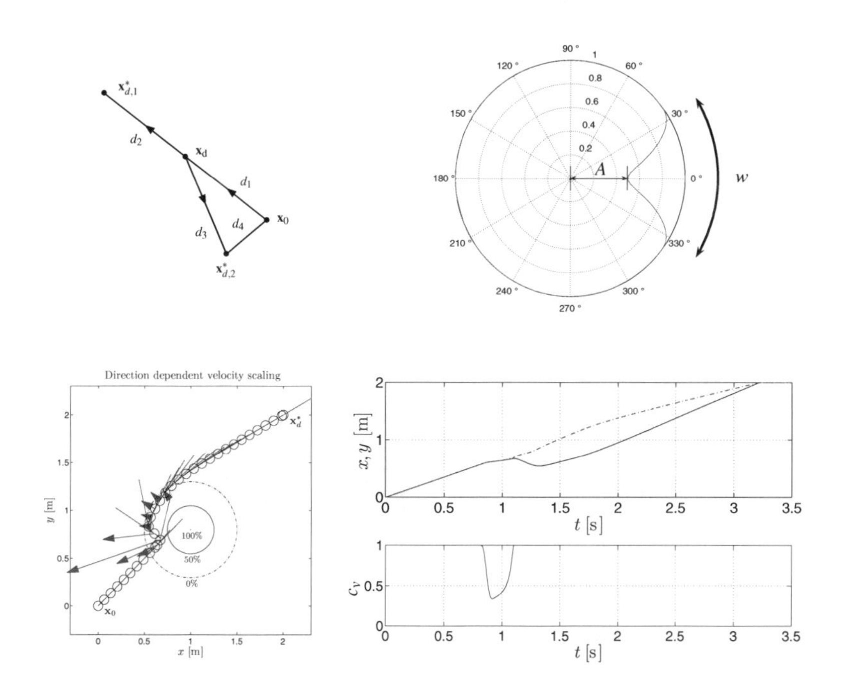

Fig. 7.8 *Upper left:* Relevant configurations for defining the distance to travel e_d. The initial configuration is denoted by $\mathbf{x}_0$, the first goal by $\mathbf{x}^*_{d,1}$ and the second one by $\mathbf{x}^*_{d,2}$. The generated trajectory is $\mathbf{x}_d$.*Upper right:* Velocity-scaling factor with respect to $\measuredangle(-\mathscr{F}_r, \dot{\mathbf{x}}_d)$. 0° means the robot is directly approaching the obstacle. *Lower left:* Direction dependent velocity scaling depicted in 2D. The left figure shows the position $\mathbf{x}_d$ at equidistant time intervals of 0.1 s (black obstacles). The red circles show the obstacle together with the force horizon for 0%, 50% and 100% maximum force. *Lower right:* trajectory of the x- and y-coordinates (with x being solid and y dashed) and the velocity scaling factor c_v.

7.2.2.3 Velocity Scaling

During (virtual) contact, the commanded velocity can be additionally shaped, similar to the method described in Chapter 3.4.2 for physical contact. Thus, due to the collision avoidance, the robot could continuously reduce speed, or even retract, while at the same time actively avoid the upcoming collision. In the most basic case the presence of external objects should lead to a lower velocity. For this purpose, the method of trajectory scaling in case of $\|\mathscr{F}_r\| > 0$ is used to slow down the motion around objects generating these virtual forces. One extension over this pure scaling of velocities in the presence of a repelling force $\|\mathscr{F}_r\| > 0$ is to scale the velocity-profile, as a function of the direction of the repelling force $\mathscr{F}_r$ and the current motion $\dot{\mathbf{x}}_{\text{ord}}$, see Fig. 7.8 (upper right). All scaling effects will be grouped in the scaling factor $c_v \in [0,1]$, which is directly multiplied with the desired track speed $\dot{x}'_d$, leading to the new direction depended input $\dot{x}''_d \in \mathbb{R}$ with

$$\dot{x}''_d = c_v \dot{x}'_d. \tag{7.13}$$

Virtual Force-Based Velocity Scaling

The angle between the repelling force $-\mathscr{F}_r$ and the commanded velocity vector $\dot{\mathbf{x}}_{\text{ord}}$ is given by

$$\phi = \arccos\left(\frac{\langle -\mathscr{F}_r, \dot{\mathbf{x}}_{\text{ord}}\rangle}{\|\mathscr{F}_r\| \|\dot{\mathbf{x}}_{\text{ord}}\|}\right), \tag{7.14}$$

with $\phi \in [0,\pi]$. (7.14) can then be used to calculate a velocity scaling factor, given the parameter for the speed-ditch width $w \in [0\ldots\pi]$ and amplitude $a \in [0,1]$. The velocity scaling factor $c_{v,\text{virt}} \in [0\ldots 1]$ is defined as

$$c_{v,\text{virt}}(\phi) = \begin{cases} 1 - a\dfrac{\cos\left(\frac{\phi\pi}{w}\right)+1}{2} & \text{if } \phi \in [-w,w], \\ 1 & \text{else,} \end{cases} \tag{7.15}$$

where $c_{v,\text{virt}} \in [0\ldots 1]$. For ensuring a smooth velocity change, a is defined as a function of $\|\mathscr{F}_r\|$,

$$a = \begin{cases} A\left(1 - \dfrac{\cos\left(\frac{\|\mathscr{F}_r\|}{f_{\max}}\pi\right)+1}{2}\right) & \text{if } \|\mathscr{F}_r\| \leq f_{\max}, \\ A & \text{else} \end{cases} \tag{7.16}$$

with $f_{\max} \in \mathbb{R}^+$ being some force saturation constant and $A \in [0\ldots 1]$. $c_{v,\text{virt}}(\phi)$ is symmetric: $c, v_{\text{virt}}(-\phi) = c, v_{\text{virt}}(\phi)$. Therefore, the restriction of (7.14) to $[0\ldots\pi]$ does not generate any conflict.

Physical Force-Based Velocity Scaling

Scaling down the velocity can be useful during physical contact. This is done by applying a monotonically decreasing scaling function $g : \mathbb{R}^n \rightarrow \mathbb{R}^+$ for obtaining the physical scaling factor $c_{v,\mathrm{ext}} \in [0,1]$

$$c_{v,\mathrm{ext}} = g(\mathscr{F}_{\mathrm{ext}}). \tag{7.17}$$

In order to incorporate physical forces, there are various behaviors that are desirable. An intuitive choice is to slow down if motion and force vector point in different directions and to accelerate if their direction is similar.

7.2.2.4 Fusion

In order to fuse both scaling factors consistently, the more conservative one is used to ensure safer behavior.

$$c_v = \min(c_{v,\mathrm{virt}}(\phi), c_{v,\mathrm{ext}}(\mathscr{F}_{\mathrm{ext}}, \ldots)) \tag{7.18}$$

This leads to a slowdown of the motion as long as the robot drives towards critical obstacles, but leaves the desired velocity untouched if bypassing or departing.

7.2.2.5 Stiffness Adaptation

As described previously, the attractor stiffness enables to change the overall attractor behavior online according to the current situation. High stiffness relates to higher convergence rate, whereas decreasing values represent an increasing decoupling from the goal configuration. This helps improve avoidance behavior, as stiffness adaptation prevents fighting between attractive and repulsive forces.

In the implementation, the information obtained from e_d, $\mathscr{F}_r$, and $\mathscr{F}_{\mathrm{ext}}$ is utilized in order to achieve intuitive behavior. If no obstacle is to be avoided the diagonal stiffness values $K_v = \mathrm{diag}\{K_{v,i}\}$ are set to high values that are in the order of the physical reflected robot stiffness as a function of e_d.

$$K_v^{\mathrm{high}} = \max\{K_v^{\mathrm{max}}(1 - e_d), K_v^{\mathrm{min}}\} \tag{7.19}$$

With this definition higher convergence rate during goal approaching phase and less spiral behavior (especially if the initial velocities are non-zero) are provided. In case avoiding behavior (due to virtual or physical forces) is desired, a relaxing behavior is activated, which enables decoupling of virtual particle and goal configuration. Figure 7.9 depicts the overall block diagram of the implemented method.

In the next subsection simulation results with a full dynamic simulation of a Cartesian impedance controlled robot are shown.

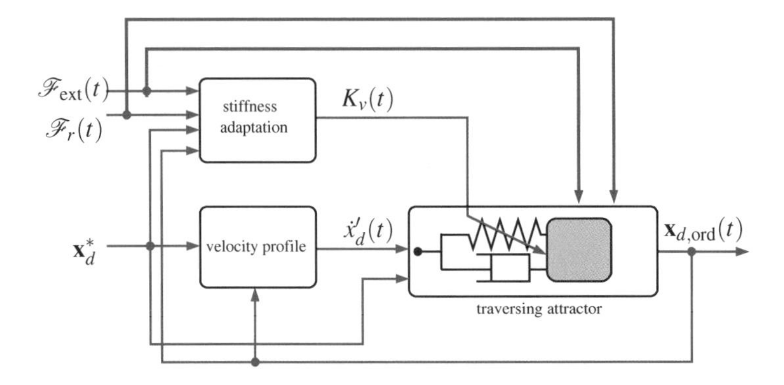

Fig. 7.9 Block diagram of the proposed method. The stiffness adaptation depends on the distance to the desired goal configuration, the current commanded pose, and the virtual and external forces. The velocity profile is based on the distance to travel and produces the desired track speed. The traversing attractor calculates with these inputs the commanded pose.

7.2.3 Simulations

The described method was implemented for a full dynamic simulation of the LWR-III in Cartesian impedance controlled mode, where only constant attractor dynamics are used for simplicity. The attractor parameters were chosen to be $K_v = \text{diag}\{1000\}$ N/m, $D_v = \text{diag}\{3.16\}$ Ns/m, and $M_v = \text{diag}\{1.0\}$ kg. At $\mathbf{x}_o = (0.3, 0.35, 0.4)^T$ m a virtual spherical object with radius $r = 0.2$ m is placed. Initially, the TCP of the robot is at $\mathbf{x}_0 = (0.05, 0.44, 0.55)^T$ m. The robot is commanded such that the orientation is kept constant and the goal configuration is at $\mathbf{x}_d^* = (0.5, 0.45, 0.35)^T$ m. Figure 7.10 (left) depicts the behavior of the motion

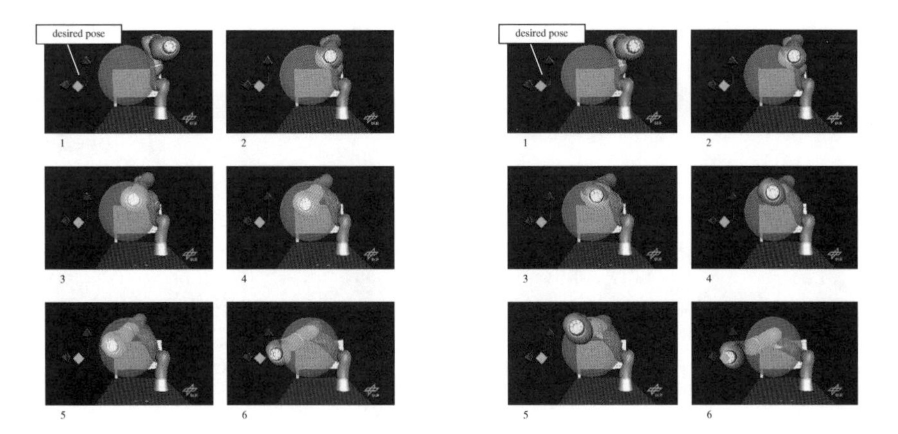

Fig. 7.10 Motion without collision avoidance (left). Motion with collision avoidance (right).

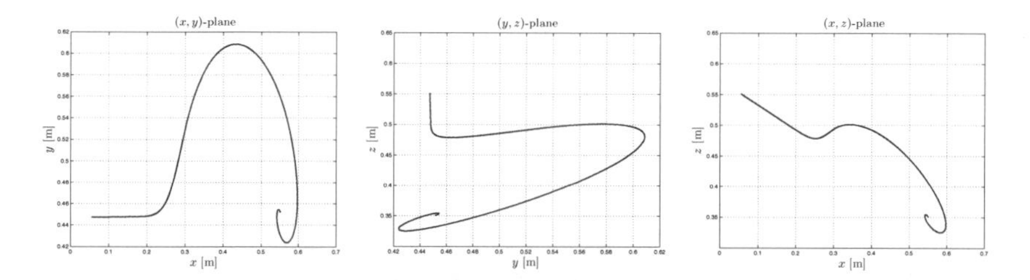

Fig. 7.11 Collision avoidance behavior in the three translational planes

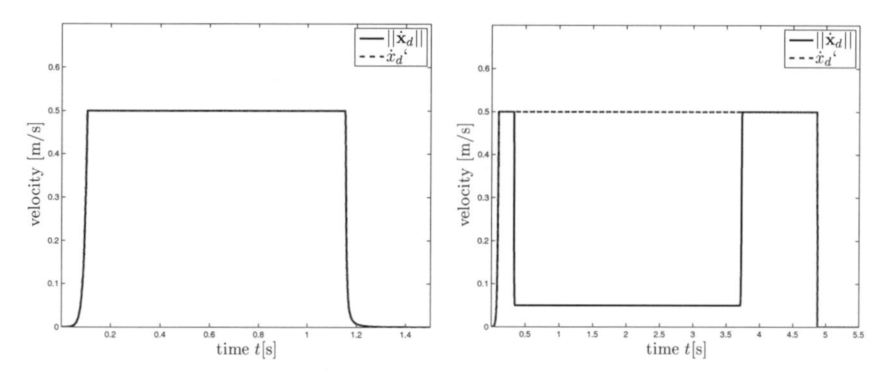

Fig. 7.12 Desired and generated velocity profile of the attractor without an external object (left), which leads to $||\dot{\mathbf{x}}_d|| = \dot{x}'_d$ and with a virtual object (right), resulting in $||\dot{\mathbf{x}}_d|| = \dot{x}''_d = k_v \dot{x}'_d$

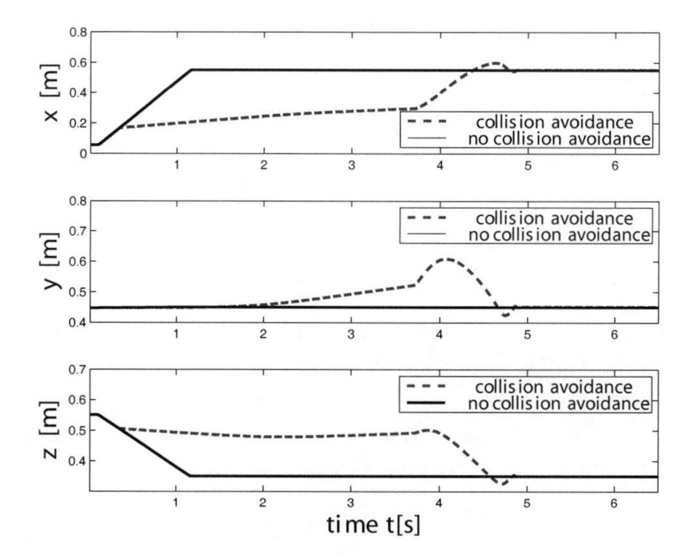

Fig. 7.13 Time courses of the generated translational motion with an unknown virtual object

generator for $\mathscr{F}_r = \mathbf{0}$. The robot is reaching its goal under the constraint of the given velocity profile, see Fig. 7.12 (left). Figure 7.10 (right) on the other hand indicates the behavior while taking the virtual object into account, with Fig. 7.13 showing the corresponding time courses. The robot deviates from its original motion path and circumvents the object accordingly. During the "contact phase", the velocity is scaled additionally (see Figure 7.12 (right)) and at the same time the virtual stiffness K_v is switched to a low value of diag$\{10\}$ N/m to provide higher obstacle avoidance performance instead of good tracking behavior[4]. Figure 7.11 depicts the generated motions in (x,y), (y,z), and (x,z) plane.

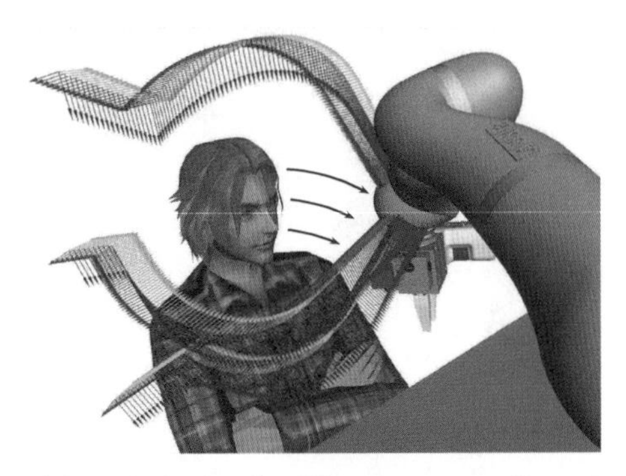

Fig. 7.14 Resulting avoidance behavior for different start configurations and the same goal configuration for a full dynamic simulation of an impedance controlled manipulator with the human moving towards the robot

Figure 7.14 shows the simulation result for different starting points and the common goal configuration of the robot. The nominal trajectory is a straight line from different starting points to a common end point. The avoidance takes place for a dynamic motion of the human towards the robot, cutting the original motion path.

After this full dynamic simulation, the experimental performance of the proposed method for the LWR-III is analyzed.

7.2.4 Experiments

In this section, experiments to examine the performance of the proposed method in various situations with static and dynamic obstacles are presented.

[4] Please note that for this simulation the more complex scaling of the attractor stiffness described in Sec. 7.2.2.5 was not used. This, however, is applied in Sec. 7.2.4 for the experimental part.

Fig. 7.15 Billiard scenario with the DLR Lightweight Robot III

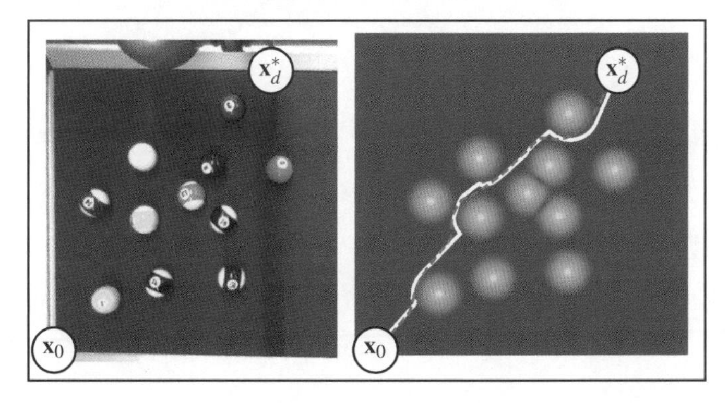

Fig. 7.16 Configuration of Billiard balls (left). 2D plot of the collision avoidance experiment with the Billiard balls (right). Please note that the radius of potentials on the right side is not the same as the one of the recognized balls. Only the position of the ball center is used for their definition.

7.2.4.1 Static Obstacles

Figure 7.15 illustrates the setup for showcasing the abilities to circumvent various static obstacles (billiard balls). The objects are manually and arbitrarily arranged on the table and then identified with an object recognition system [7]. Their positions are used to define the artificial repulsive potential fields. In Figure 7.16 (left) the scene view from above is shown, where the robot reached its target configuration. Figure 7.16 (right) depicts the commanded motion (solid) and the real path of the robot (dashed). v_d was chosen to be 0.2 m/s. The slight deviation is caused by the

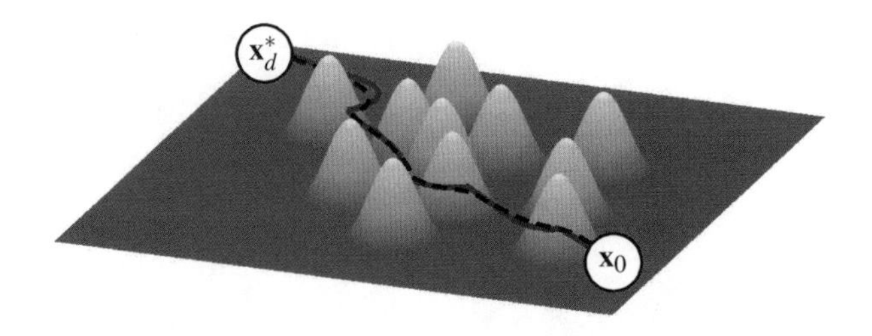

Fig. 7.17 3D plot of the collision avoidance experiment with the Billiard balls. Please note this is a 2D experiment visualized in 3D. Therefore (7.10) is only defined in 2D and the visualized height corresponds to the magnitude of the repulsive elements (which is sinusoidally shaped).

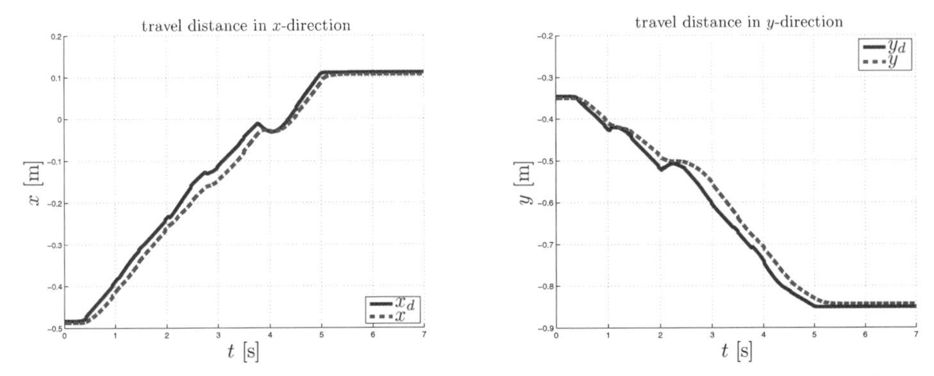

Fig. 7.18 Time courses of the avoidance in x-direction (left) and y-direction (right) as a function of time. The plot shows the desired trajectory $\mathbf{x}_d$ and the real motion of the robot $\mathbf{x}$.

use of Cartesian impedance control since no feed-forward compensation was used. Figure 7.17 denotes the 3D visualization and Figure 7.18 the timely behavior of the robot.

7.2.4.2 Dynamic Obstacles

The performance of the method is evaluated for three distinct dynamic situations. Figure 7.19 depicts the situation for the first experiment. The DLR 3D Modeler [9] is mounted on the robot in order to use the integrated laser scanner for acquiring proximity data. In the second experiment an ART tracking system is used for passively tracking the human wrist pose. In the third scenario the estimated external forces provided by (3.96) were utilized as the repulsive input. This experiment shows how the method can cope with robot-human collisions and unexpected rigid impacts with the environment, see Fig 7.22.

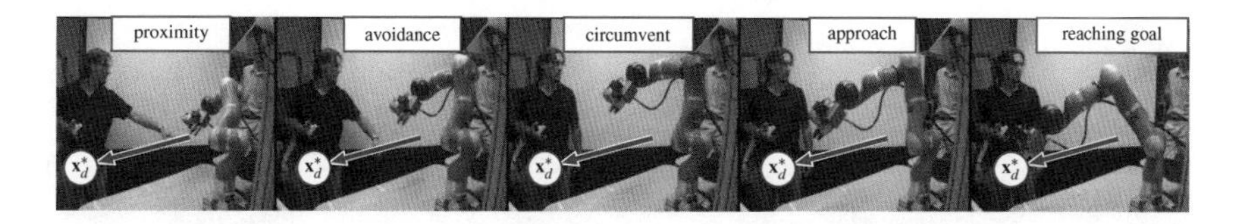

Fig. 7.19 Collision avoidance behavior with a proximity sensor. The laser scanner mounted in the DLR 3D Modeller is used. The device is mounted on the TCP and gives proximity information for an opening angle of 270 ^{o}. The original motion path in absence of a disturbance is indicated by the arrow and the goal configuration is denoted by $\mathbf{x}_d^*$.

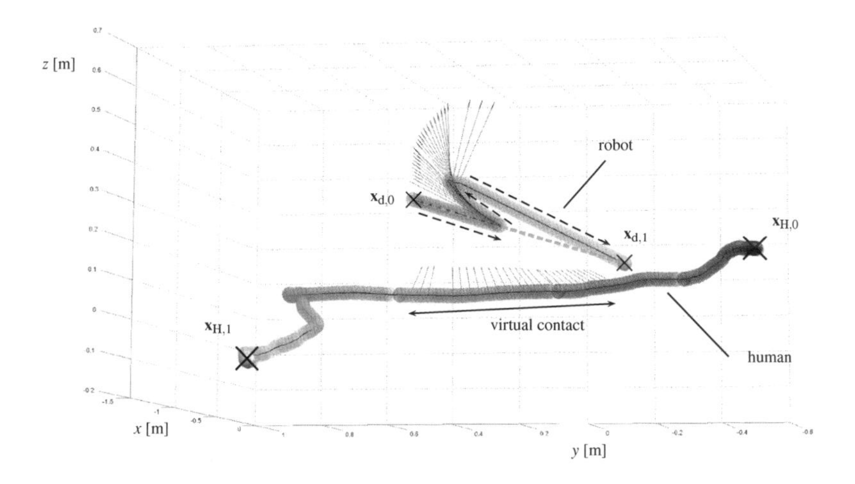

Fig. 7.20 Dynamic collision avoidance with an ART-tracking system

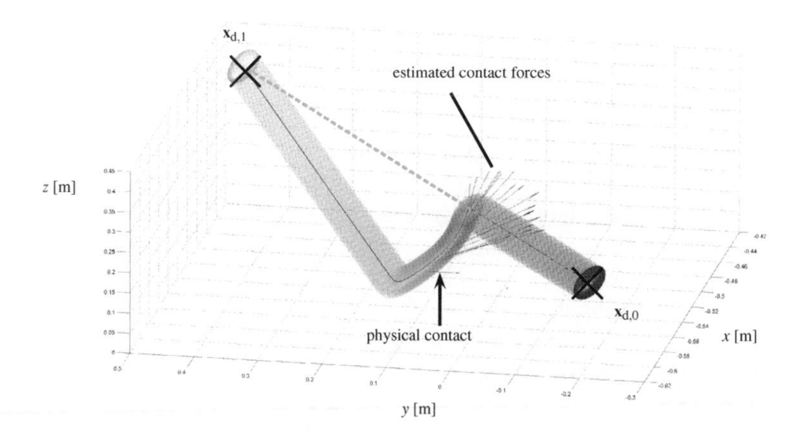

Fig. 7.21 The plot depicts the behavior for pushing the robot. After the collision the robot quickly recovers from the contact and finds its way into the final goal.

Fig. 7.22 The robot is pushed by the human and thus deviates its trajectory. Since the robot has no knowledge about the position of the table, the robot collides into the table. However, it smoothly and quickly recovers from this rigid contact due to the deformation of the path and the used impedance control. Finally it reaches the desired goal configuration $\mathbf{x}_d^*$. The external forces are obtained with the nonlinear observer based on the generalized momentum.

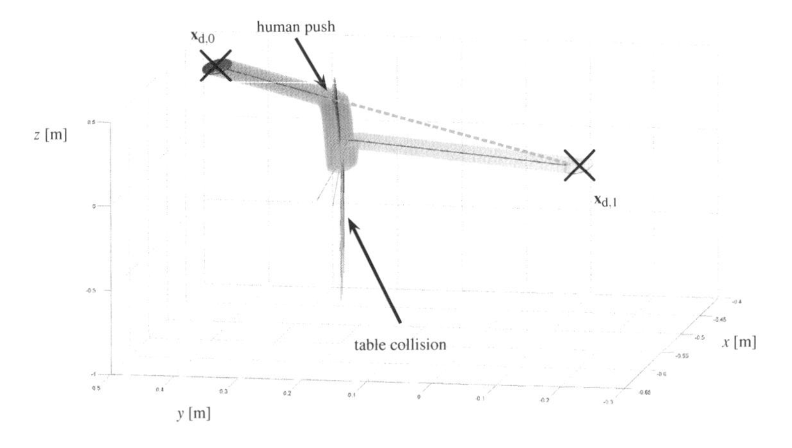

Fig. 7.23 The plot depicts the behavior for pushing the robot harder, which results in a second collision with the table. Even though the robot has no prior knowledge of the table it quickly recovers from the second contact and finds it way into the final goal.

The measurement results of the second experiment are given in Fig. 7.20. The robot is commanded to reach the desired goal configuration $\mathbf{x}_d^*$. As soon as the human holds his arm into the workspace and blocks the initial motion path, the robot circumvents the hand and reaches the goal. The original desired motion is depicted and the generated virtual forces are shown along the human motion path and the resulting robot trajectory. The human moves from the right side to the left, while the robot intends to reach the right configuration. As soon as the robot is affected by virtual forces it starts deviating from the path. After the human surpasses it, it moves again towards the goal and terminates there.

From Fig. 7.21 and Fig. 7.23 one can see how the method can cope with external forces in the same way as with virtual ones. The human pushes the robot while it is moving and the desired motion is deformed such that the robot is deviated from its path, see Figure 7.21. In Figure 7.23 the robot collides with the table after being pushed by the human into the unknown object. Then, the contact information is used to recover from this second collision. Finally, the robot reaches its goal position.

Especially for the table impact one can see how the Cartesian impedance control, the external force estimation, and the collision avoidance work together to recover from this unexpected rigid contact and how the robot still reaches its goal.

7.3 Summary

In this chapter, two methods for reactive motion generation were outlined. The first one is a straightforward extension of the trajectory scaling method developed in Chapter 3 to the case of virtual forces for collision prevention. The second one is based on an intuitive physical interpretation, namely an impedance-like motion generation. The method is well suited to serve in between global motion planning and control to provide well defined and safe behavior even for unexpected virtual and physical contact. It is designed to serve as a relief for both sides and provides a safe motion in complex environments, taking into account both proximity to objects, and external forces. The method allows to command arbitrary velocity profiles to the robot and provides collision avoidance behavior at the same time. Even during circumvention of obstacles the track speed can be commanded such that no unexpected velocity or acceleration jumps occur.

References

[1] ART: advanced realtime tracking GmbH: Your expert for infrared optical tracking systems. advanced realtime tracking GmbH (2010), http://www.ar-tracking.de/
[2] Aström, K., Wittenmark, B.: Adaptive Control. Dover Books on Engineering (2008)
[3] Camacho, E., Bordons, C.: Model Predictive Control. In: Advanced Textbooks in Control and Signal Processing. Springer (2007)
[4] Choset, H., Lynch, K., Hutchinson, S., Kantor, G., Burgard, W., Thrun, S., Kavraki, L.: Principles of Robot Motion: Theory, Algroithms, and Implementation. MIT Press, Cambridge (2005)
[5] Haddadin, S., Parusel, S., Vogel, J., Belder, R., Rokahr, T., Albu-Schäffer, A., Hirzinger, G.: Holistic design and analysis for the human-friendly robotic co-worker. In: IEEE/RSJ Int. Conf. on Intelligent Robots and Systems (IROS 2010), Taipeh, Taiwan, pp. 4735–4742 (2010)
[6] Khatib, O.: Real-time obstacle avoidance for manipulators and mobile robots. The Int. J. of Robotics Research 5, 90–98 (1985)
[7] Parusel, S.: Playing billard with an anthropomorphic robot arm. Master's thesis, FH Kempten & German Aerospace Center (DLR) (2009)
[8] Siciliano, B., Khatib, O. (eds.): Springer Handbook of Robotics. Springer (2008)
[9] Suppa, M., Kielhöfer, S., Langwald, J., Hacker, F., Strobl, K.H., Hirzinger, G.: The 3D-Modeller: A multi-purpose vision platform. In: IEEE Int. Conf. on Robotics and Automation (ICRA 2007), Rome, Italy, pp. 781–787 (2007)

Chapter 8
Towards the Robotic Co-worker

Various human-friendly motion control methods were presented and analyzed. These are independently useful tools for numerous applications as they open up entirely new robot behaviors. However, due to their complex interrelationship in this chapter it is discussed how to integrate the presented methods into a more general hybrid state-based control architecture. Even though the focus is on robotic co-workers, the elaborated schemes are also applicable to service robots. The implementation of such a sensor-based robotic co-worker that brings robots closer to humans in industrial settings and achieve close cooperation is currently a challenging goal in robotics. Pioneering examples of intimate collaboration between human and robot, whose origin can be found in [10], are Intelligent Assist Devices (IADs), as the *skill assist* described in [18]. In 1983 a method was proposed at DLR for allowing immediate "programming by touch" of a robot through a force/torque-sensor-ball [8], see Fig. 8.1 (left). Despite being a common vision in robotics the robotic co-worker has not become reality yet, as there are various open questions still to be answered. Apart from the control and safety aspects, the architectural level also poses significant challenges.

In this chapter, a solid architectural concept and a prototype realization of a co-worker scenario are developed in order to demonstrate that state of the art technology is now mature enough to reach this aim. The ideas are supported by addressing the industrially relevant bin-picking problem with the LWR-III, which is equipped with a time-of-flight camera for object recognition and the DLR 3D-Modeller (DLR-3DMo) for generating accurate environment models. The chapter describes the application of the control schemes from Chapter 3 in combination with robust computer vision algorithms, which leads to a reliable solution for the chosen problem. Strategies are devised for safe interaction with the human during task execution, state depending robot behavior, and the appropriate mechanisms to realize robustness in partially unstructured environments. The theoretical basis as well as requirements regarding task execution and safe interaction are elaborated which rely mainly on sensor-based reaction strategies. The concept requires flexibility of the system in terms of sensor integration and programming. This flexibility

S. Haddadin: *Towards Safe Robots*,
Springer Tracts in Advanced Robotics 90,
DOI: 10.1007/978-3-642-40308-8_8, © Springer-Verlag Berlin Heidelberg 2014

is currently not available in the state-of-the-art first generation industrial robots, designed mainly to position objects or tools in six dimensions.

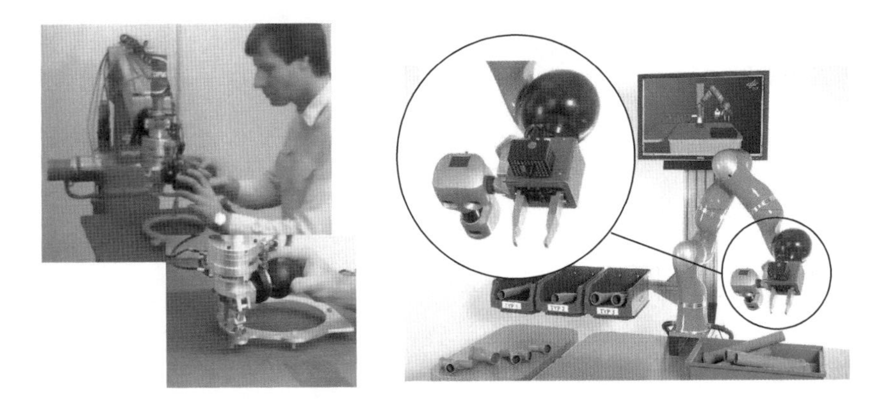

Fig. 8.1 The concept of sensor programming was developed at DLR in 1983 for teaching robot paths and forces/torques simultaneously (left). The DLR Co-Worker consisting of the DLR Lightweight Robot III, the DLR-3DMo, and a Time-of-Flight Camera (ToF-camera) (right).

Therefore, for the new generation of industrial robot, a fundamental change in concept is required to enable the implementation of the robotic co-worker. This new approach is derived from the fusion of robots with innovative and robust control schemes ("soft-robotics" features) with exteroceptive sensing such as 3D vision modalities for safely perceiving the environment of the robot. Together with additional sensing capabilities for surveillance, such technology will open entirely new application fields and manufacturing approaches. In order to develop and evaluate the proposed concept, the DLR Co-Worker was constructed as a demonstration platform, see Fig. 8.1 (right).

Complementary sensor fusion[1] plays a key role in achieving the desired performance through the combination of complementary input information. As demonstrated in [13] a prioritized and sequential use of vision and force sensor based control leads to robust, fast, and efficient task completion using the appropriate sensor information depending on the particular situation.

Presently, industrial robot applications require complete knowledge of the process and environment. This approach is prone to errors due to model inaccuracies. The central approach is to use intelligent sensor-based reaction strategies to overcome the weaknesses of purely model-based techniques. Thus, the sensor noise and limited robot positioning accuracy can be overcome. The robot task is described in high-level functions encapsulated in the states of hybrid automata. The state transitions are based on the decisions made by using sensor inputs. This enables the robot to react to "unexpected" events not foreseen by the programmer. These events are

[1] Please note the difference of complementary from competitive sensor fusion.

induced by the human behavior, which cannot be completely modeled analytically. Furthermore, the human is encouraged in the experiments to physically interact with the robot as a modality of "communication" to provide task-relevant information. This also improves the fault tolerance functionality of the task since only absolute worst-case contacts trigger a complete emergency stop in contrast to approaches for current robots, where opening the fences leads to an immediate robot stop.

The presented concept for realizing the robotic co-worker is fundamentally different from classical industrial robots. None of the components are supposed to be intrinsically fail safe, but the appropriate combination of all components makes the system more robust, safe, and reliable. Multiple sensor information of the robot and external sensing is used for increasing the error tolerance and fault recovery rate. Finally, the stage of a highly flexible state-based programming concept for various applications is reached. The associated task description allows for novel switching strategies between control modes, sensory reaction strategies, and error handling.

This chapter is organized as follows. In Sec. 8.1 the general functional modes required for a robotic co-worker are described. Then, the interaction concept is outlined in Sec. 8.2, followed by the elaboration of the task description. Finally, the developed concepts are applied to a robotic bin picking scenario with user interaction as a case study in order to demonstrate their practical relevance and implementation in Sec. 8.3.

8.1 Functional Modes

Currently, industrial settings incorporate, in most cases, simple sequences of tasks whose execution orders are static, sometimes allowing limited binary branching. Fault tolerance during task execution is, apart from certain basic counterexamples[2], usually not an issue due to the well designed environment. Furthermore,

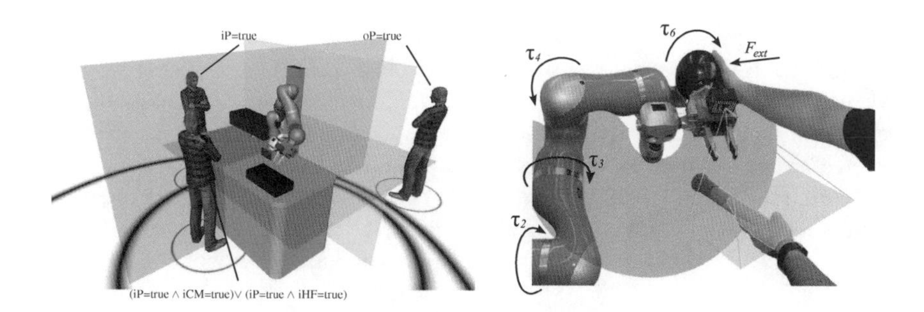

Fig. 8.2 Proximity and task partition (left) and modalities for multi-sensor Human-Robot Interaction in the DLR Co-Worker (right)

[2] Checking for a successful grasp is e.g. commonly used.

Human-Robot Interaction is not yet safely and effectively implemented. Its legal foundations are to a large extent nonexistent at the current stage. In industrial settings a fault immediately leads to a complete stop of the manufacturing process, i.e. robust behavior in a (semi-)unstructured environment has not been addressed until now. In this monograph, an integrated and flexible approach is proposed to carry out the desired task in a robust yet efficient manner. This approach is able to distinguish between different fault stages, which stop the entire process and lower the efficiency only in the absolute worst-case. Flexible jumps within execution steps are part of the concept and do not require special treatment. In order to optimally combine human and robot capabilities, the robot must be able to quickly adapt to the human intention during task execution for both achieving safe interaction and high productivity. Thus, the measured human state is the dominant transition between the proposed functional modes.

Estimating the human state is a broad topic of research and has been addressed in recent work [9]. The focus is often on estimating the affective state of humans, which is however of secondary interest during an industrial process. The more relevant information is the physical state that the human currently occupies, and the estimation of the human attention, so that a clear set of sufficient behaviors can be selected and activated. This leads to robust and reliable overall performance. In order to keep the discussion focused, attention estimation or gesture recognition is not considered and, instead, the focus is on considering the human state.

Following selection of physical states were compiled to provide sufficient coverage for cases relevant to the present study, see Fig. 8.2 (left).

- **oP**: out of perception
- **iP**: in perception
- **iCM**: in collaborative mode
- **iHF**: in human-friendly zone

oP denotes that the human is out of the perceptional ranges of the robot and therefore not part of the running application. **iP** indicates that the human is in the measurement range of the robot. Thus its presence has to be part of the robot control. **iCM** and **iHF** indicate whether a collaborative or human-friendly behavior must be ensured. Each physical state is subdivided, depending on the task. However, only when **iCM** $= true$ should the collaborative intention be taken into account: This leads to a complex physical interaction task. In this chapter, the "hand-over and receive" process is used as an example, see Fig. 8.2 (right).

The human state is primarily used to switch between different functional modes of the robot which in turn are associated with fault behaviors. As shown in Fig. 8.3 it is distinguished between four major functional modes of the robot in a co-worker scenario:

1. *Autonomous task execution:* autonomous mode in human absence
2. *Human-friendly behavior:* autonomous mode in human presence
3. *Co-Worker behavior:* cooperation with human in the loop
4. *Fault reaction behavior:* safe fault behavior with and without human in the loop

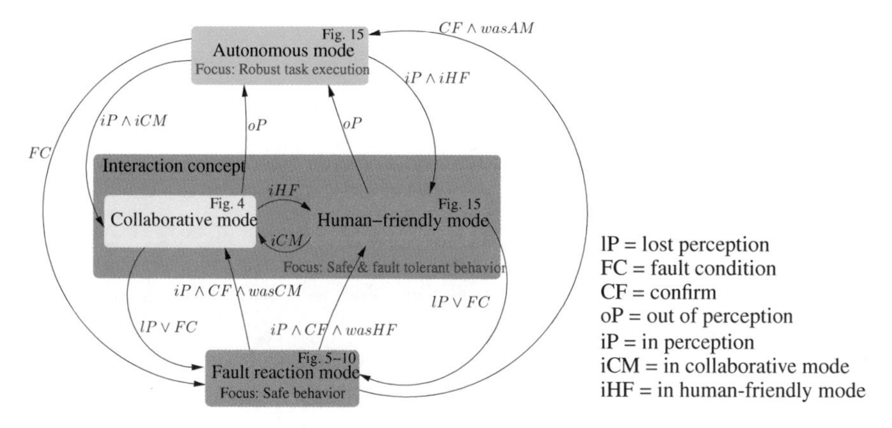

Fig. 8.3 Functional modes for the DLR Co-Worker

Their interrelationship and transition conditions provide high flexibility in the application design. In the *first* functional mode the robot is autonomously fulfilling its given task without considering the human presence. The task is carried out under certain optimality criteria, such as cycle time, in order to maximize the productivity. In the *second* and *third* modes, a concise partition of the task space is needed which subdivides the given workspace of the robot into regions of interaction. These incorporate the "hand-over" schemes as described in Sec. 8.2.2 and "human-friendly" behavior, whose core elements are reactive collision avoidance and self-collision avoidance schemes. In the *third* mode interaction tasks are carried out. These tasks have to be specified or generated for fulfilling a common desired goal, involving a synergy of human and robot capabilities in an efficient manner. These two modes form an integrated interaction concept, allowing seamless switching between each other. The *fourth* mode defines the fault reaction behavior, addressing the appropriate and safe state-dependent fault reaction of the robot. It incorporates both the robustness concepts during autonomous reaction, as well as human-safe behavior. Since each mode possesses an underlying safety concept, it will be described later in more detail.

8.2 Interaction Concept

In this section, the developed interaction schemes are described. First, the proposed task space partition is outlined, followed by the interaction layer. Then, the different collision avoidance techniques from Chapter 7, as well as physical collision detection and reaction methods for safe pHRI from Chapter 3 are grouped. Finally, the resulting safety architecture which structures the different schemes is presented.

8.2.1 Proximity and Task Partition

In case humans are in close proximity to robots in current industrial installations, the robots reside inside safety cages in order to prevent any physical contact and thus minimize the risk for humans. However, when humans and robots collaborate, such a plant design is no longer an option. The human location must be taken into account in the control scheme and in higher level planning of the robot as an integral part of the system design. The previously introduced physical human states have to be mapped into a topology as the one shown in Fig. 8.2 (left), where the four distinctive classes are indicated. They should be established with respect to the task and the robot workspace for assessing whether the human does not have to be taken into account and therefore the robot still behaves autonomously regardless of the iP state. In case the human does not enter the robot workspace, it is not necessary to degrade the productivity of the robot. In this sense, the functional mode of the robot changes only if the human clearly enters the workspace of the robot (indicated by the inner circle). If the human has entered the robot workspace a distinction between human-friendly behavior (on the right side of the table in Fig. 8.2 (left)) and the cooperative mode (and their respective submodes) is required (on the left side of the table in Fig. 8.2 (left)). If perception is lost while $iP = \text{true}$, the robot assumes a severe error condition, stops and waits for further instruction. If the presence of the human was not detected (a worst-case from a safety point of view) various safe control schemes ensure the safety of the human during possible unforeseen collisions.

Defining these regions is part of the application design and definition phase. Furthermore, switching zones are introduced, which are boundary volumina of predefined thickness between task partitions (see Sec. 8.2.3 for details).

8.2.2 Interaction Layer

Interaction between robot and human is a delicate task, which requires multi-sensory information. Furthermore, robust as well as safe control schemes are called for to enable intuitive behavior. The main physical collaboration schemes are "joint manipulation" and "hand-over and receive". "Parallel execution" may be part of a task, but usually without physical interaction. Some work has been carried out on exchanging objects between human and robot based on reaching gestures [3]. In [2] the concept of interaction history was used to achieve cooperative assembly.

Figure 8.2 (right) shows a "hand-over" and "receive" example with the DLR Co-Worker. Its central component is the LWR-III with soft-robotics features. As a default its high-performance Cartesian and joint level impedance control are used and it is only switched to other schemes, such as position control, if necessary. Due to its internal joint torque sensors, the robot is well suited for realizing various important features such as load loss detection and online load identification without additional force sensing in the wrist. Collision detection and reaction, depending on the potential physical severity of the impact as well as on the current state of the robot and the

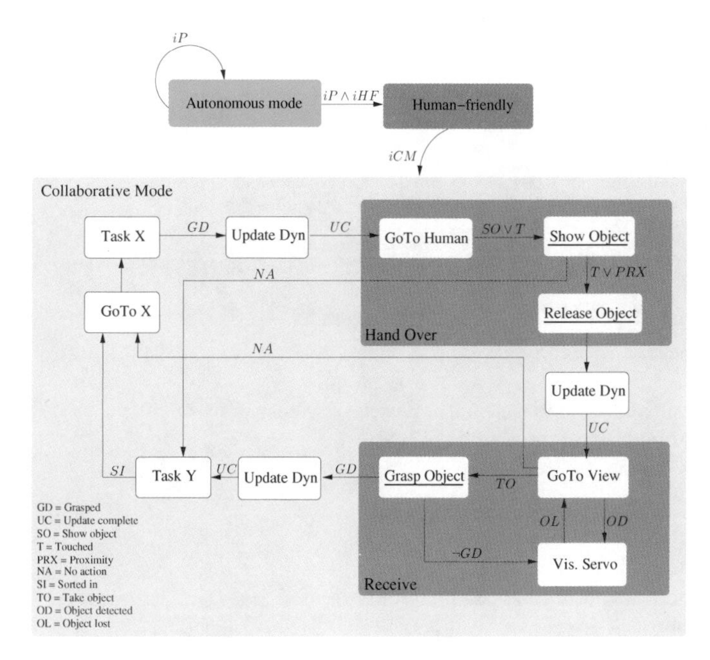

Fig. 8.4 Example for "hand-over" and "receive". Underlined states incorporate explicit physical interaction.

application, are central features used for detecting and isolating contacts of different intensity along the entire robot structure. By being able to distinguish different contact types, fault tolerant and situation suited behavior is possible.

Virtual walls are utilized for avoiding collisions with the environment. In order to realize an effective reactive behavior, it is important to change the stiffness, velocity, disturbance residuals (see Sec. 8.2.3), trajectory generators, collision severity reaction strategies, and robot control parameters on the fly within the lower level control cycles (here 1 ms), even during motion or state execution. With the combination of exteroceptive sensing, capabilities of object recognition, tool surveillance, and human proximity detection (shown in Fig. 8.2 (right)), interaction processes such as the aforementioned "hand-over" and "receive" can be achieved, see Fig. 8.4. "Receiving" or "handing-over" the object is simply triggered by touching the robot at any location along its structure or by using the proximity information from the exteroceptive sensors.

8.2.3 Absolute Task Preserving Reaction

In this monograph, following control points/structure pairs are used (for detailed explanations see Chapter 8):

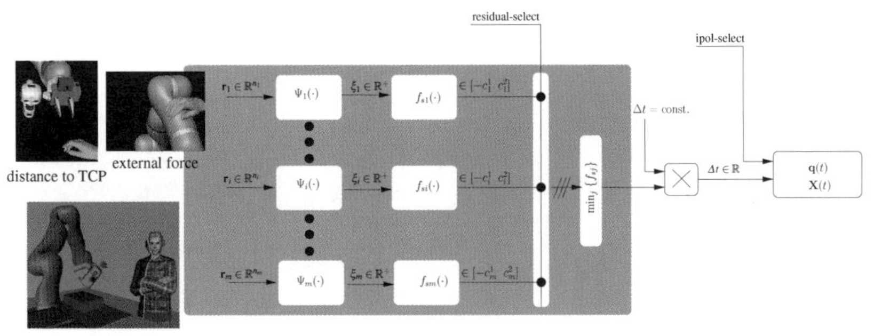

Fig. 8.5 Residual fusion for integrated trajectory scaling. Ψ_i is a normalization function and f_{si} a sigmoid function for time scaling [7, 6].

1. R1: Human-scenario proximity
2. R2: Human-TCP proximity
3. R3: Human-switching zones proximity
4. R4: TCP-table proximity
5. R5: TCP-hang-in proximity
6. R6: Elbow-workspace proximity

These proximity pairs were chosen due to their importance to the implementation presented later. The first three (R1-R3) are used for generating residuals for trajectory scaling, while the last three signals (R4-R6) are used for calculating virtual forces acting on an additional torque control input. R1 is taking into account the distance of the human to the robot workspace. The distance between human and robot TCP is important due to the fact that gripper and grasped objects are often characterized by sharp edges. Human-switching zones are boundary surfaces that separate different task workspaces and the related robot behavior depending on the human position. Since in the vicinity of the switching surfaces human behavior is not necessarily unambiguously classifiable, it is of large benefit to use this information (R3) as a possible residual. R4-R6 are chosen for showcasing collision avoidance during Cartesian impedance control and torque control with gravity compensation. They can be used to prevent the TCP from colliding with the table or the elbow with other objects in the environment without altering the desired motion path.

While the robot is in human-friendly mode, its intention is to fulfill the desired task efficiently, despite human presence. In order to accomplish this, it is necessary to equip the robot with reactive motion generators that take into account the human proximity and prevent collisions if possible without inefficient task abortion.

Trajectory scaling preserves the original motion path and at the same time provides compliant behavior by influencing the time generator of the desired trajectory, see Chapter 3.4.2. In this approach, physical contact residuals such as the estimated external joint torque or the external contact wrench are used, together with proximity-based residual signals such as the human-robot proximity, the human-switching zones proximity, and the human-workspace proximity. The usefulness of the approach becomes apparent when considering cases where humans are moving close to switching zones. If the robot would simply use binary switching information about the current state of the human, undesired oscillating behavior would occur due to the imprecise motions and decisions of the human. By using the human proximity to this border as a residual the robot always slows down and stops until the human clearly decides his next action. This way, the user receives elaborate visual feedback, indicating that the robot is aware of his presence and waits for further action.

The fusion of the different residuals is shown in Fig. 8.5 for several aforementioned signals. This concept allows to bring quantities of different physical interpretation together and use them in a unified way for trajectory scaling. Each residual is normalized[3] and then nonlinearly shaped to be an intuitive time scaling factor. Depending on the current state, the user can choose suitable residuals accordingly during application design.

8.2.4 Task Relaxing Reaction

Apart form task preserving reaction as described in the previous subsection, reactive real-time reaction with task relaxation is an important element for dealing with dynamic environments as well. For this the method introduced in Chapter 7 is used.

8.2.5 Dealing with Physical Collisions

The approaches introduced and derived in Chapter 3 provide the possibility to divide the impact severity into several classes, using a disturbance observer. This method for detecting contacts is also able to give an accurate estimation of the external joint torques τ_{ext}, which in turn can be used to classify collisions with the environment according to their "severity level". This allows to react differently to particular *collision severity stages*, leading to a *collision severity based behavior*. Apart from this nominal contact detection, the developed algorithms are also able to detect malfunctions of the joint torque sensors, based on model inconsistencies interpreted as a collision.

8.2.6 Safety Architecture

Apart from gaining insights into the mechanisms behind safe pHRI and isolated tools, it is critical to determine how to apply the knowledge and methodologies in a consistent manner. Schemes to utilize these features appropriately were developed

[3] Please note that an appropriate handling is referred to as e.g. projecting external forces to the velocity direction of the robot or similar transformations.

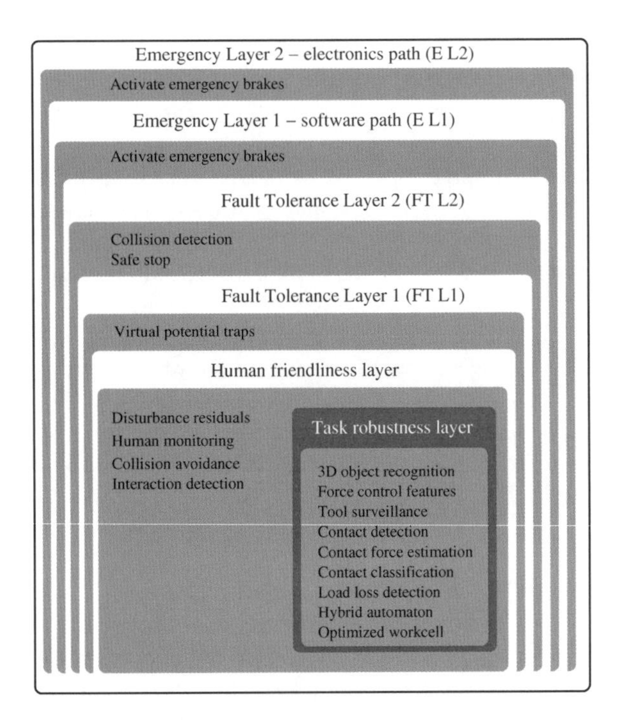

Fig. 8.6 Safety architecture of the DLR Co-Worker. Only the first two stages are user specific.

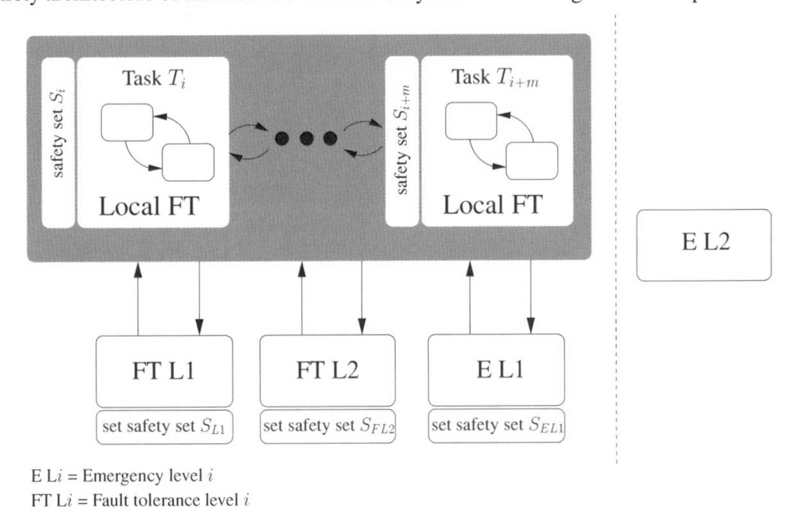

Fig. 8.7 Safety background of the DLR Co-Worker

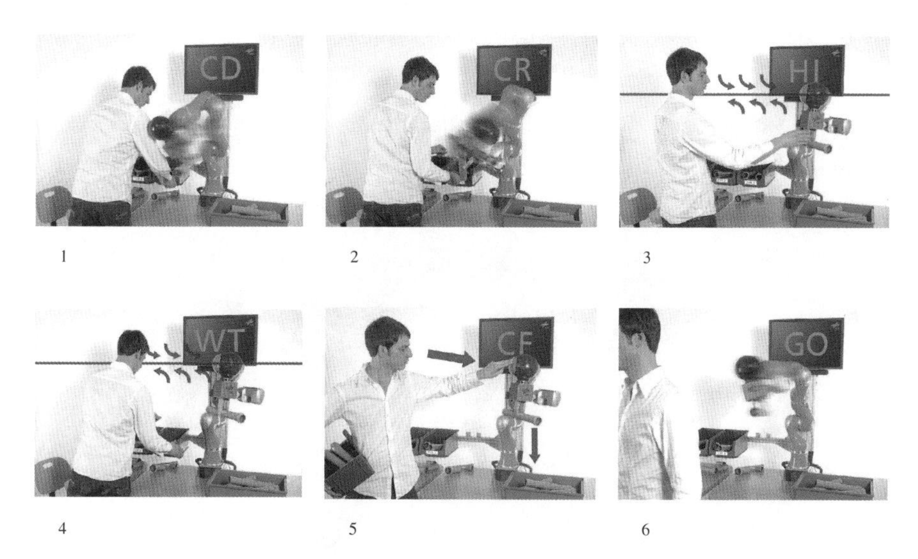

Fig. 8.8 Safe physical Human-Robot Interaction. Detecting and recovering from a collision in FT L1. It was assumed that the human was not perceived to have entered the workspace.

in order to maximize task performance under the constraint of achieving sufficient human-friendly behavior, see Fig. 8.6. Each feature is shown at the according hierarchical level where it is introduced and made available in the respective layer of the process.

Figure 8.7 outlines how the fault management and emergency components are embedded as underlying components for each task. Every task has certain low-severity-fault tolerant components to make it robust against external disturbances in general and prevent unnecessary task abortion. Each of them activates their distinct safety set S_j which is compatible with a particular goal, see Fig. 8.9 for details.

Figure 8.8 shows an example of an *unexpected collision* between a worker and a human ①, leading to a collision in layer FT L1. The robot switches to a compliant behavior ② after the collision is detected (CD). Due to the Collision Reaction (CR), the robot can be freely moved in space. This could lead to secondary collisions with the environment. Therefore, nonlinear virtual walls were designed (Fig. 8.2 (left)) to prevent physical collisions of the robot and secure the sensitive parts as the ToF-camera and the 3DMo. Moreover, the human can simply grab the robot anywhere along the structure and hang it like a tool into a predefined arbitrarily shaped virtual potential trap[4] (HI) ③, which smoothly drags it in and keeps it trapped. The human

[4] A current implementation generates an attractive region associated with a vertical virtual force. The human may now "hang-in" the freely movable robot into this "virtual trap". While being trapped by this potential the robot is free to move in horizontal direction but is relatively firmly confined in vertical direction. If a human pushes or pulls on the robot such that a certain "confirmation force" is detected this signal is used as volitional confirmation to re-enter the previously aborted process.

can then complete his task, ④, while the robot waits (WT) for further action. After completion is confirmed (CF) in ⑤ the robot continues ⑥ with the interrupted task (GO). If no confirmation arrives, the robot stays in its constrained passive behavior until either a confirmation for continuation occurs, or a human dragged him out of the hang-in field. Figure 8.9 shows how such behavior is triggered in a hybrid automaton and displays the safety sets involved in this process.

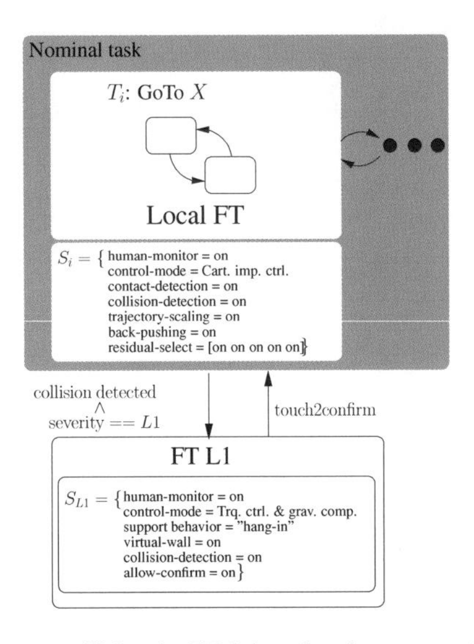

Fig. 8.9 Safe reaction to a collision in FT L1 under the assumption that the human was not perceived to have entered the workspace. A simple and convenient behavior is triggered, which can be realized by intuitive use of well designed state-dependent control scheme selection.

8.3 Interactive Bin Picking

In this section, the focus is on describing solutions for an industrially relevant autonomous task, which combines computer vision techniques with soft-robotics features. Furthermore, it should be embedded into an interaction scenario with the human. To demonstrate the performance of the system during autonomous task execution, the classical bin picking problem is addressed, which is a classical benchmark since the mid-1980s. However, such problems have remained difficult to be solved effectively. This is confirmed in different literature, as exemplified below:

> "Even though an abundance of approaches has been presented a cost-effective standard solution has not been established yet."
>
> *Handbook of Robotics 2008* [12]

In this monograph, environmental modeling, robust and fast object recognition, as well as quick and robust grasping strategies are combined in order to solve the given task. The setup depicted in Fig. 8.1 (right) serves as the demonstration platform. It is further used for realizing a scenario where the human assembles parts which are supplied by the robot and, after a "hand-over" and "receive" cycle, sorted into a depot by the robot, see Fig. 8.15. This fully sensor-based concept is entirely embedded in the proposed safe interaction framework. The intention of this application is to augment human capabilities with the assistance of the robot and achieve seamless cooperation between each other.

8.3.1 Vision Concept

The LWR-III is equipped with two exteroceptive sensors: the DLR 3D-Modeller and a time-of-flight camera so that their complementary features can be used within this scenario.

8.3.1.1 DLR 3D-Modeller

System: The DLR-3DMo is a multi-purpose vision platform [16], which is equipped with two digital cameras, a miniaturized rotating laser scanner and two line laser modules, see Fig. 8.1 (right). The DLR-3DMo implements three range sensing techniques:

1. laser-range scanning [5]
2. laser-stripe profiling [14]
3. stereo vision

Fig. 8.10 Generated 3D model from a series of sweep scans over the filled bin

These techniques are applicable to a number of vision tasks, such as the generation of photo realistic 3D models, object tracking, collision detection, and autonomous exploration [15].

Implementation: The laser-range scanner used for determining obstacles and free regions, provides range data coupled with a confidence value. The proposed application employs the rotating laser range scanner for two tasks. First, the wide scan angle of 270 degrees enables nearly complete surveillance of the working range around the gripper. Secondly, the measured distance data provides information about occupation of the space between the jaws of the gripper and indicates whether a target object is located there.

The laser-stripe profiler is used for modeling the environment and can be used for the localization of the bin or accurate modeling of the entire workcell, see Fig. 8.10. The shown model was generated with a series of sweep motions of the LWR-III across the scenario. The main purpose of the laser-stripe profiler is to acquire accurate data for model generation, in contrast to the safety functionality of the laser-range scanner.

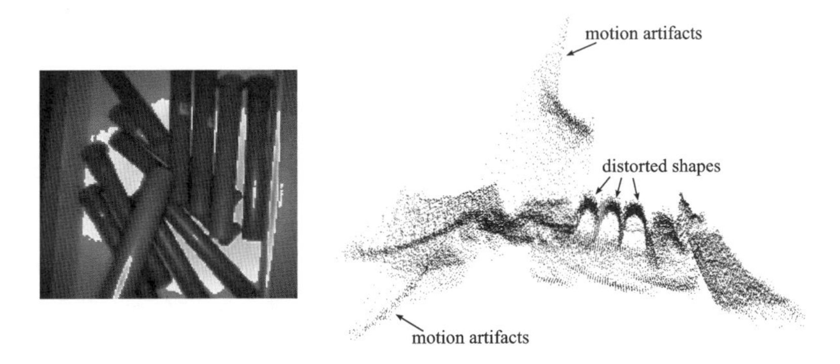

Fig. 8.11 Amplitude and depth data from view into the bin (left) showing large signal noise (right)

8.3.1.2 Time-of-Flight Camera

System: The Time-Of-Flight (ToF) camera Swissranger SR 3000, mounted on the robot, has a resolution of 176×144 pixels. An important feature of this device is the ability to capture $2\frac{1}{2}$D depth images at ≈ 25 Hz. Unlike stereo sensors, ToF-cameras can measure untextured surfaces because the measurement principle does not depend on corresponding features. Furthermore, due to the active illumination, ToF-cameras are robust against ambient illumination changes. These properties enable the recently established use in the robotics domain for tracking, object detection, pose estimation, and collision avoidance. Nonetheless, the performance of distance measurements with ToF-cameras is still limited by a number of systematic and non-systematic error sources, which would turn out to be challenges for further processing.

Figure 8.11 highlights the non-systematic errors such as noise, artifacts from moving objects, and distorted shapes due to multiple reflections. While noise can be handled by appropriate filtering, the other errors mentioned here are system-inherent. The systematic distance-related error can be corrected by a calibration step down to 3 mm, see [4].

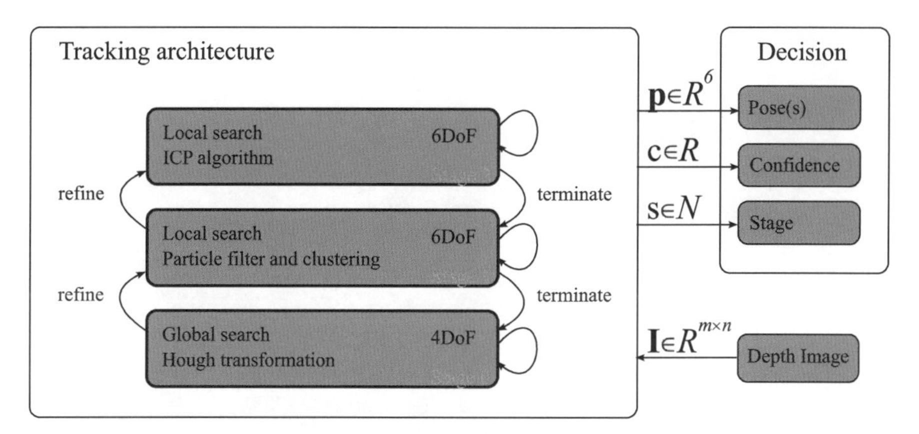

Fig. 8.12 Multi-stage tracking architecture based on [11]

Implementation: Generally, the high sampling rate of the ToF-camera guarantees fast object localization and robust object tracking performance based on a three staged tracking architecture, see Figure 8.12. In each stage a different algorithm processes an incoming depth image to provide a list of pose hypotheses for the potential object, which is additionally tagged with a confidence value. The stages are continuously monitored and executed according to suitable termination criteria or reentered for refinement.

The first stage is a global search, consisting of edge filtering and a Hough transformation for identifying lines as initial hypotheses for the tube location. In the second stage these hypotheses are locally consolidated and clustered by a particle filter. Third, an Iterative Closest Point Algorithm (ICP) provides an accurate pose estimation of the target object at a frame rate of $\approx$ 25 Hz. Both ICP and particle filter directly process 3D data and a 3D model of the target. The 3D model is represented by a point set with corresponding normals. This can be either generated from CAD models or surface reconstruction. The target object can be localized and tracked with an accuracy of $\approx$ 7 mm.

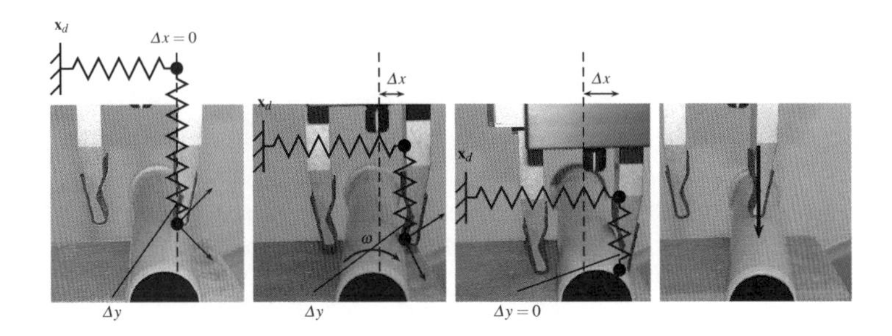

Fig. 8.13 Compliant grasping strategy

8.3.2 Soft-Robotics Control for Grasping

The soft-robotics features of the LWR-III greatly provide powerful tools to realize such a complex task as bin picking. Cartesian impedance control [1] is used as a key element for robust grasping despite the aforementioned recognition uncertainties. The impedance behavior of the robot is adjusted according to the current situation in order to achieve maximal robustness. Furthermore, the previously introduced strategies for fault detection are used to recognize grasp failures or unexpected collisions with the environment based on force estimation. In addition there are virtual walls preventing collisions with the static environment. The robustness of grasping against errors in object localization and errors in positioning due to the used impedance control is of great importance for this application. The grasping strategy shown in Fig. 8.13 successfully copes with possible translational deviations in the range of 55 mm before the grasp fails. Due to the compliant behavior of the robot the gripper-object and object-ground friction, the object is rotated into a firm grasp. The last image shows a case expected to be a failure. However, due to the rotational stiffness implemented along the axis perpendicular to the image plane grasping can still be achieved.

8.3.3 Autonomous Task Execution

Figure 8.14 depicts the autonomous bin picking task automaton, which merges the presented concepts into a high-level task description. The application is comprised of object recognition, grasping, and sort-in phases[5]. If the bin is depleted, the robot waits for further supply. Fault tolerant behavior is realized by introducing various branching possibilities for each state execution. In case of a failure, the robot

[5] The initial view and sort-in frames are taught in torque control mode with gravity compensation. This enables the user to freely move the robot to a desired configuration and save the pose in the application session.

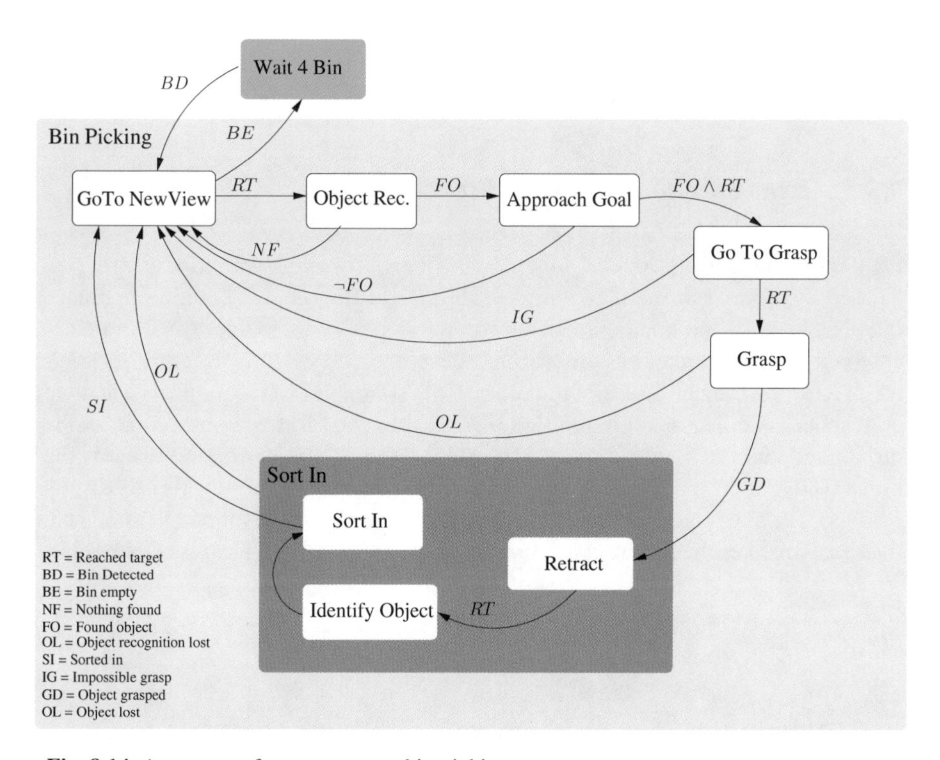

Fig. 8.14 Automaton for autonomous bin picking

recovers by monitoring conditions like object recognition dropouts, load losses, or impossibility of grasps.

8.3.4 Evaluation of Grasping Success

The efficiency and robustness of the approach was tested in a series of autonomous trials. For this evaluation the bin (Fig. 8.10) was replenished after each successful grasp in order to have a filled bin and independent trials. On average, the robot needed 6.4 s for one grasping process, which comprises of object detection from an arbitrary viewing position, approaching and grasping, unbagging, and moving back to the initial viewing position. The robot was able to grasp an object in every cycle for 80 trials. The overall cycle success rate of 100 % was achieved. This result was only achievable due to the fault tolerance capabilities of the system along the entire process, such as the detection of a physical impossibility of a planned grasp, of the non-successful grasp (overall 3 times), losing an object in tracking, or localization without finding any result of requested quality. The last fault mainly only occurred when the searched objects were partially in the field of view, so that the robot had to move to a new view position. All of these failure modes where detected or realized by the system and induced a restart of the grasping process. Consequently, the number of average views to recognize an object was $N_{\text{view}} = 2.2$.

8.3.5 Extension to Interactive Bin-Picking

Figure 8.15 depicts the concept for an interactive bin-picking scenario, merging interaction features and the autonomy capabilities of the robot. The initial entrance into the scene by the human is not shown, but is part of the demonstrator, i.e. it is assumed that the human has entered the scene, the "way into interaction" is completed, and the human is part of the process. ① shows the view into the bin and the corresponding object recognition (OR). Then, the robot grasps an object out of the bin ② and identifies it according to its weight, followed by a motion towards the human (GH) in ③. The "hand-over" ④ then takes place, after which the robot waits (WT) for the human to complete his process ⑤. As soon as the human has finished, the robot receives the object in a visual servo loop (VS) in ⑥. The classified object

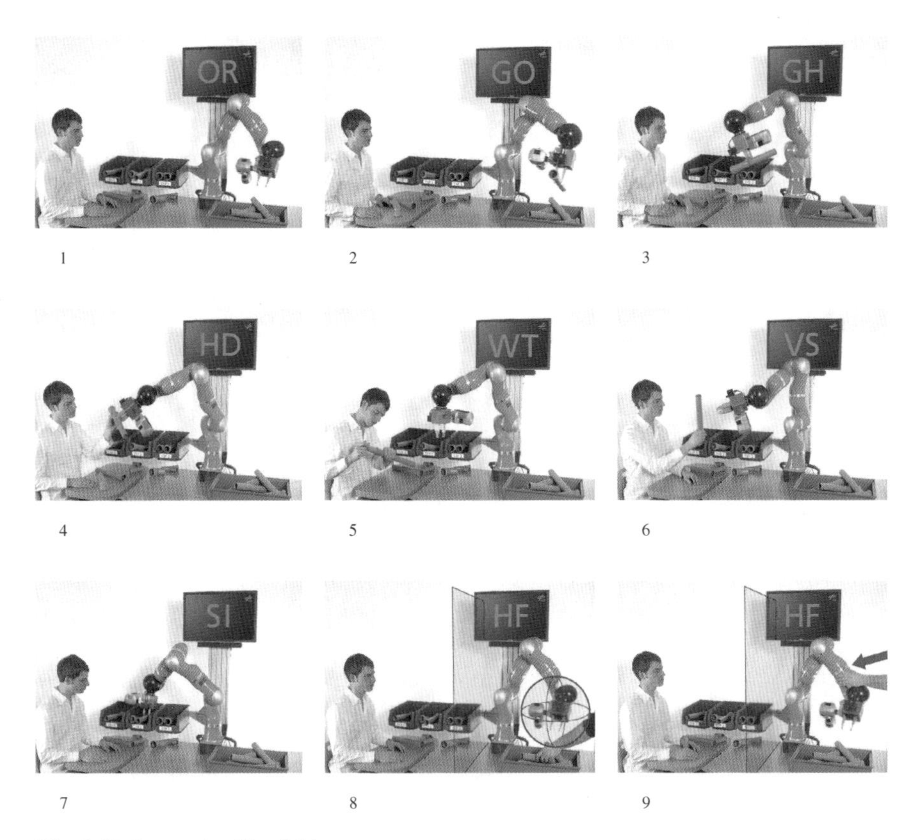

Fig. 8.15 Interactive bin picking

is sorted into (SI) one of the trays ⑦ and the robot goes back to ①. ⑧ and ⑨ show how human-friendly (HF) behavior is an integrated part of the co-worker design even in the presence of multiple humans. In ⑧ and ⑨ the tool surveillance and the physical contact during task execution are shown, respectively.

In summary, the system described here presents a versatile and robust solution with standard components for achieving safe and effective human-robot collaboration and a solution for the bin picking problem. Various explicitly non-trained test subjects were able to intuitively use the system.

Recently, the proposed concepts were integrated into a new human-friendly control architecture for the LWR-III. Its basic structure is depicted in Fig. 8.16 and shows the four central entities for robot control:

1. Task Control Unit (TCU)
2. Robot Control Unit (RCU)

    a. Safety Control Unit (SCU)
    b. Motion Control Unit (MCU)

The TCU is the general state-based control entity for gathering non-real-time data and provides the correct nominal behavior changes on an abstract level to the RCU. The RCU runs in hard real-time and assigns control, motion generation, and safety methods. The SCU serves as an underlying safety layer within the RCU that combines all low-level safety behaviors and activates them consistently. The MCU manages the correct switching and activation of motion generators and controllers.

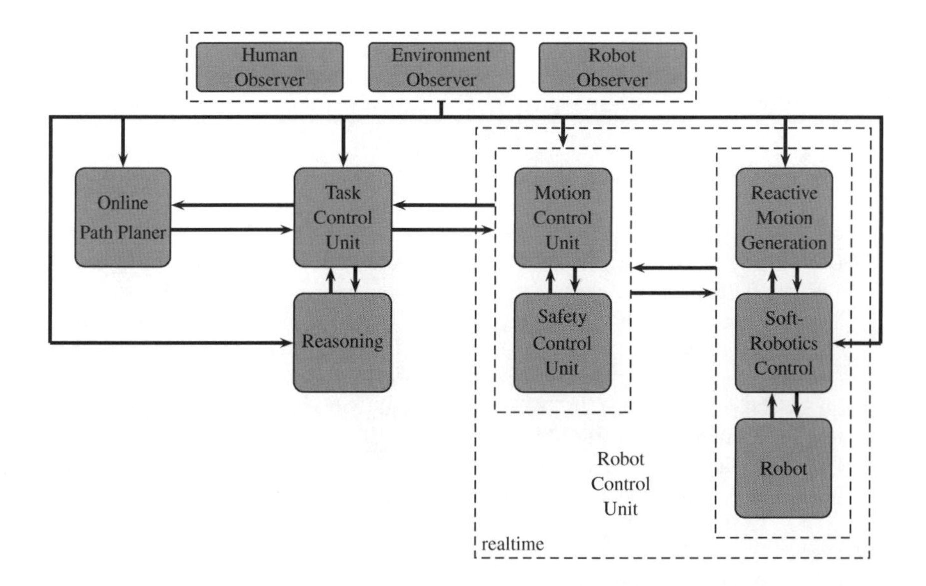

Fig. 8.16 Overview of the LWR-III architecture for human-friendly behavior

This novel concept enabled various applications requiring to a large extent pHRI as e.g. the first continuously brain controlled robot [17].

8.4 Summary

In this chapter, a general concept for the robotic co-worker was proposed and a prototype demonstration based on commercially available technology, namely an LWR-III, ToF camera, electro-mechanical gripper, and passive tracking system was developed for validation. An integrated solution was outlined for combining soft-robotics schemes with multi-sensor vision schemes. Flexible hybrid automata can robustly and safely control the modalities of the co-worker in a partially known environment and especially handle the complexity as well as the necessary branching factor during the execution of the tasks. Based on the results in safe physical Human-Robot Interaction elaborated in this monograph, effective combination of various control and motion schemes with vision sensing capabilities for the robot was achieved. This effectively accomplishes the task in a manner which is safe for the human. Furthermore, exteroceptive sensing is used in combination with compliance control for implementing industrially relevant autonomous tasks. The fusion of these concepts leads to high fault tolerance, proven by the results of the presented bin picking application. The use of multi-sensor information enabled to combine the proposed interaction and robust autonomy concepts needed for the *robotic Co-worker*.

References

[1] Albu-Schäffer, A., Ott, C., Hirzinger, G.: A unified passivity-based control framework for position, torque and impedance control of flexible joint robots. The Int. J. of Robotics Research 26, 23–39 (2007)
[2] Dominey, P.F., Metta, G., Natale, L., Nori, F.: Anticipation and initiative in dialog and behavior during cooperative human-humanoid interaction. In: IEEE-RAS International Conference on Humanoid Robots (HUMANOIDS 2008), Daejeon, Korea, pp. 693–699 (2008)
[3] Edsinger, A., Kemp, C.: Human-robot interaction for cooperative manipulation: Handing objects to one another. In: IEEE International Symposium on Robot & Human Interactive Communication (RO-MAN 2007), Jeju Island, Korea, pp. 1167–1172 (2007)
[4] Fuchs, S., Hirzinger, G.: Extrinsic and depth calibration of ToF-cameras. In: IEEE Conference on Computer Vision and Pattern Recognition (CVPR 2008), Anchorage, USA, pp. 1–6 (2008)
[5] Hacker, F., Dietrich, J., Hirzinger, G.: A laser-triangulation based miniaturized 2-D range-scanner as integral part of a multisensory robot-gripper. In: EOS Topical Meeting on Optoelectronic Distance/Displacement Measurements and Applications, Nantes France (1997)

[6] Haddadin, S.: Towards the human-friendly robotic co-worker. Master's thesis, Technical University of Munich (TUM) & German Aerospace Center (DLR) (2009)
[7] Haddadin, S., Albu-Schäffer, A., Luca, A.D., Hirzinger, G.: Collision detection & reaction: A contribution to safe physical human-robot interaction. In: IEEE/RSJ Int. Conf. on Intelligent Robots and Systems (IROS 2008), Nice, France, pp. 3356–3363 (2008)
[8] Hirzinger, G., Heindl, J.: Sensor programming - a new way for teaching a robot paths and forces. In: International Conference on Robot Vision and Sensory Controls (RoViSeC3), Cambridge, Massachusetts, USA (1993)
[9] Kulic, D., Croft, E.: Affective state estimation for human-robot interaction. IEEE Transactions on Robotics 23(5), 991–1000 (2007)
[10] Moshner, R.: From handiman to hardiman. Trans. Soc. Autom. Eng. 16, 588–597 (1967)
[11] Sepp, W., Fuchs, S., Hirzinger, G.: Hierarchical featureless tracking for position-based 6-DoF visual servoing. In: IEEE/RSJ Int. Conf. on Intelligent Robots and Systems (IROS 2006), Beijing, China, pp. 4310–4315 (2006)
[12] Siciliano, B., Khatib, O. (eds.): Springer Handbook of Robotics. Springer (2008)
[13] Stemmer, A., Albu-Schäffer, A., Hirzinger, G.: An analytical method for the planning of robust assembly tasks of complex shaped planar parts. In: Int. Conf. on Robotics and Automation (ICRA 2007), Rome, Italy, pp. 317–323 (2007)
[14] Strobl, K.H., Wahl, E., Sepp, W., Bodenmüller, T., Seara, J., Suppa, M., Hirzinger, G.: The DLR hand-guided device: The laser-stripe profiler. In: Int. Conf. on Robotics and Automation (ICRA 2004), New Orleans, USA, pp. 1927–1932 (2004)
[15] Suppa, M.: Autonomous robot work cell exploration using multisensory eye-in-hand systems. Ph.D. thesis, Gottfried Wilhelm Leibniz Universität Hannover (2007)
[16] Suppa, M., Kielhöfer, S., Langwald, J., Hacker, F., Strobl, K.H., Hirzinger, G.: The 3D-Modeller: A multi-purpose vision platform. In: IEEE Int. Conf. on Robotics and Automation (ICRA 2007), Rome, Italy, pp. 781–787 (2007)
[17] Vogel, J., Haddadin, S., Simeral, J.D., Stavisky, S.D., Bacher, D., Hochberg, L.R., Donoghue, J.P., van der Smagt, P.: Continuous control of the DLR Lightweight Robot III by a human with tetraplegia using the BrainGate2 neural interface system. In: International Symposium on Experimental Robotics (ISER 2010), Dehli, India (2010)
[18] Yamada, Y., Konosu, H., Morizono, T., Umetani, Y.: Proposal of skill-assist: a system of assisting human workers by reflecting their skills in positioning tasks. In: IEEE International Conference on Systems, Man, and Cybernetics (SMC 1999), Tokyo, Japan, pp. 11–16 (1999)

Chapter 9
Competitive Robotics

The monograph dealt to a large extent with currently open problems, which are important for both robotics industry and standardization organizations. In the present chapter topics are discussed, which are relevant in the more distant future while at the same time being tightly interrelated with a very recent topic of robotics research:

(variable) intrinsic joint elasticity

However, this is only one aspect among many others in the context of what is called *Competitive Robotics*.

Competitive Robotics deals with human-robot games that involve intentional physical contact of human and robot being opponents.

The most prominent example of Competitive Robotics is the RoboCup [23] with the goal: winning against the human world soccer champion team by the year 2050. This implies real tackles and fouls between humans and robots, raising safety concerns for the robots and even more important for the human players, similar to the questions that were already discussed in the context of pHRI.

The first contribution of this chapter is to shed light on the pHRI aspects of such a hypothetical human-robot match. Therefore, two matches from the (2006) FIFA World Cup in Germany are used as examples and are analyzed with respect to scenes with physical interaction. These interactions are related to results in pHRI and sports science by speculating what would have happened if one of the opponents was a robot. The most important finding is that elastic joints are needed to reduce the impact joint torques during collisions. The second part of the analysis focuses on the robot's robustness and safety. How can it withstand the impact of kicking the ball or even fouls? And finally, it is discussed how joint elasticity can be used to achieve the kick velocity of human soccer players. The discussion includes experiments with traditional robots with little elasticity, experiments using a joint with large elasticity, and theoretical result on optimal control of an elastic joint.

Overall, this chapter analyzes the possibilities of a future vision. However, all the conclusions are based on actual simulations, experiments, derivations, or findings from sports science, forensics, and pHRI. Furthermore, this chapter lays the ground

S. Haddadin: *Towards Safe Robots*,
Springer Tracts in Advanced Robotics 90,
DOI: 10.1007/978-3-642-40308-8_9, © Springer-Verlag Berlin Heidelberg 2014

work for numerous findings about variable stiffness actuation, extensively discussed in Chapter 10.

The RoboCup 2050 Challenge

Soon after establishing the RoboCup competition in 1997, the RoboCup Federation proclaimed an ambitious long term goal, see Fig. 9.1.

> "By mid-21st century, a team of fully autonomous humanoid robot soccer players shall win the soccer game, comply with the official rule of the FIFA, against the winner of the most recent World Cup."
>
> *H. Kitano and M. Asada* [23]

Soccer is a contact sport and injuries of players are frequent [26]. The FIFA rules state explicitly, that

> "Football is a competitive sport and physical contact between players is a normal and acceptable part of the game. [...]"
>
> *Laws of the game, 2006* [10]

A soccer match between humans and robots implies physical Human-Robot Interaction including tackles and fouls between the participants. In order to come closer to that vision, an evaluation of the fundamental requirements and challenges the human presence would bring into such a match is crucial and remains an open issue. This not only makes sense from the perspective of ensuring human safety but also of defining requirements a robot has to fulfill in order to withstand the enormous strains posed by such a real soccer match. These problems can only be approached and tackled by treating the robotic and biomechanical aspects as complementary.

In the domains of industrial assistance and service robotics, robots are and will be designed to cause absolutely no harm to any human. Presumably, such a robot could never win a sports game. However, it is demand that a human-robot match should not be more dangerous than a regular soccer match.

> "A competitive robot may not be more dangerous than a human being."

Hence, it is focused on situations, where a robot is expected to potentially cause more injury than a human player.

This chapter is organized as follows. Section 9.1 introduces some preliminaries. Section 9.2 discusses the safety of humans in the context of human-robot soccer and analyzes potentially dangerous situations. Section 9.3 investigates how to protect robot joints from external loads, leading to the necessity of introducing joint compliance for protection. This intrinsic compliance is useful for increasing velocity performance with appropriately designed trajectories, which lead to efficient elastic energy storage and release. The details are discussed in Sec. 9.4.

Fig. 9.1 The RoboCup 2050 Challenge

9.1 Preliminaries

In this chapter, it is focused on the benefits of elastic joints for safety and kicking performance. Nevertheless, in a soccer scenario this would also imply to walk and run with these joints. So the state of the art in this field is briefly reviewed. Since numerous conclusions made throughout this chapter are interlinked with VIA also a brief overview on the concept is given. Furthermore, a short comment is made on the dynamic models used for kicking analysis in some of the presented simulations.

Compliance for Walking and Running

Current large and medium scale anthropometric humanoids as H6, H7 [32], P2 [15], ASIMO [16], JOHNNIE, LOLA [27], WABIAN-2 [34], KHR-2 [22], HRP, HRP-2 [21], and SAIKA [41] represent major achievements over the last years. In these systems, locomotion is mostly realized with stiff actuation in combination with rigid high-geared transmission mechanisms. Due to the lack of an appropriate storage mechanism, the entire energy is lost during decelerating while walking and running and has to be continuously injected by active actuation. The same holds for the robots in the RoboCup domain, where usually no deliberately introduced compliance is used.

However, some realizations already exist, which successfully used intrinsically compliant joint designs for biped walking. In WL-14 [47, 48], a sophisticated nonlinear spring mechanism was used for stiffness adjustment. More recently in Lucy [43], a biped that is able to walk in the sagittal plane, approaches were made to utilize adjustable passive compliance for high energy efficiency during walking. The robot Flame [17] uses constant compliance (Series Elastic Actuation) in the hip, knee, and ankle pitch joint. HRP-2LR [19] is equipped with a compliant toe in both feet, each having a constant rotational spring. The authors predicted via simulation a running speed of 3 km/h with this device compared to 0.58 km/h achieved with HRP-2LT that has no such compliant toes. The authors already demonstrated hopping with both feet.

Apart from these first realizations in the field of biped walking, there is clear evidence in biomechanics that intrinsically compliant actuation is critical to terrestrial locomotion [31]. So to summarize, running with elastic joints seems to be difficult but possible and probably of long-term benefit.

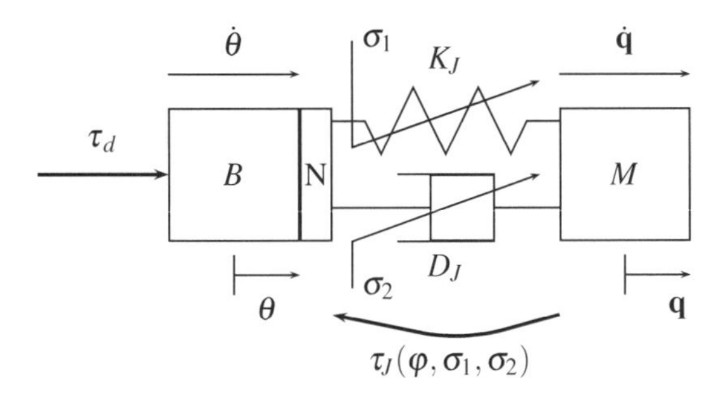

Fig. 9.2 1-DoF model of a VIA joint

Variable Impedance Actuation

The principle of Variable Stiffness Actuation is truly human-inspired in the sense that it intends to approach the impedance adjustment capabilities of the human musculoskeletal system. In humans all muscles work in pairs, namely the agonist and the antagonist. Since human muscles are only able to pull via contracting this arrangement is needed for moving in both directions (one muscle pulling). A contraction of both muscles at the same time changes the joint stiffness due to the nonlinear elastic properties of the tendon-muscle complex (If both pull asymmetric a combination of motion and stiffness change is achieved). A well known example for such a muscle pair is the biceps brachii and the triceps brachii. There are numerous concepts for transferring this design idea to robotic actuation. However, they show the characteristic of an intrinsically variable impedance element between actuator and link, see Fig 9.2. The elastic joint torque $\tau_J(\varphi, \sigma_1, \sigma_2)$ between motor and link is in general a function of the elastic deflection $\varphi = \theta - q$ as well as of the stiffness and damper actuation variables σ_1, σ_2. This model can be seen as the extension of the flexible joint model introduced in Chapter 3. A major difference is that the deflection can no longer be considered as small. Passive deflection may occur in considerable ranges of the joint space.

A Simulation Model for a Humanoid Soccer Robot Leg

Simulated and real experiments in this chapter primarily refer to the LWR-III and the DLR Variable Stiffness Joint (DLR VS-Joint), a prototype developed for the new intrinsically compliant DLR hand-arm system [11, 1], see Chapter 10. This joint is a representative of intrinsically compliant devices and all major conclusions made in this chapter related to joint elasticity are of general character. Although the LWR-III is designed as an arm, it has inertial and geometric properties comparable to a human leg $\left(\frac{LWR-III}{Leg} \approx 1.2\right)$ [7, 13]. So it is used as a "model" for the leg of a future humanoid soccer robot throughout this chapter, while not claiming that the design is feasible for a leg in general. DLR has recently developed a biped [36] based on the LWR-III technology. With 130 ^{o}/s, its maximum joint velocity is still much lower than that of a human soccer player at 1375 ^{o}/s [33]. Hence in simulations a hypothetical, faster LWR-III as a model is often considered.

9.2 Safety of the Human

This section is concerned with typical physical interaction in soccer. It mainly covers fouls in human soccer after a short overview of collisions in robot soccer. These are classified into different categories and discussed from a pHRI perspective. Afterwards, a simulation and experimental analysis of impacts is presented. In particular elbow checks as a major injury source are considered.

9.2.1 Physical Interaction in Humanoid Robot Soccer

Most RoboCup Soccer leagues, including the Humanoid League, already base their rules roughly on the official FIFA laws of the game. Thus, physical interaction and fouls are specified together with the resulting consequences [25]. However, the level of detail is much lower than in the original rules, which even include *Additional Instructions and Guidelines for Referees* [10] to distinguish types of physical interaction explicitly.

Even with 20 degrees of freedom, current humanoid soccer robots are not able to perform sophisticated movements comparable to humans. Thus, the RoboCup Humanoid league only differentiates between having physical contact (independent of the involved body parts) or not. In general, physical contact is allowed but should be minimized. Prolonged contact must be avoided and leads to an intervention of the referee. The rules of other robot soccer leagues are similar, but might specify different periods and intensities of contact.

This indifference between the kinds of contacts becomes obvious when examining matches in the Humanoid Kid-Size league, especially the 2008 final between *Nimbro* and *Team Osaka*. Within this eventful 3 vs. 3 match, many physical interactions occurred. In contrast to the variety of interactions in human soccer, which are described in the following section, only one reoccurring pattern can be observed: robots have contact, lose their balance, and fall over. The intensity of the impact with the floor is disproportionately higher than any contact with any robot trunk or limb.

Due to the crudeness of the current state of the art different kinds of physical interactions (active or passive) to prevent damages have not been addressed in the RoboCup community so far.

9.2.2 Physical Interaction in Human Soccer

In this subsection, possible physical interaction occurring in soccer are separated into different classes and their injury potential for the human and the robot is discussed. A set of scenes from the 2006 FIFA world championship serves as examples. The final (Italy vs. France) as well as one of the most physical[1] matches of the tournament (Portugal vs. the Netherlands) were chosen for the analysis. Table 9.1 and Tab. 9.2 show the analyzed and classified scenes and which players were involved. To investigate possible injury mechanisms, frequently involved body parts must be identified. According to [28], adult soccer injury spreads almost over the whole body, but especially concerns the limbs (arm 15 %, hand 9 %, ankle 32 %, and knee 26 %), the back (5 %), and the head (11 %), whereas the rest of the torso seems to be in less danger. Injury causes were analyzed in [24], indicating that collisions with opponents (22.4 %) or the ball (20.3 %), incidents while being in motion (17.1 %) or after falling down (8.2 %) are most frequent. In this chapter, these dominant injury sources and mechanisms are focused on.

9.2.3 Tripping and Getting Tripped Up

Tripping at high speed over the opponent's legs has a relatively high injury potential and is a commonly observed action. It is not necessarily an intended foul, but can be a legal tackle which aims at the ball. Roughly, tripping someone up in soccer can be divided into three categories:

- Hitting the opponent's feet intentionally by a sliding tackle (Fig. 9.3a, b[2]; T1, T2).
- Hitting the opponent's feet or legs unintentionally while chasing the ball (Fig. 9.3c; T3, T4).

[1] 16 yellow cards (including four second cautions) denoted the maximum value of the entire tournament.

[2] To avoid any copyright conflicts, the most significant situations are sketched.

Table 9.1 Scenes of the FIFA World Cup 2006 for the match Italy vs. France (final)

Scene	Time	Description	Figure
T1	30:37	Costinha skids to Cocu's feet and overthrows him (yellow card).	9.3 b
T2	72:03	Heitinga runs fast while his leg is thrusted by sliding Deco (yellow card).	9.3 a
T3	86:33	Ooijer trips Petit up.	9.3 c
T4	94:52	Van Bronckhorst trips Tiago up (yellow card).	
I1	92:50	Ricado and Kuyt both jump to reach a ball, colliding in the air.	
L1	06:52	Bouhlarouz hits Ronaldo's thigh with his boot while his leg is half-elongated (yellow card).	
L2	38:20	Costinha tries to play the ball but hits Ooijer's shin.	
L3	41:50	Robben and Valente both approach a high ball, Valente jumps with elongated leg and hits Robben's chest.	9.5 a
L4	62:00	Bouhlarouz approaches Figo from the side to gain access to the ball. While running, he hits Figo with his elbow in the face.	9.5 c
L5	79:40	Kuijt skids towards Ricardo and hits the goalkeeper's shank with his boots causing a minor injury.	9.6 a
L6	87:38	Simão steps on the goalkeeper van der Saar.	
P1	14:48	Kuyt and Carvalho run leaning against each other in parallel, Kuyt falls.	
P2	61:50	Van der Vaart and Figo chase the ball and push against each other.	9.6 b

Table 9.2 Scenes of the FIFA World Cup 2006 for the match Portugal vs. Netherlands

Scene	Time	Description	Figure
T5	04:22	Zambrotta hits Vieira's supporting leg and Vieira falls badly (yellow card).	
I2	00:35	Cannavaro and Henry collide with their trunks, Henry falls.	9.4 a
I3	34:03	Materazzi moves forward, Ribery backward, both collide.	
I4	65:23	Ribery and Grosso jump, Grosso lands on Ribery's back.	
I5	71:39	Camoranesi is running and gets blocked by standing Abidal, Camoranesi falls.	
I6	79:04	Cannavaro and Zidane jump. Cannavaro jumps higher and drags Zidane to the ground.	
I7	E01:13	Makelele and Gattuso jump. Makelele jumps higher and lands on Gattuso's back.	9.4 b
I8	E04:16	Gattuso rushes into Malouda while approaching the ball.	
L7	10:47	Sagnol runs into Grosso, their knees collide (yellow card).	9.5 b
L8	31:53	Ribery steps on Zambrotta's ankle joint.	
L9	72:03	Toni kicks the ball in a 180° rotation and hits Thuram's knee with his shin.	
L10	E04:16	Malouda hits Gattuso's face with his lower arm.	
P3	35:07	Thuram and Toni run with entangled arms.	
P4	45:10	Zidane pushes Gattuso.	
P5	74:57	Malouda jumps to head the ball and gets pushed by Zambrotta.	
P6	E02:17	Malouda and Cannavaro run parallel and push each other. In the end, both fall.	

- Directly attacking the opponent's legs (T5) without any chance of playing the ball.

This kind of interaction usually causes two mechanisms of injury: fractures of lower and upper extremities, ankle or knee injuries by direct contact [28], and indirect ones from resulting tumble. Soft covering of the robotic leg can decrease this injury potential dramatically and protects the robot structure. Because tripping can be sudden with little time to actively react, an overall compliant covering of the robot may be required. This is because the robot could fall in a more or less arbitrary direction with an undefined impacting zone. Passive compliance in the joints appears to be an effective countermeasure to intrinsically decouple impacting masses and decrease potential danger.

A necessary action the robot has to perform is minimizing impact forces on its body, similar to humans, by preshaping its limbs. This prevents singular configurations during tackling and therefore protects both human and robot.

Fig. 9.3 Typical tripping scenes: a, b) A player slides to the ball and touches his opponent's legs (T2,T1). c) A player trips his opponent up (T3).

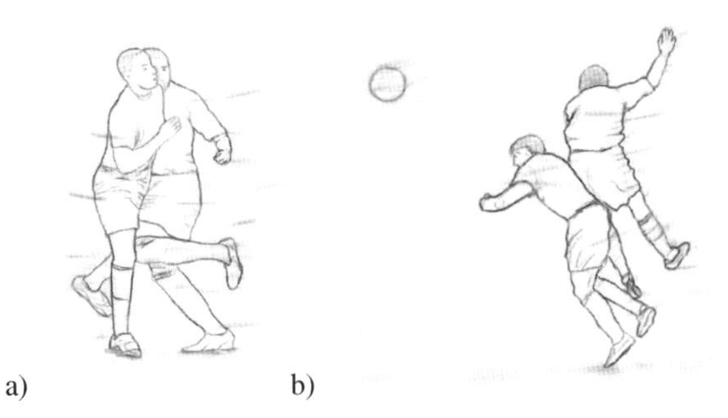

Fig. 9.4 Trunk impacts: a) Two players run into each other (I2). b) During a header, two players collide. Afterwards, one of them falls upon his opponent (I7).

9.2.4 Trunk and Head Impacts

Trunk and head impacts occur frequently and they are usually caused by

- Two players colliding while running towards each other (Fig. 9.4a; I2, I3, I8)
- One player body-checking the other player (I5, I8)
- Two players jumping back to back at each other when fighting for a header (Fig. 9.4b; I1, I6, I7)
- One player falling on the other one who is lying on the ground (I4, I7)

This particularly limits the robot's weight as kinetic energy is an indirect indicator of head injury according to [30, 40]. Therefore, the robot's weight has to be similar to the professional soccer players. This was also stated by Burkhard et al.: "The robots should have heights and weights comparable to the human ones (at least for safety reasons) [...]" [5]. According to [8], the average weight of the *FIFA Worldcup 2002* participants was 75.91 ± 6.38 kg. For much higher robot masses, the situation of a human clamped on the ground by a robot that outweighs him, poses significant danger to the limbs, chest, and other body parts. The weight of current humanoids, such as ASIMO (54 kg), HRP-2/3 (58/65 kg), WABIAN (64.5 kg) or HUBO (57 kg) is generally less than the ones of an average soccer player but all of them are smaller.

Apart from limiting the robot weight, its body surface should be padded to avoid human injuries from sharp edges, resulting in fractions, lacerations or cuts which already occur during blunt impacts [28]. The spinal column and facial bones are very sensitive parts of the human body, having relatively low fracture forces [29], which necessitates compliant properties of the robot's back, see Fig. 9.4b. Nevertheless, one should keep in mind that headers require a hard contact surface to accelerate the ball fast enough and therefore use a thinner coating for the head. Hard, elastic materials such as rubber, polyurethane or silicone are some possible choices for the coating. Further aspects concerning weight and height are discussed in Sec. 9.2.8.

9.2.5 Limb Impacts

Dangerous impacts caused by limbs, i.e. colliding legs or arms with the opponent's body can be roughly divided into

- Elbow checks (intended or unintended) to the other's face (Fig. 9.5c; L4, L10)
- A player sliding into or stepping on another player who is on the ground (Fig.9.6a; L5, L6)
- A leg hitting the opponent's trunk (Fig. 9.5a; L3)
- Legs or feet of two players colliding (Fig. 9.5b; L1, L2, L7, L8, L9)

The first class of impacts can be reduced to subhuman injury level by padding the robot's elbow. The other ones are caused by the boot which is the same for robots and humans. The enormous velocity of the kicking foot (see Sec. 9.4.2) can be fatal, so the robot must detect the absence of the human head absolutely reliably in order to protect it.

Impacts with parts of the goalkeeper other than the head are not clearly separable from the third injury source, where passive compliance in the joints is crucial to decouple the impact area from the rest of the robot[3]. This protects both, the human and the robot from being injured/damaged. In other words, passive joint compliance enhances safety for both, the human and the robot. This mechanism has limitations

[3] Please note that it is not referred to the immediate impact but the following contact dynamics, where the joint stiffness plays an important role indeed.

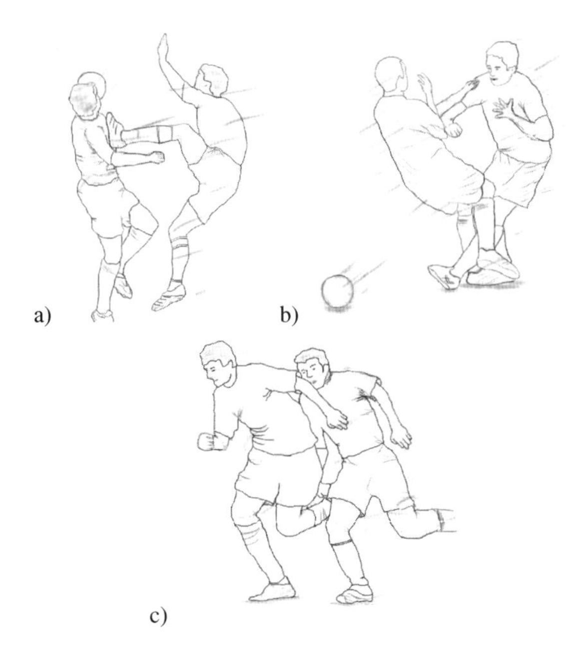

Fig. 9.5 Different situations of limb impacts: a) A high foot hits the opponent's chest instead of the ball (L3). b) Two players collide and hit each other's knee (L7). c) A player pushes his elbow into a chasing opponent's face (L4).

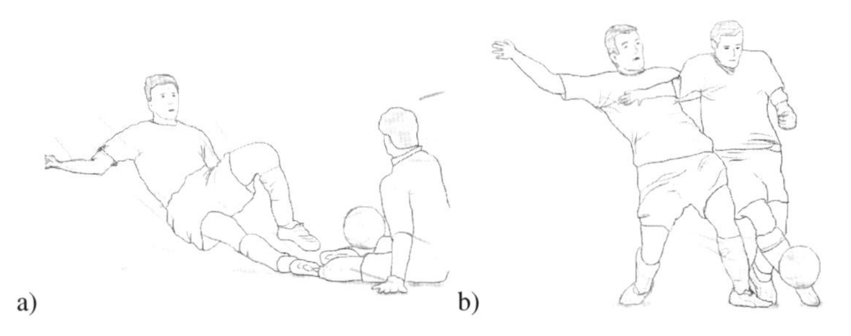

Fig. 9.6 a) A striker tries to reach the ball and slides into the sitting goalkeeper (L5). b) A typical situation with two players pushing each other while chasing the ball (P2).

as well: in an outstretched singular configuration, joint compliance has no effect and the Cartesian reflected inertia is vastly increased. As for humans, this configuration has to be avoided during such an impact under any circumstances to prevent both parties from damage.

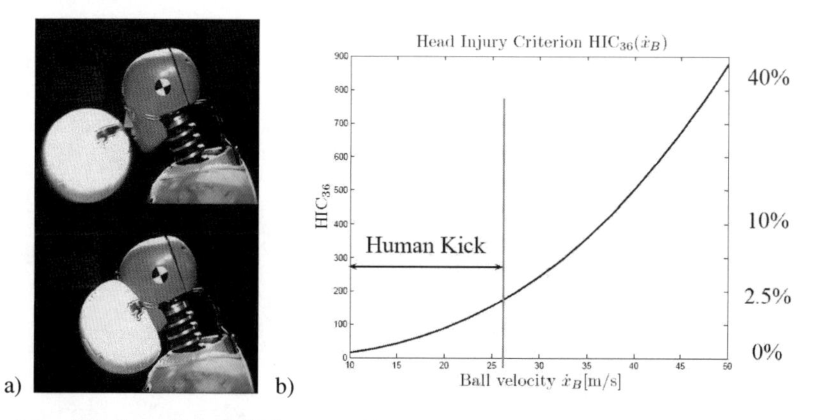

Fig. 9.7 a) Hitting a Hybrid III dummy with a soccer ball. The impact is almost fully defined by the properties of the ball. The elasticity of the head can be neglected. Courtesy of the German Automobile Club (ADAC). b) The HIC as a function of impact velocity and resulting probability of *serious* (AIS = 3) injury.

9.2.6 Being Hit by the Ball

Being hit by a fast moving soccer ball can be a painful experience. In order to analyze such an impact, a one-dimensional simulation was carried out. The human head is modeled as a simple mass and the ball as a mass-spring system[4], justified by high-speed camera recordings, see Fig. 9.7a. Injury severity is expressed by HIC, following the extended Prasad/Mertz curves[5] for the conversion to probability of injury. In Figure 9.7b, the resulting Head Injury Criterion is plotted against impact velocity, and the probability of *serious* injury for different impact velocities is indicated. It shows that kicks, carried out by humans do not pose a serious threat, whereas increasing ball speed by only 50% would be already significantly more dangerous. These observations strongly suggest avoiding an approach to counterbalance lack of robot intelligence by simple power, i.e. no "brute force" solution in robot-soccer. In addition to the potential threat posed to human heads by faster impacts, the joints of the robot can suffer damage from such fast kicks. This type of loading is mostly the same as if the robot kicks the ball and is discussed in Sec. 9.4.2.

9.2.7 Secondary Impacts

A situation more unlikely to happen but still worth mentioning are secondary impacts such as the ones during heading duels, where one of the players clashes against

[4] Because no adequate damping models are available, this effect is neglected.

[5] There exist various mappings to injury probability and interpretations of the HIC leading to different numerical values. However, *one* of them is used to show its extreme velocity dependency.

the goalpost. Additionally, a player could be pushed against the boards next to the field[6]. These secondary impacts are potentially dangerous to both human and robot, so the robot should have sufficient understanding to avoid such situations if possible. In order to protect itself from being damaged, padding and compliant joints appear to be an adequate countermeasure.

9.2.8 Further Aspects

Besides the interactions described in Sec. 9.2.2, of which most are fouls or tackles, several other comparatively light contacts occur in soccer. In almost all matches, situations in which two players in parallel run to a ball and mutually obstruct each other could be observed (Fig. 9.6b; P1, P2, P6). Light pushes (P4, P5) without any consequences happen as well as entangling arms in crowded situations (P3). This raises the question whether a soccer robot would benefit from a touch sensitive skin.

Another aspect not fully discussed in this chapter is the possible necessity of specialized team role robots. Because of the varying loading of players in different positions, having different types of players is beneficial. For example, goalkeepers seldomly sprint but often dive and fall on the whole body when defending a ball, whereas field players are posed to sustained loads and duels. According to [28], injury severity and mechanisms highly depend on the position of the player, pointing out that goalkeeper have been shown to have more head, face, neck, and upper extremity injuries than lower extremity injuries. Another reason to design different player types is that because of their inertial properties, massive and hence slow players cannot fulfill the role of a fast and flexible playmaker. It can often be observed in real world soccer that manipulability of the body is more important than simple speed and strength. According to [44] the average height of players is different between striker ($\approx$ 176 cm), defender ($\approx$ 185 cm) and goalkeeper ($\approx$ 190 cm), clearly indicating the necessary specialization for each position. An obvious reason for this difference in height are headers, or for a goal keeper reaching the kick in terms of reach [44]. Furthermore, there is the natural advantage of heterogeneity and diversity within the team.

In the following, soft-tissue injuries and injuries caused by elbow checks are outlined and how they can be reduced. Under certain circumstances it is even possible to limit them to lower levels than presumably caused by humans.

9.2.9 Injuries from Blunt Impacts with Soft-Tissue

In order to further analyze the benefit of intrinsic joint compliance, the blunt soft-tissue impact of a rigid robot joint with the lower abdominal area will be evaluated.

---

[6] In new soccer arenas, tracks are usually left out so that this is definitely not too unlikely.

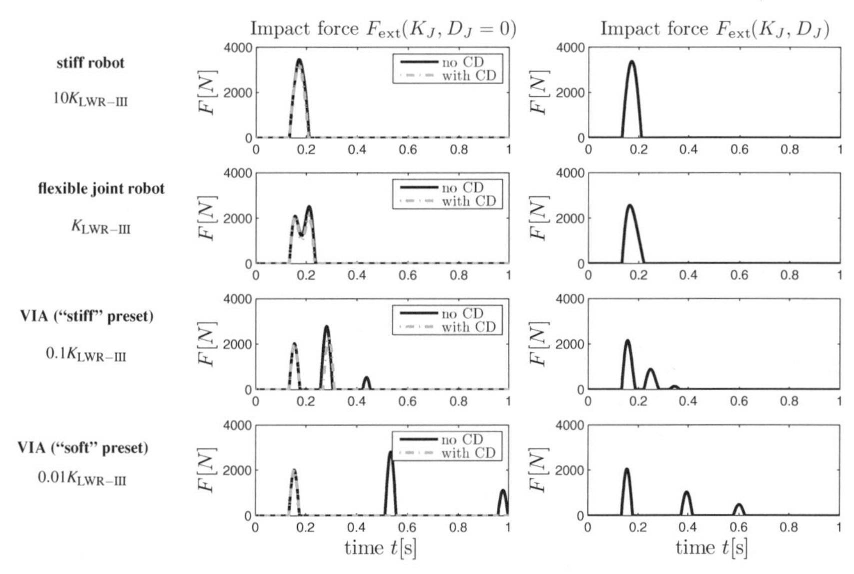

Fig. 9.8 Impacting the abdomen at 7.5 m/s with a robot. The inertial parameters of the robot are the reflected ones of the LWR-III and the joint stiffness is chosen to be $10, 1, 0.1, 0.01$ times the one of the LWR-III. In other words, the compliance varies from low to very high. The left plots show a robot without joint damping with and without CD. If a collision is detected, the robot reacts by braking with full available motor torque. The right plots show the behavior of a critically damped link for the same impact conditions.

Then, it is outlined how decreasing the stiffness results in significantly improved safety characteristics.

A main benefit of intrinsic joint compliance is that it gives a physical collision detection mechanism more time to detect and react to the collision since it decouples motor and link inertia. Before presenting the impact results, a short assessment of abdominal injury will be given to introduce a relevant injury severity index for the abdomen.

The abdomen is located between the thorax and the pelvis. A large amount of literature exists on abdominal injury describing various different injury criteria with an overview given in [18]. For simplicity, the side force criterion that is part of the EuroNCAP crash test is used. It states that the contact force must be

$$F_{ext} \leq 2.5\ \text{kN}. \tag{9.1}$$

This criterion will be used with a mass-spring system as a simple model of the lower abdomen. The spring stiffness of $K_{Abd} = 20\ \text{kN/m}$ can be estimated from data published in [6]. It will be assumed that the impact involves only the torso with a weight of 34 kg [7].

A kick with a hypothetical, faster version of the LWR-III at 7.5 m/s is simulated, which is above any velocity common in Human-Robot Interaction but reasonable for a soccer game. The reflected inertia of the motors and links are 13 kg and 4 kg. In the following analysis, the joint stiffness is varied from very rigid to fully compliant[7]. It is shown how collision detection together with intrinsic joint compliance significantly reduces the potential injury risk during a robot-human impact.

In Figure 9.8, the contact force of a typical instep kick into the abdomen is depicted with and without collision detection (left column), while on the right column the effect of joint damping is depicted. In current variable stiffness joints, physical joint damping is usually undesired [46] because it introduces hysteresis and possibly non-linear behavior. However, human joints clearly are damped and therefore some properties related to damped joints are shown as well.

For a very stiff robot, such as a typical industrial robot, the impact force results from an immediate impact of both link *and* motor inertia acting as one interconnected mass. The force limit of the abdomen is exceeded and therefore such an impact poses a severe threat to the human. In case of a flexible joint robot as the LWR-III, the joint stiffness is already low enough to partially decouple link and motor inertia. The latter becomes significant approximately 50 ms after the link impact. This reduces the maximum force and gives a collision detection mechanism time to react. Due to the low link inertia, the first force peak is below the tolerance force of the lower abdomen. For even lower joint stiffness (VIA "stiff" preset and VIA "soft" preset), both components are more decoupled and the delay of the second peak increases (caused by the much slower increasing joint force). This property would give a less sensitive collision detection scheme sufficient time to react.

In order to show how effectively collision detection and reaction could reduce the impact forces caused by the contribution of the motor, a collision detection and reaction is analyzed in Fig. 9.8. The robot reacts to the detected impact by braking with maximum motor torque as soon as a the collision is observed. For a very compliant robot, there is only the first impact peak remaining. However, for a joint stiffness comparable to the one of the LWR-III, the height of the larger second peak can be diminished to a similar level as the first one.

Introducing joint damping D_J has an interesting influence on the impact characteristics. For a flexible joint robot, motor and link inertia show less decoupling than for the undamped case. However, the maximum value of the force is attenuated compared to the stiff robot. For a VIA system, the damping leads to a larger joint force which decreases the effect of the motor inertia during the second peak. This way, the potential threat to the abdomen is fully eliminated even without any collision detection mechanism.

9.2.10 Analysis of Elbow Checks

According to [3], in professional football 41 % of head injuries result from collisions with the elbow, arm, or hand of the opponent. In the following simulation, results

[7] The problem of impacting in pretensioned state is not part of this analysis.

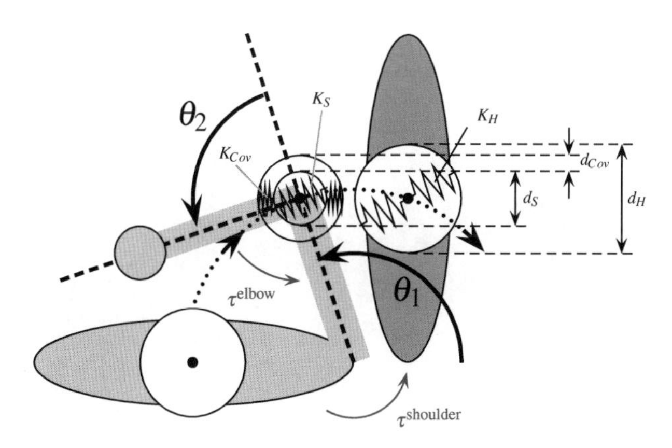

Fig. 9.9 Two-dimensional modeling view from above of an elbow check. The left player hits the right player with his elbow on the head. The elbow is adjusted such that it produces the worst-case impact force for each setting.

point out how dangerous elbow checks generally are. Furthermore, it will be shown that this threat can be reduced to lower levels than presumably caused by humans and even facial fractures can be prevented at all.

Figure 9.9 depicts the model. The human head is represented as a mass-spring system, with a head mass of 4 kg [7], a contact stiffness of $K_H = 10^5$ N/m (maxilla, i.e. upper jaw [12]), and a fracture force of 660 N [2, 29, 9]. The arm/robot that is carrying out the elbow check is represented as a 2R rigid body system [] with inertial parameters of the human arm [7]. The hand mass is assumed to be rigidly attached to the lower arm. The contact stiffness K_S of the robot structure is modeled as the human elbow stiffness which is $K_S = 7 \times 10^5$ N/m during quasi-static bending [20].

In [45], elbow to head impacts were evaluated with human soccer volunteers and a HIII dummy. Impact velocities were 1.7–4.6 m/s. Hence an impact velocity of 3 m/s was chosen and assumed here that the involved players have no relative velocity during the incident. Also the worst elbow angle of $\theta_2 = \frac{\pi}{2}$ was chosen, see Fig. 9.9. The maximum human shoulder and elbow torques according to [14] are

$$(|\tau_{\max}^{\text{shoulder}}|, |\tau_{\max}^{\text{elbow}}|) = (80, 60) \text{ Nm}. \tag{9.2}$$

These are calculated by analyzing baseball pitches during a throw. In order to show the improvement adequate covering could have, the influence of covering thickness and material type on the contact force are analyzed in Fig. 9.10. The elasticity modulus E_{cov} of the covering was chosen to range up to rather hard rubber and its thickness increases up to $d_{\text{cov}} = 0.15$ m.

Without any countermeasure the contact force easily exceeds the fracture tolerance of the human maxilla, see Fig. 9.10. On the other hand, with an ideal collision detection and reaction scheme according to Chapter 3, it is possible to reduce impact forces significantly, even without any covering ($d_{\text{Cov}} = 0$ m) by ≈ 150 N. The feasi-

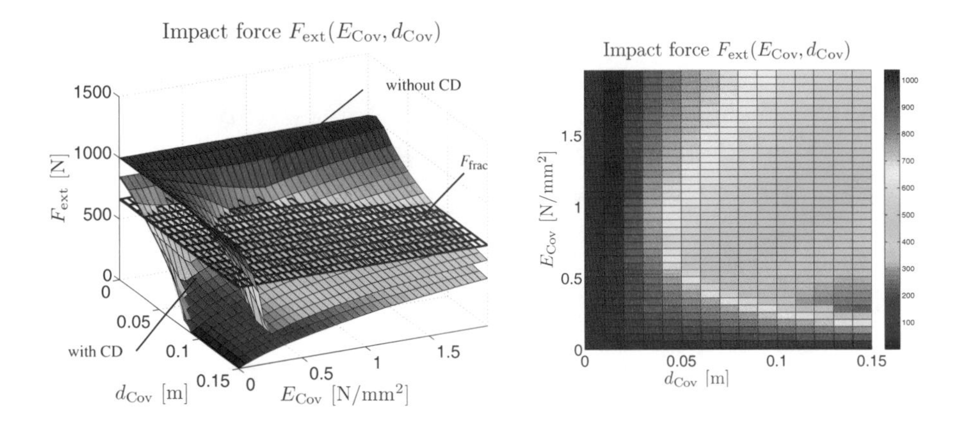

Fig. 9.10 Impact force as a function of covering elasticity modulus and thickness for an elbow check with the maxilla (upper jaw) at 3 m/s. CD indicates whether a collision detection and reaction scheme is activated or not. The reaction consists basically of rapidly "fleeing" from external forces (left). It becomes clear that (without CD) for each specified covering thickness $d_{Cov} \geq 4$ cm should be a different optimal material which is able to provide impact forces below the fracture tolerance (right).

ble reaction torque is bounded by (9.2). Compliant covering is the second effective approach to reduce dynamic impact forces. Particularly interesting is that for each covering thickness an optimal value for the elasticity modulus exists, see Fig. 9.10 (right).

In the simulation, it appears that a good collision detection and reaction scheme is almost as effective in reducing impact forces as providing thick covering. In reality, this is not only limited by the maximum available torque (considered by (9.2)) but also by the full motor dynamics and the corresponding non-ideal motor torques (joint torques in the flexible case). Furthermore, detection delays and system latencies need to be considered which additionally lower the absolute effectiveness of collision detection.

9.3 Robot Joint Protection

In this section, a trend in physical Human-Robot Interaction is discussed that led to the development of novel joint designs incorporating mechanical joint compliance [38] or even VSA. As mentioned in Chapter 2, various control schemes to realize compliance by means of active control are described in the literature. However, motion in sport happens at extreme joint velocities, e.g. 1375 ^{o}/s for instep-kicking [33] or even 6900–9800 ^{o}/s during a baseball pitch [14]. At such velocities, it seems unrealistic to achieve compliance by control, since results in Chapter 3 indicate a limit at much lower velocities for a state of the art robot as the LWR-III. One reason for this is actuator saturation. In this section, it is focused on the situation of an external impact. For a stiff joint, the motor has to immediately stop at impact, leading to an extreme torque that can damage the gears, see Chapter 3. Since the torque is much higher than what the motor can generate, this problem cannot be solved by control but only by mechanical compliance in the joint, which relaxes the requirements posed by motor and gear box.

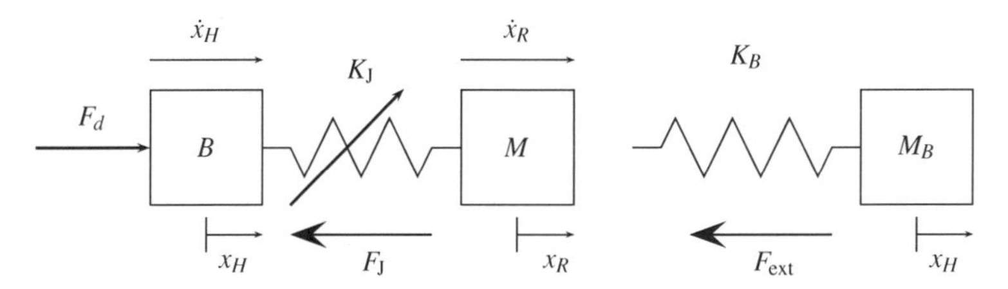

Fig. 9.11 One-dimensional model of kicking a soccer ball with a variable stiffness robot. The robot is modeled as a mass-spring-mass system, representing the motor mass, joint stiffness, and link mass with $B = 13$ kg, $M = 4$ kg, and $K_J \in \{130, 1300, 13000\}$ N/m. The ball is modeled as a mass-spring element with $M_B = 0.45$ kg, and $K_{\text{Ball}} = 43.7$ kN/m. B, M were selected to be the reflected inertias in case of a typical stretched out collision configuration with the LWR-III.

9.3.1 Joint Stiffness and Kicking Force

In order to visualize the effect of joint elasticity on the joint torque, a one-dimensional example is simulated, see Fig. 9.11. It outlines the dramatic decrease of joint torque during an impact with a soccer ball at $\dot{x}_R \in \{2, 4, 10\}$ m/s for a variable stiffness joint. In Fig. 9.12, the impact forces are given, showing that even with reduced joint stiffness they basically stay the same at different kicking velocities[8]. This is again due to the decoupling of link and motor inertia happening already at a moderately high stiffness.

Concerning the load on the joint, one can see that although the contact force F_{ext} stays the same, the joint force F_J decreases dramatically for a joint stiffness reduced by one or two orders of magnitude compared to the LWR-III. A full-robot simulation of this phenomenon is documented in [13]. So one can say that more elasticity helps protecting both robot and human. However, for the human a benefit can be seen only up to the point where motor and link become practically decoupled.

[8] Please note again that impact force refers to the right hand force acting on the link side inertia and joint torque to the elastic joint torque between motor and link.

Now an experimental evaluation of a new variable stiffness joint prototype [46] is discussed with the aim of quantifying the achievable gain in joint protection during kicking a soccer ball.

9.3.2 Kicking a Soccer Ball with the VS-Joint

There are generally two main approaches to realize variable joint compliance. The first one is the biologically motivated antagonistic concept using its two actuators for both position and stiffness adjustment. The second one is to assign one actuator mainly for positioning and the other one for changing the joint stiffness. However, most conclusions made in this chapter can be generalized to both types. The prototype used in this chapter is of the second type and its basic concept is visualized in Fig. 9.13. In Chapter 10 the classification and design of intrinsically compliant joints are discussed in detail. The positioning motor of the DLR VS-Joint is connected to the link via a Harmonic Drive gear. Mechanical compliance is introduced by a mechanism which forms a flexible rotational support between the Harmonic Drive and the joint base. In case of a compliant deflection of the joint due to an external torque, the entire Harmonic Drive gear rotates relative to the base. At the same time the positioning motor does not change its position.

The effect of joint stiffness on the resulting joint torque of the DLR VS-Joint prototype is investigated during impact loading with a soccer ball. When kicking or throwing a ball against the link, it is difficult to reproduce impact position and velocity. Therefore, instead of kicking the ball, the entire setup is moved along a trajectory and hits the soccer ball at a constant velocity. This was achieved by mounting the setup upside down on the TCP of a KUKA Robocoaster, see Fig. 9.14. This robot weighs 2500 kg and can therefore be treated as a velocity source during the following analysis[9]. In this setup, the maximum horizontal velocity is achieved by moving the Robocoaster in an "outstretched" configuration at maximum velocity in its first joint. A wooden shoe-tree in a standard football shoe is attached to the tip of the joint lever. The joint torque $\tau_J \in \mathbb{R}$ is measured ($\tau_{\text{msr}} \in \mathbb{R}$) with a strain gauge torque sensor at the base of the link lever. Furthermore, the joint motor position $\theta \in \mathbb{R}$ and the link lever position $q \in \mathbb{R}$ are measured by rotational encoders. The difference between both is the passive joint deflection $\varphi := \theta - q$. The impact configuration was an instep kick, see Sec. 9.4.3.

The impact tests were carried out at four different impact velocities and with three parameterizations of the torque-deflection function[10], see Fig. 9.15. Two stiffness setups are realized via the passively compliant VS-Joint. The most compliant as well as the stiffest configuration ($\sigma = 0$ and $\sigma = \sigma_{\text{max}}$) are chosen. Depending on the joint deflection, the corresponding stiffness ranges from 0 Nm/rad to 2120 Nm/rad in the

[9] The robot is basically a KR500 as the one used in Chapter 5.

[10] The joint stiffness $K_J(\varphi, \sigma^*) = \frac{\partial \tau_J(\varphi,\sigma)}{\partial \varphi}$ for some stiffness preset σ^* is a highly non-linear function as can be observed in Fig. 9.15.

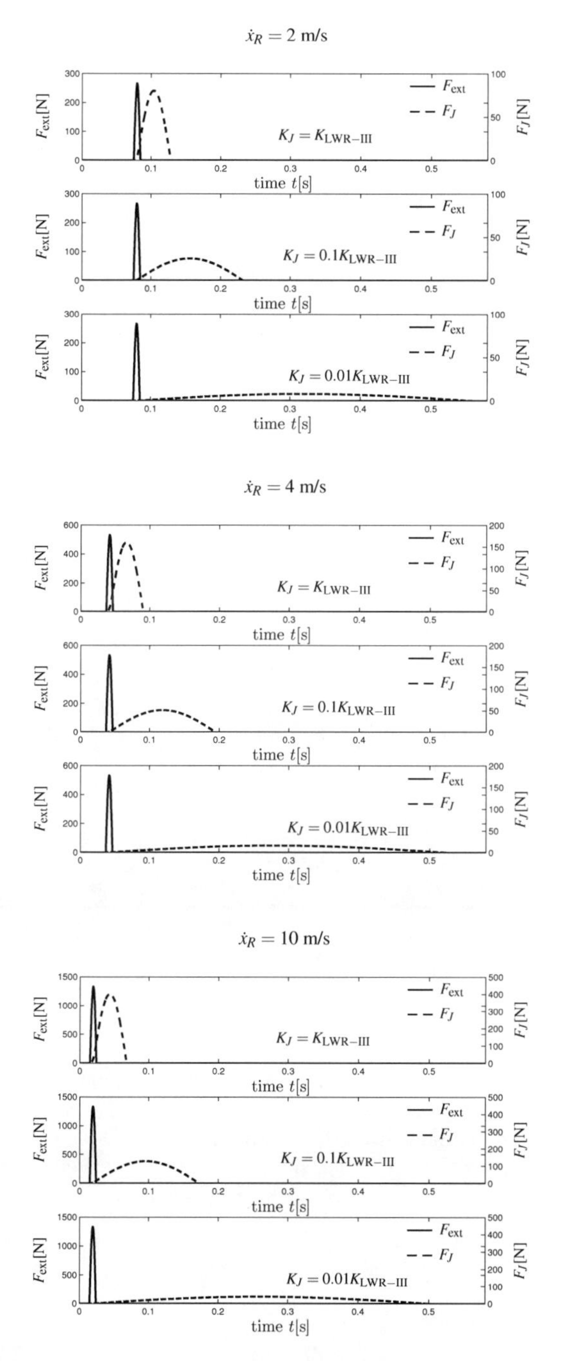

Fig. 9.12 Simulation describing the effect of stiffness reduction on impact force and spring force for a kicking velocity of 2 m/s, 4 m/s, 10 m/s. The solid line indicates the contact force and the dashed line the spring force. The spring force decreases in magnitude and increases in duration when reducing the spring stiffness, whereas the contact force basically stays the same for each particular impact velocity.

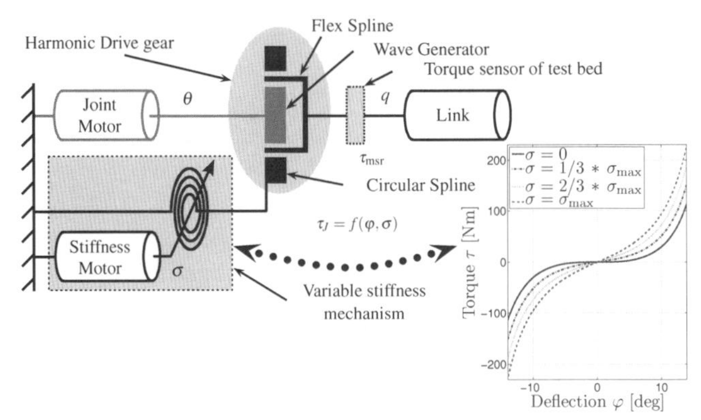

Fig. 9.13 Principle of variable stiffness joint mechanics. The circular spline of the Harmonic Drive gear is supported by the VS-Joint mechanism.

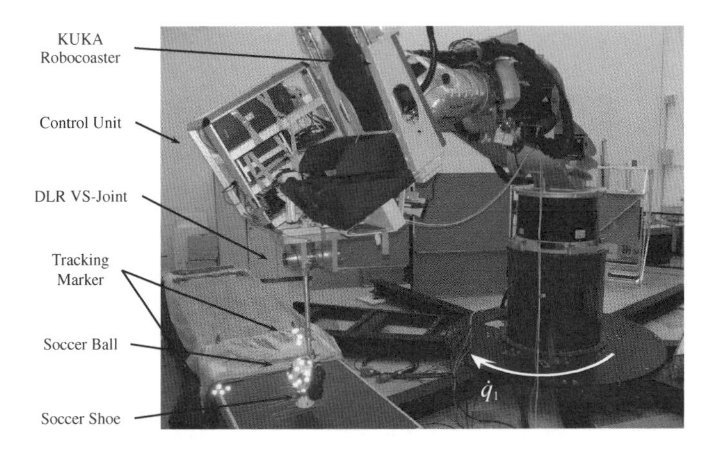

Fig. 9.14 Test setup for hitting the VS-Joint with a soccer ball. The testbed for the VS-Joint is mounted upside down on a KUKA KR500/Robocoaster. The entire joint testbed is moved horizontally with a constant Cartesian velocity of up to 3.7 m/s by the KR500. The link hits the resting ball in non-pretensioned state with an attached foot that is equipped with a standard soccer shoe, see Fig. 9.18. This allows to investigate the effect of the resting joint being hit by a ball in a controlled and reproducible environment.

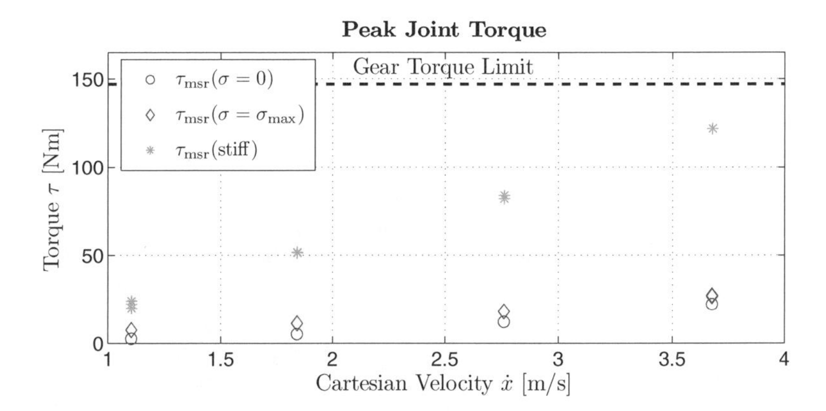

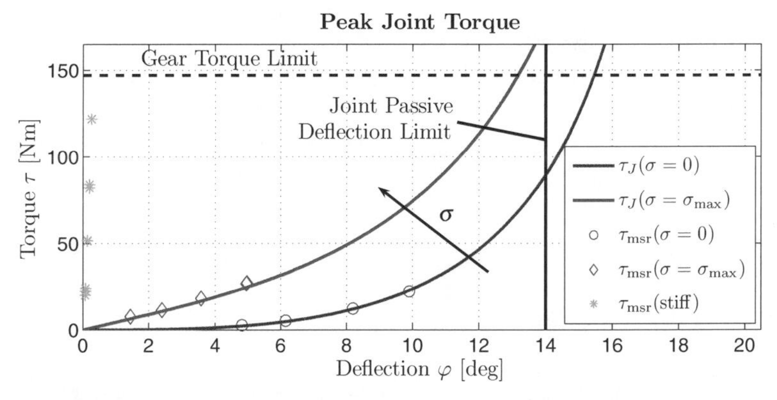

Fig. 9.15 Peak joint torque during impacts with a soccer ball and the VS-Joint. The impact velocity ranges up to the maximum velocity of the KR500/Robocoaster. Three different stiffness setups are examined: VS-Joint at low stiffness preset, VS-Joint at high stiffness preset, and an extremely stiff joint without deliberate elasticity (upper). Peak joint torque during impacts of a soccer ball on the soccer foot mounted on the joint. Higher impact velocities result in larger peak torque and passive joint deflection. At the same speed a soft joint stiffness preset ($\sigma = 0$) causes significantly lower joint torque but higher joint deflection. Therefore, a very soft joint faces a higher risk of running into the deflection limits. Please note that this depends on the joint design, which in this case has a constant deflection limit. In general, φ_{max} can be dependent of σ, but in any case, both maximum torque and $\varphi_{max}(\sigma)$ need to be avoided. For a very stiff joint, the gear torque limit poses an upper bound for the maximum impact velocity. At most two trials were carried out for each velocity and stiffness configuration (lower).

compliant and from 315 Nm/rad to 3150 Nm/rad in the stiffest configuration. In the third setup, a mechanical shortcut is inserted into the testbed instead of the VS-Joint mechanism, leading to a rather stiff intrinsic behavior with $\approx$ 30000 Nm/rad. The numerical value is in the range of the LWR-III elasticity in the first joint which is $\approx$ 20000 Nm/rad.

Both, increasing impact speed and joint stiffness result in higher peak joint torques as visualized in Fig. 9.15 (upper). The maximum peak torque limit of the joint gear is almost reached with the stiff joint at an impact velocity of ≈ 3.7 m/s, whereas the compliant VS-Joint is still far in the safe torque region.

During the impact, a certain amount of kinetic energy is transferred to the joint. Apart from parasitic effects such as friction and damping, the complete transferred energy is stored as potential energy in the joint spring. Increasing impact velocity naturally enlarges the amount of transferred energy. This, in turn, results in increased joint deflection during the impact, see Fig. 9.15 (lower). If the compliant joint has a maximal passive deflection angle, this poses a second safety limit to the joint. Therefore, a trade-off must be made: On the one hand, lower stiffness results in lower peak torques but higher joint deflections and one may run into joint limits. On the other hand, higher stiffness causes higher peak torques and may damage the gears or the structure of the joint itself. The stiffness has to be chosen such that both limits are avoided, if possible.

The preceding evaluation outlined how joint elasticity can effectively reduce high impact joint torques and the related risk of joint damage. In the following, the ability of a VSA to use its inherent physical elasticity as an energy storage and release mechanism is investigated. This feature is especially powerful for achieving very high link speeds, which are necessary for kicking a soccer ball strong enough.

9.4 Robot Performance Improvement

For future soccer robots, kicking a ball at human speed level is a major requirement in order to be serious opponents to their human counterparts (Fig. 9.16, left). This section discusses, how joint elasticity can be used to close the large gap in joint

Fig. 9.16 Kicking a soccer ball at high impact speed (left). A football kick with a KUKA KR500 weighing 2500 kg at maximum velocity. The reflected inertia during such an impact is 1870 kg (right).

velocity between current robots and human soccer players [33]. A general argument in favor of intrinsic joint compliance is its ability to store and release energy

1. for decreasing the energy consumption of the system or
2. to increase peak power output.

The former has received larger attention especially for biped walking [47, 48, 43]. The focus lies on the latter as it allows to considerably increase the link speed [39, 37, 35, 13, 46] above motor speed level.

9.4.1 Kicking in RoboCup

For comparing the results presented in this chapter with the performance of current soccer robots, a short overview of the state of the art regarding ball manipulation abilities in RoboCup is given in the following.

Currently the largest and most powerful (by means of joint torque) humanoid soccer robots play in the Humanoid Teen Size League. In this league, an orange beach handball (size 2; 18 cm diameter, weighing 294 g) is used [25]. The robots have to manipulate the ball using their legs. In most cases, a humanoid leg is constructed as a sequence of six joints which allow – in addition to kicking – omnidirectional walking patterns. The 2007 world champion, team NimbRo from Freiburg, Germany [4], powered these joints with Dynamixel RX-64 servo motors (as several other teams do), which have a holding torque of 6.4 Nm and a maximum velocity of about 360 ^{o}/s (specification from manufacturer) without load. By coupling pairs of these motors in several joints of their robot *Robotina*, the torque is doubled. The knees of this robot are additionally supported by torsional springs. Robotina is able to kick the standard ball at a velocity of about 2 m/s but cannot lift it from the ground significantly.

9.4.2 Required Joint Velocity

In the following, the joint velocity necessary for kicking a ball with the LWR-III at a speed comparable to a human instep kick is calculated. According to [26], the velocity of the ball can be expressed sufficiently accurately by

$$\dot{x}_B = \dot{x}_F \frac{m_F(1+e)}{m_F + m_B}, \tag{9.3}$$

where m_F is the effective striking mass of the foot and $m_B = 0.45$ kg is the ball mass. The coefficient of restitution is $e \approx 0.5$. In [26] the ratio $\frac{m_F}{m_F+m_B}$ is described to be typically 0.8. Since the LWR-III has in outstretched position a reflected inertia of ≈ 4 kg along the impact direction (thus more than twice as large as the human foot), the velocity of the robot's end needs to be $\approx 0.75\dot{x}_B$, leading with 16 m/s $\leq \dot{x}_B \leq 27$ m/s for real kicks to

$$12\ \mathrm{m/s} \leq \dot{x}_F \leq 20.25\ \mathrm{m/s}. \tag{9.4}$$

This corresponds to a joint velocity of 414 ^{o}/s to 700 ^{o}/s, much higher than the maximal joint velocity of the LWR-III (130 ^{o}/s). Due to the smaller reflected inertia of a human foot, humans kick at even higher joint velocities of up to 1375 ^{o}/s for knee extension and with joint torques up to 280 Nm [33]. Kicking a soccer ball at the maximum nominal joint velocity of the LWR-III leads to a ball velocity of $\approx$ 4.5 m/s, i.e. six times slower than required. Even with such a low velocity, the joint torques already become critical (80 % of maximum nominal torque) [13]. This is confirmed by observations made during robot-dummy impacts presented in Chapter 5, where the exceedance of maximum nominal joint torques was observed already at impact velocities of $\approx$ 1 m/s.

9.4.2.1 Kicking with a Heavy-Duty Industrial Robot

In order to show the performance limits of classical actuation by an intuitive experiment, a soccer ball was kicked with a KUKA KR500, one of the world's largest robots (500 kg payload) weighing almost 2500 kg. Maximum joint velocity results in an impact at 3.7 m/s, see Fig. 9.16 (right). Still, the ball hits the ground after a flight of only $\approx$ 2 m. This example gives a good feeling about the large gap in joint velocity between current robots and the RoboCup 2050 challenge requirements and especially supports the claim that increasing robot mass does not significantly enhance kicking performance.

9.4.3 Kicking a Ball with an Elastic Joint

Asimo, currently one of the fastest biped humanoid robots, or the successful robots of Humanoid Team NimbRo kicking a soccer ball reveal a large gap in the kicking performance between current humanoid robots and humans. In this part of the chapter, it is shown how much higher kicking performance is achievable with a single elastic joint. This experiment is not meant as an assessment but to show the potential of elastic joints. The DLR VS-Joint is equipped with an adjustable passive elastic element which serves as an energy storage and release mechanism, see Fig. 9.13. It allows to significantly increase the link speed as pointed out and partially analyzed in [39, 35, 13, 37]. In order to show that the proposed increase in kicking performance is not only achievable for a particular type of kicking, experiments with five basic kicking techniques were conducted, see Fig. 9.17.

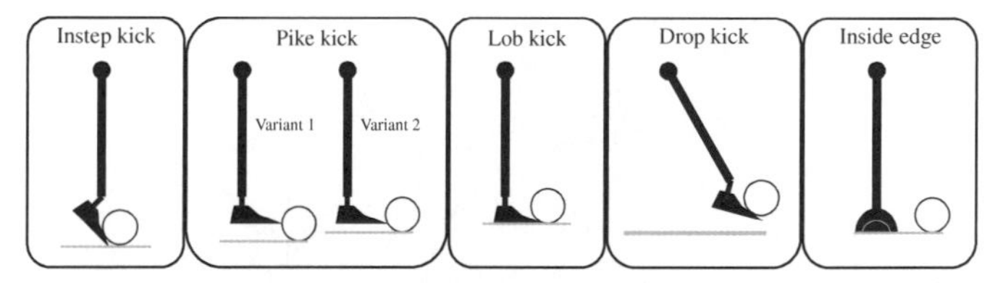

Fig. 9.17 Kicking techniques investigated in this monograph. Only the drop kick allows a foot position below the ball.

9.4.3.1 Kicking Test Setup

In the following, the most common kicking techniques used in soccer are evaluated: *instep*, *pike*, *lob*, and *drop kick* as well as *inside edge pass*. These techniques require appropriate foot angle setups, see Fig. 9.18. For this reason, the foot angle can be changed in two axes. The first axis is concentric to the joint lever. Its angle ϕ_1 is set to $0\,^o$ for all techniques except for the inside edge pass where it is set to $-90\,^o$. The second axis is rotated by $90\,^o$ relative to the first axis and is parallel to the joint axis in case of $\phi_1 = 0$. The angle ϕ_2 of the second axis is changed according to the kick technique. The inertia of the lever and foot is $J \approx 0.57\ \mathrm{kg\,m^2}$ and slightly depending on the foot orientation. The height h_B of the ball can be changed to adjust the position of the contact point between ball and foot. An ART passive marker tracking system was used to track the position of the link S_l and of the ball relative to a world coordinate system S_w. This is done by two 6-DoF markers mounted to the link and to the table, respectively. The coordinate system S_f was identified with the tracking system for each foot position relative to S_l. Furthermore, the surface of the shoe was sampled by grid points relative to S_f. This allows to calculate the contact normal $\mathbf{n}_C$ between the foot and the ball out of the tracking data. The trajectory of the ball is also measured by the tracking system.

9.4.3.2 Kicking Trajectory

The link velocity of a stiff joint is limited by the velocity of the driving motor. In a flexible joint, the potential energy stored in the system can be used to accelerate the link relative to the driving motor. Additionally, potential energy can be inserted by the stiffness adjuster of the variable stiffness joint.

In the experiments presented in this section, a simple strike out trajectory is used, see Fig. 9.19. A motor position ramp accelerates the link backwards to increase its kinetic energy. Then the motor reverts its motion which in turn leads to a transformation of the kinetic link energy into potential energy stored in the VS-Joint spring. The stiffness adjuster starts moving with maximum velocity to the stiffest configuration, increasing the potential energy of the system. The next step is to accelerate the motor up to its maximum velocity, adding kinetic energy to the VS-Joint (this topic is theoretically addressed in Sec. 9.4.4 and Chapter 10.7 in more depth). As

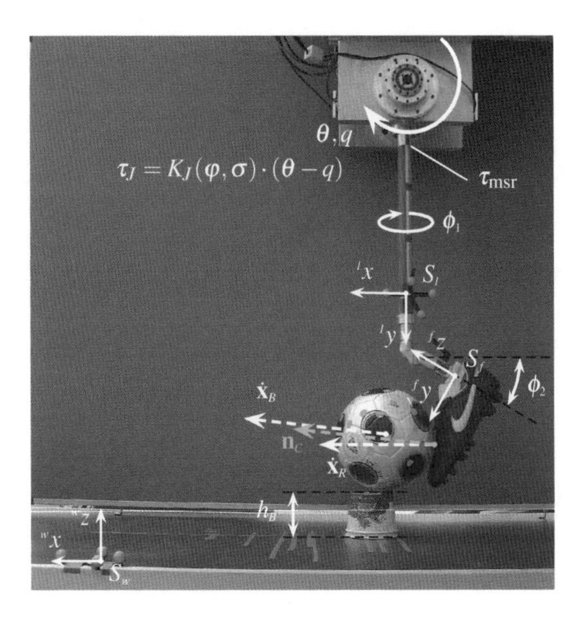

Fig. 9.18 Test setup for kicking a ball depicted for an instep kick. The testbed for the DLR VS-Joint is mounted upside down. The angle ϕ_2 between the foot and the limb (joint lever) is altered by a hinge. The height of the ball h_B is adjusted by the number of piled cups underneath and adjusted according to the investigated kicking technique, see Fig. 9.17. The normal on the contact point between foot and ball is denoted as $\mathbf{n}_C$.

soon as the link starts to catch up with the motor, its velocity increases up to the motor maximum velocity plus a term correlating to the stored potential energy.

9.4.3.3 Experimental Results

With the VS-Joint prototype it is possible to achieve a maximum link velocity of $\dot{q} = 490\ ^o$/s at a motor velocity of $\dot{\theta} = 200\ ^o$/s. This is a speedup of 2.45 compared to the rigid case. All subsequently presented tests with the VS-Joint were carried out at this maximum joint velocity, leading to Cartesian kicking velocities of up to 6.65 m/s (depending on the configuration of the foot). In Table 9.3 the results for the stiff joint and the VS-Joint are given, showing the large increase in kicking performance with the latter. The tests were repeated several times and the resulting ranges for the external force F_{ext}, the kicking range x_{kick}, and the ball velocity $\dot{x}_B$ are given accordingly.

An *instep kick* is characterized by large ball velocities which reached up to 7.5 m/s in the experiments, depending on the angle ϕ_2 between foot and limb (link lever). The impact force is calculated using the dynamic joint model, the torque sensor signal, and the link position signal. Compared to Fig. 9.12, the impact force is smaller. This has two main causes: First, the signal is heavily filtered to obtain the

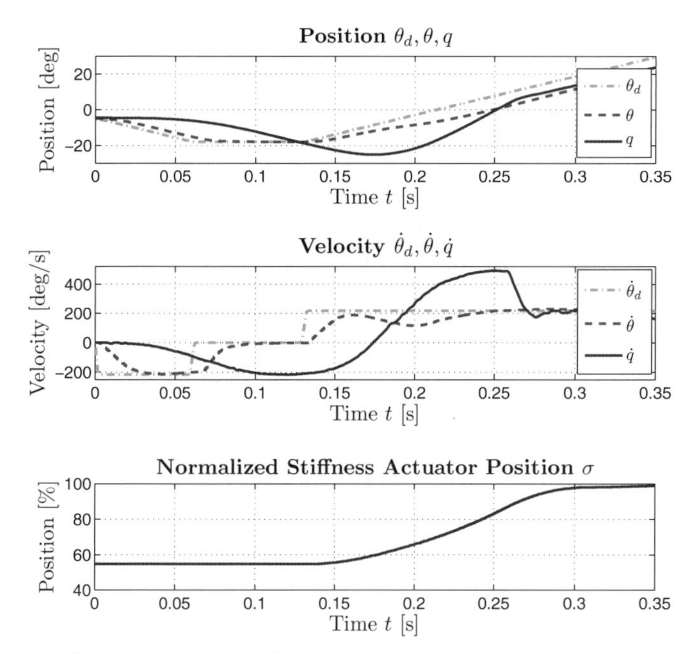

Fig. 9.19 A strike out trajectory of the joint motor in combination with an increase of the stiffness preset are used to gain maximum link velocity

Table 9.3 Results for the different kicks investigated given for the VS-Joint and for the entirely stiff joint

Type	Ball	Variant	# trials	ϕ_2 [^{o}]	$\dot{q}$ [^{o}/s]	$\dot{x}_R$ [m/s]	F_{ext} [N]	x_{kick} [m]	$\dot{x}_B$ [m/s]	h_{kick} [m]
Instep kick	Football	Stiff	1	30	228	3.05	144	–	–	–
Instep kick	Football	VSA	3	30	498	6.65	343 – 359	–	6.6 – 7.5	–
Instep kick	Football	VSA	4	45	490	6.56	387 – 473	–	6.0 – 7.0	–
Instep kick	Football	VSA	3	60	490	6.50	503 – 591	3.40 – 3.65	5.7 – 6.0	–
Pike kick	Football	Stiff 90 ^{o}	1	90	231	3.09	141	0.60	3.0	–
Pike kick	Football	VSA 90 ^{o}	3	90	489	6.20	447 – 503	2.90 – 3.50	8.0 – 10.0	–
Pike kick	Football	Stiff 45 ^{o}	1	90	226	3.02	111	1.43	5.0	–
Pike kick	Football	VSA 45 ^{o}	3	90	489	6.20	548 – 640	3.20 – 3.40	5.5 – 7.7	–
Lob kick	Football	Stiff	1	90	228	3.04	96	–	1.9	0.65
Lob kick	Football	VSA	3	90	488	6.00	374 – 390	–	3.9	0.84
Drop kick	Football	Stiff	1	30	229	3.06	172	1.60	–	–
Drop kick	Football	VSA	3	30	475	6.35	354 – 483	3.80 – 4.05	–	–
Drop kick	Handball	VSA	3	30	477	6.37	389 – 419	3.40 – 3.70	–	–
Drop kick	RoboCup	VSA	4	30	476	6.36	163 – 203	5.90 – 6.30	–	–

link acceleration and second, the radial force component cannot be calculated from the torque signal.

For $\phi_2 = 30\,^o$ and $\phi_2 = 45\,^o$, it is not meaningful to measure x_{kick} since the ball practically does not lift.

Kicking with the pike is mainly varied by the position at which the ball is struck. In this monograph, only the vertical variation is evaluated, because horizontal variation causes spin and is left for future work. Two impact positions are investigated, which were chosen to be perpendicular to the ball surface ($90\,^o$ contact) and hitting the ball at an angle of $45\,^o$ ($45\,^o$ contact), see Fig. 9.17. The impact forces were generally higher compared to the instep kick and the kicking ranges are large as well. This seems mainly to be caused by the rigid contact at the pike.

The lob is basically a pike kick striking the ball at a contact point that is as low as possible, generating a smooth parabolic trajectory, lower ball velocities and contact forces. The main idea behind a lob is to kick the ball beyond the opponent (often the goalkeeper in a direct one to one situation). So one has to lift the ball rapidly very high. The robot was able to kick the ball such that it lifted 0.82 m at a horizontal traveling distance of 0.6 m.

In order to compare the *drop kicks*, the kicking range was measured with three different balls. Apart from the soccer ball, an indoor handball and a plastic RoboCup ball, used in the Standard Platform League, were evaluated. Each ball was hit such that it was contacted at a $45\,^o$ angle. The ball velocities were lower than for the other kicks but at the same time it was possible to shoot up to a distance of 4 m with a football and more than 6 m with the RoboCup ball. The handball was not a beach handball as used in the Humanoid soccer league but an indoor version which is heavier (0.45 kg). It has basically the same weight as a soccer ball but different contact characteristics which is presumably due to the different requirements from the sport itself (kicking vs. dribbling and throwing).

For the *inside edge pass*, the entire foot was rotated to $\phi_1 = -90\,^o$ and ϕ_2 was set to $90\,^o$. The robot was able to kick the ball with the inside edge of the shoe. With this type of kick it is possible to kick the soccer ball the fastest[11] so that it reached maximum velocities of 7.8–9.8 m/s.

While evaluating such a kick in terms of the physical parameters, as done so far in this chapter, is straightforward, evaluating the effectiveness of a kick can be difficult since it depends on the game situation whether it was a success or failure.

After this evaluation of the kicking performance with different techniques, a remarkable observation can be made when comparing the drop kick of a stiff with a VS-Joint by means of speed, kicking range, and impact joint torque. Although the impact speed with a VS-Joint more than doubles and the kicking range can be more than three times higher compared to a stiff joint, the impact joint torque during the observed kicks is only 10 Nm for the VS-Joint in contrast to 85 Nm for the stiff joint. This shows that performance can be increased along with effective joint protection.

[11] This is presumably due to the fact that the surface stiffness of the foot-show complex is the largest at this point. Furthermore, the structural compliances are also lower than for the other kicking configurations.

Fig. 9.20 Comparing the kicking abilities of a 5 year old boy with the DLR VS-Joint prototype. Position and velocity of foot and ball were tracked.

9.4.3.4 Comparison with a Human Child Kick

It is not possible to shoot close to professional level or even comparable to an adult human kick with a single-joint-setup. However, in order to compare performance as a show-case to a real human, a 5 year old boy kicked the soccer ball lying on the ground and on the same height as used for the instep kick, see Fig. 9.20. The leg length of the child is shorter (0.54 m) compared to the prototype link length but he was allowed to kick as hard as possible without any restrictions on the used degrees of freedom of leg and body.

The boy achieved ball velocities of 5 – 6 m/s, comparable to the setup. The kicking length range was 1.5–4.2 m depending on the "quality" of the kick. The foot velocity was relatively constant 10–13 m/s at the time instant of the kick, leading to the conclusion that the reflected inertia is significantly lower than for the robotic setup.

To sum up, in all evaluated cases good kicking performance was obtained and the benefit of the intrinsic joint elasticity was verified. It seems promising to further evaluate the n-DoF case in the future.

9.4.4 Optimal Control for Kicking with an Elastic Joint

In this section it is analyzed theoretically, how much velocity can be gained from using (constant) joint elasticity and what the price is. Therefore, a standard elastic joint model [42] with the motor acting as a pure velocity source is considered. At this point no geometric constraints or non-linear elasticity are considered as it would not contribute to a better understanding of the main idea. In Chapter 10.7 the entire problem is treated from a more general point of view and more complicated models are analyzed. The considered model is

$$\dot{\theta}(t) = u(t), \quad |u(t)| \leq u_{\max} \tag{9.5}$$

$$\ddot{q}(t) = \frac{K_J}{M}(\theta - q) \tag{9.6}$$

$$q(0) = \dot{q}(0) = \theta(0) = \dot{\theta}(0) = 0, \tag{9.7}$$

where $q \in \mathbb{R}$ is the joint position, $\theta \in \mathbb{R}$ the motor position, $K_J \in \mathbb{R}^+$ the joint stiffness, $B \in \mathbb{R}^+$ the link inertia, and $u \in \mathbb{R}$ the control command. Without damping, a mass-spring system can be excited to arbitrarily large oscillations. However, these need time to build up. So the question is asked what is the largest joint velocity that can be achieved within a time t_f, leading to an optimal control problem. To address this, the closed solution of (9.5)-(9.7) is considered.

$$\theta(t_f) = \int_0^{t_f} u(t)dt \tag{9.8}$$

$$q(t_f) = \int_0^{t_f} u(t)\big(1 - \cos(\omega(t_f - t))\big)dt, \tag{9.9}$$

with $\omega = \sqrt{\frac{K_J}{M}}$. It can be verified by taking derivatives of (9.9):

$$\begin{aligned}
\dot{q}(t_f) &= \underbrace{u(t_f)(1 - \cos(0))}_{0} + \omega \int_0^{t_f} u(t)\sin(\omega(t_f - t))dt \\
\ddot{q}(t_f) &= \underbrace{u(t_f)\sin(0)}_{0} + \omega^2 \int_0^{t_f} u(t)\cos(\omega(t_f - t))dt \\
&= \omega^2(\theta - q)
\end{aligned} \tag{9.10}$$

t_f is assumed to be fixed, i.e. the goal is to maximize the joint velocity at a known point in time. Then the integrand of $\dot{q}(t_f)$ in the first equation of (9.10) can be maximized for every t independently by setting $u(t) = u_{\max}\,\mathrm{sgn}\sin(\omega(t_f - t))$ leading to the overall maximum

$$\max_u \dot{q}(t_f) = u_{\max}\omega \int_0^{t_f} |\sin(\omega(t_f - t))|dt \tag{9.11}$$

$$= u_{\max} \int_0^{\omega t_f} |\sin(x)|dx \tag{9.12}$$

$$= u_{\max}\big(2n + 1 - \cos(\omega t_f - n\pi)\big), \tag{9.13}$$

with $n = \lfloor \frac{\omega t_f}{\pi} \rfloor$. The last equation is obtained by splitting (9.12) at multiples of π according to the sign of $\sin(x)$, see Fig. 9.21. Even for $\omega t_f = \pi$, i.e. half a cycle of the spring-mass eigenfrequency, the joint velocity can already be doubled. This is achieved by simply commanding maximum motor velocity, i.e. without any back

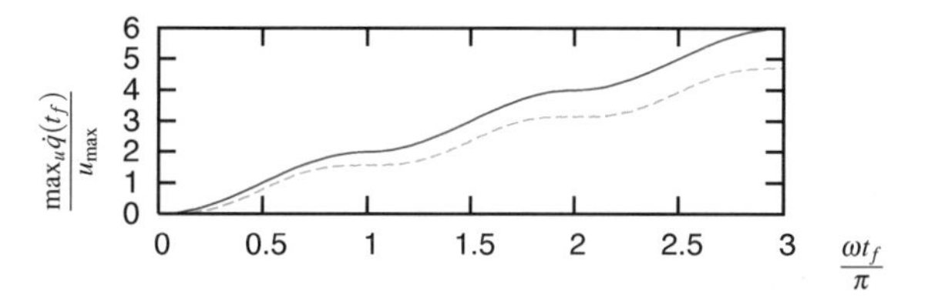

Fig. 9.21 The speedup achievable in time t_f. The X-axis indicates the time $\frac{\omega t_f}{\pi}$ in half-cycles of the spring-mass eigenfrequency. The Y-axis indicates the achievable joint velocity $\frac{\max_u \dot{q}(t_f)}{u_{\max}}$ relative to the motor velocity. The continuous line depicts optimal bang-bang control, the dashed line shows sinusoidal control.

and forth motion. For $\omega t_f = 2\pi$, i.e. a full cycle or going one time back and forth, the joint velocity can be quadrupled. Using more than a full cycle seems unrealistic for soccer as an application.

These results refer to an idealized setting. In reality, the system would involve motor inertia, friction, damping, and torque limits. Damping and friction on the link side reduce the obtainable velocity. However, they are mainly built-up over many cycles, so they create no severe problem. Friction on the motor side only increases the torque required, hence effectively reducing any torque limit. Motor inertia prevents ideal bang-bang control which would require infinite acceleration $\ddot{\theta}$. To analyze this effects, a rather conservative sinusoidal control is evaluate now.

$$u(t) = u_{\max} \sin(\omega(t_f - t)). \tag{9.14}$$

$$\dot{q}(t_f) = u_{\max}\omega \int_0^{t_f} \sin^2(\omega(t_f - t))dt \tag{9.15}$$

$$= u_{\max} \left(\frac{\omega t_f}{2} - \frac{\sin(2\omega t_f)}{4} \right) \tag{9.16}$$

As Fig. 9.21 (dashed line) shows, the speedup reduces from 2 and 4 to $\frac{\pi}{2}$ and π respectively but is still substantial.

Torque limits have an important effect that can be seen from the energy balance. A motor with limited velocity and torque can only generate limited power and hence energy can only be build up $\sim t_f$ and velocity only $\sim \sqrt{t_f}$. As both control policies discussed above result in a linear built-up of velocity, they will at some point exceed the motor's torque limit. A detailed analysis of this problem is provided in Chapter 10.7.

When comparing these theoretical results in Fig. 9.21 to the practical ones shown in Fig. 9.19, some caution is needed. The experiments there show a back-and-forth motion, roughly corresponding to $\frac{\omega t_f}{\pi} = 2$. So a factor of 4 could be achieved with an ideal velocity source, or $\pi \approx 3.14$ with sinusoidal control. In the experiments, only a factor of 2.45 has been achieved. However, $\dot{\theta}$ in Fig. 9.19 is far from being sinusoidal, let alone from an ideal step trajectory. To the rough extend, the experiments correspond to the theory expected from the simple model (9.5)-(9.7).

Another problem arising from the elasticity can be seen in (9.13) near $t_f = 0$, see Fig. 9.21. The term $1 - \cos(\omega t_f)$ has 0 derivative there. In this time, no velocity increase can be obtained. This is the well known problem that elasticity in the joints reduces joint dynamics. Overall, there are some other problems in using elasticity to increase velocity as e.g. the elastic deflection limit, how to adjust the joint stiffness for the VIA case, and the limited motor dynamics. However, for Competitive Robotics, the obtainable gains far outweigh these problems.

9.5 Summary

In this chapter, safety and performance challenges imposed by the RoboCup 2050 vision of a human-robot soccer match were analyzed. An attempt for a pHRI view on human-robot soccer was taken. For this scenes from real soccer matches were selected and discussed what could have happened if one of the teams consisted of robots instead of humans. The interaction scenarios were grouped and solutions for resolving or attenuating the corresponding safety problems were pointed out. A key finding is the necessity of a new actuation paradigm, including elasticity (i.e. mechanical compliance) in the robot joints. This contributes to three important challenges of human-robot soccer:

Safety of the Human

Joint elasticity decouples motor and link inertia. Hence, someone hit by the robot feels only the impact of the link at first. The impact of the motor inertia transmitted through joint stiffness is delayed and less severe than for the rapid case. It can be further reduced by a collision detection mechanism[12]. Even though this has been shown in Chapter 5 to be already the case for the LWR-III, the next two aspects necessitate to further increase the decoupling effect. Therefore, the deliberate introduction of strongly elastic elements becomes necessary.

Protection of the Robot

The decoupling effect described above also protects the robot during an impact, as it gives the motor more time to react e.g. by decelerating. This reduces the peak gear torque, avoiding gear damage. Figuratively speaking, if a stiff robot bangs its fist on a table, it could hurt its shoulder. Joint elasticity prevents this. The benefit for the robot is therefore even higher than for the human.

[12] For this strategy to be effective, singularities must be avoided.

Robot Performance

Stiffness elements can store and release energy. Thereby allowing the robot to increase the link velocity to a multiple of the maximum motor velocity. This makes motion control, in particular walking, much more difficult, but helps to close the large gap in peak joint speed performance between humans and robots, for the case that inertial, payload, and structural properties should be in a similar range for robots and humans.

As a further contribution the effectiveness of padding by means of a biomechanical worst-case injury study was evaluated, leading to a discussion about its desired mechanical characteristics.

The results of the present chapter are used in Chapter 10 for further investigation of the aforementioned three challenges. However, the focus in Chapter 10 is on physical Human-Robot Interaction and extending the optimal control methodology.

References

[1] Albu-Schäffer, A., Eiberger, O., Grebenstein, M., Haddadin, S., Ott, C., Wimböck, T., Wolf, S., Hirzinger, G.: Soft robotics: From torque feedback controlled lightweight robots to intrinsically compliant systems. IEEE Robotics and Automation Mag.: Special Issue on Adaptable Compliance/Variable Stiffness for Robotic Applications 15(3), 20–30 (2008)

[2] Allsop, D., Warner, C., Wille, M., Schneider, D., Nahum, A.: Facial impact response - a comparison of the Hybrid III dummy and human cadaver. SAE Paper No.881719, Proc. 32th Stapp Car Crash Conf., pp. 781–797 (1988)

[3] Andersen, T.E., Árnason, A., Engebretsen, L., Bahr, R.: Mechanisms of head injuries in elite football. British Journal Sports Medicine 38, 690–696 (2004)

[4] Behnke, S., Schreiber, M., Stückler, J., Schulz, H., Böhnert, M., Meier, K.: NimbRo teensize 2008 team description. In: RoboCup 2008: Robot Soccer World Cup XII Preproceedings (2008)

[5] Burkhard, H., Duhaut, D., Fujita, M., Lima, P., Murphy, R., Rojas, R.: The road to RoboCup 2050. IEEE Robotics and Automation Mag. 9(2), 31–38 (2002)

[6] Cavanaugh, J., Nyquist, G., Goldberg, S., King, A.: Lower abdominal impact tolerance and response. SAE Paper No.861878, Proc. 30th Stapp Car Crash Conf. (1986)

[7] Chandler, R., Clauser, C., McConville, J., Reynolds, H., Young, J.: Investigation of inertial properties of the human body. Tech. Rep. DOT HS-801 430, Aerospace Medical Research Laboratory (1975)

[8] Durašković, R., Joksimović, A., Joksimović, S.: Weight-height parameters of the 2002 world football championship participants. Physical Education and Sport 2, 13–24 (2004)

[9] European Commission Framework: Improved frontal impact protection through a world frontal impact dummy. Project No. GRD1 1999-10559 (2003)

[10] Fédération Internationale de Football Association: Laws of the game 2006 (2006)

[11] Grebenstein, M., van der Smagt, P.: Antagonism for a highly anthropomorphic hand-arm system. Advanded Robotics 22(1), 39–55 (2008)
[12] Haddadin, S., Albu-Schäffer, A., Hirzinger, G.: The role of the robot mass and velocity in physical human-robot interaction - part I: Unconstrained blunt impacts. In: IEEE Int. Conf. on Robotics and Automation (ICRA 2008), Pasadena, USA, pp. 1331–1338 (2008)
[13] Haddadin, S., Laue, T., Frese, U., Hirzinger, G.: Foul 2050: Thoughts on physical interaction in human-robot soccer. In: IEEE/RSJ Int. Conf. on Intelligent Robots and Systems (IROS 2007), San Diego, USA, pp. 3243–3250 (2007)
[14] Herman, I.: Physics of the Human Body. Springer (2007)
[15] Hirai, K., Hirose, M., Haikawa, Y., Takenaka, T.: The development of honda humanoid robot. In: IEEE Int. Conf. on Robotics and Automation (ICRA 1998), Leuven, Belgium, pp. 1321–1326 (1998)
[16] Hirose, M., Haikawa, Y., Takenaka, T., Hirai, K.: Development of humanoid robot ASIMO. In: IEEE/RSJ Int. Conf. on Intelligent Robots and Systems (IROS 2001): Workshop 2, Maui, USA (2001)
[17] Hobbelen, D., de Boer, T., Wisse, M.: System overview of bipedal robots flame and tulip: Tailor-made for limit cycle walking. In: IEEE/RSJ Int. Conf. on Intelligent Robots and Systems (IROS 2008), Nice, France, pp. 2486–2491 (2008)
[18] Johannsen, H., Schindler, V.: Review of the abdomen injury criteria. Tech. Rep. AP-SP51-0039-B, Institut National de Reserche sur les Transports et leur Sécurité (2005)
[19] Kajita, S., Kaneko, K., Morisawa, M., Nakaoka, S., Hirukawa, H.: ZMP-based biped running enhanced by toe springs. In: Int. Conf. on Robotics and Automation (ICRA 2007), Rome, Italy, pp. 3363–3369 (2007)
[20] Kallieris, D., Rizzetti, A., Mattern, R., Jost, S., Priemer, P., Unger, M.: Response and vulnerability of the upper arm through side air bag deployment. SAE Transactions 120, 143–152 (2004)
[21] Kaneko, K., Kanehiro, F., Kajita, S., Hirukawa, H., Kawasaki, T., Hirata, M., Akachi, K., Isozumi, T.: Humanoid robot HRP-2. In: Int. Conf. on Robotics and Automation (ICRA 2004), New Orleans, USA, pp. 1083–1090 (2004)
[22] Kim, J.Y., Park, I.W., Lee, J., Kim, M., Cho, B., Oh, J.H.: System design and dynamic walking of humanoid robot KHR-2. In: Int. Conf. on Robotics and Automation (ICRA 2005), Barcelona, Spain, pp. 1443–1448 (2005)
[23] Kitano, H., Asada, M.: RoboCup humanoid challenge: That's one small step for a robot, one giant leap for mankind. In: IEEE/RSJ Int. Conf. on Intelligent Robots and Systems (IROS 1998), Victoria, Canada, pp. 419–424 (1998)
[24] Knobloch, K., Rossner, D., Jagodzinski, M., Zeichen, J., Gössling, T., Martin-Schmitt, S., Richter, M., Krettek, C.: Prävention von Schulsportverletzungen - Analyse von Ballsportarten bei 2234 Verletzungen. Sportverletzungen Sportschäden 19, 82–88 (2005) (German)
[25] Kulvanit, P., von Stryk, O.: Robocup soccer humanoid league rules and setup for the 2008 competition in Suzhou, China (2008)
[26] Lees, A., Nolan, L.: The biomechanics of soccer: A review. Journal of Sport Sciences 16(3), 211–234 (1998)
[27] Lohmeier, S., Buschmann, T., Ulbrich, H., Pfeiffer, F.: Modular joint design for a performance enhanced humanoid robot. In: IEEE Int. Conf. on Robotics and Automation (ICRA 2006), Orlando, USA, pp. 88–93 (2006)

[28] McGrath, A., Ozanne-Smith, J.: Heading injuries out of soccer: A review of the literature. Tech. Rep. 125, Monash University Accident Research Center (1997)
[29] Melvin, J.: Human tolerance to impact conditions as related to motor vehicle design. SAE J885 APR80 (1980)
[30] Newman, J., Shewchenko, N., Welbourne, E.: A proposed new biomechanical head injury assessment function - the maximum power index. Stapp Car Crash Journal, SAE paper 2000-01-SC16 44, 215–247 (2000)
[31] Nigg, B., MacIntosh, B., Mester, J. (eds.): Biomechanics and biology of movement. Human Kinetics Pub. Inc., Champaign(2000)
[32] Nishiwaki, K., Sugihara, T., Kagami, S., Kanehiro, F., Inaba, M., Inoue, H.: Design and development of research platform for perception-action integration in humanoid robot: H6. In: IEEE/RSJ Int. Conf. on Intelligent Robots and Systems (IROS 2000), Takamatsu, Japan, pp. 1559–1564 (2000)
[33] Nunome, H., Asai, T., Ikegamiand, Y., Sakurai, S.: Three-dimensional kinetic analysis of side-foot and instep soccer kicks. Medicine & Science in Sports & Excercise 34(12), 2028–2036 (2002)
[34] Ogura, Y., Aikawa, H., Shimomura, K., Kondo, H., Morishima, A., Lim, H.O., Takanishi, A.: Development of a new humanoid robot WABIAN-2. In: IEEE Int. Conf. on Robotics and Automation (ICRA 2006), Orlando, USA, pp. 76–81 (2006)
[35] Okada, M., Ban, S., Nakamura, Y.: Skill of compliance with controlled charging/discharging of kinetic energy. In: IEEE Int. Conf. on Robotics and Automation (ICRA 2002), Washington, USA, pp. 2455–2460 (2002)
[36] Ott, C., Baumgärtner, C., Mayr, J., Fuchs, M., Burger, R., Lee, D., Eiberger, O., Albu-Schäffer, A., Grebenstein, M., Hirzinger, G.: Development of a biped robot with torque controlled joints. In: IEEE-RAS International Conference on Humanoid Robots (HUMANOIDS 2010), Nashville, USA (2010)
[37] Paluska, D., Herr, H.: The effect of series elasticity on actuator power and work output: Implications for robotic and prosthetic joint design. Robotics and Autonomous Systems 54, 667–673 (2006)
[38] Pratt, G., Williamson, M.: Series elastics actuators. In: IEEE/RSJ Int. Conf. on Intelligent Robots and Systems 1995 (IROS 1995), Victoria, Canada, pp. 399–406 (1995)
[39] Schempf, H., Kraeuter, C., Blackwell, M.: ROBOLEG: A robotic soccer-ball kicking leg. In: IEEE Int. Conf. on Robotics and Automation (ICRA 1995), Nagoya, Aichi, Japan, vol. 2, pp. 1314–1318 (1995)
[40] Shewchenko, N., Withnall, N., Keown, M., Gittens, R., Dvorak, J.: Heading in football. part 1: Development of biomechanical methods to investigate head response. British Journal Sports Medicine 39, 10–25 (2005)
[41] Shirata, S., Konno, A., Uchiyama, M.: Design and development of a light-weight biped humanoid robot Saika-4. In: IEEE/RSJ Int. Conf. on Intelligent Robots and Systems (IROS 2004), Sendai, Japan, pp. 148–153 (2004)
[42] Spong, M.: Modeling and control of elastic joint robots. IEEE Journal of Robotics and Automation, 291–300 (1987)
[43] Vanderborght, B., Verrelst, B., Ham, R.V., Damme, M.V., Lefeber, D., Duran, B., Beyl, P.: Exploiting natural dynamics to reduce energy consumption by controlling the compliance of soft actuators. The Int. J. of Robotics Research 25(4), 343–358 (2006)
[44] Wesson, J.: The Science of Soccer. IOP Publishing Ltd., Dirac House (2002)
[45] Withnall, C., Shewchenko, N., Gittens, R., Dvorak, J.: Biomechanical investigation of head impacts in football. British Journal of Sports Medicine 39, 49–57 (2005)

[46] Wolf, S., Hirzinger, G.: A new variable stiffness design: Matching requirements of the next robot generation. In: IEEE Int. Conf. on Robotics and Automation (ICRA 2008), Pasadena, USA, pp. 1741–1746 (2008)
[47] Yamaguchi, J., Inoue, S., Nishino, D., Takanishi, A.: Development of a bipedal humanoid robot having antagonistic driven joints and three DOF trunk. In: IEEE/RSJ Int. Conf. on Intelligent Robots and Systems (IROS 1998), Victoria, B.C., Canada, pp. 96–101 (1998)
[48] Yamaguchi, J., Nishino, D., Takanishi, A.: Realization of dynamic biped walking varying joint stiffness using antagonistic driven joints. In: IEEE Int. Conf. on Robotics and Automation (ICRA 1998), Leuven, Belgium, pp. 2022–2029 (1998)

Chapter 10
Intrinsic Joint Compliance

Human-friendly robots are usually characterized either by active compliance control or intrinsically compliant behavior. Active compliance control has already reached a mature stage and recently went to market. Intrinsic compliance on the other hand is currently investigated in several large European projects and other research projects worldwide. Due to the significant increase in mechanical design complexity, the additional degrees of freedom needed for adjusting stiffness and related questions regarding control, there are still several open issues to be addressed in order to validate the VIA concept. DLR is currently developing an integrated hand-arm system [1, 13], which will be fully equipped with variable stiffness actuation, see Fig. 10.1.

In this chapter, first general design considerations for intrinsically compliant joints are presented, leading to a new design concept, the Quasi-Antagonistic Joint (QA-Joint). The approach has an elastically coupled drive unit with variable stiffness achieved via superposition of antagonistic torque/displacement characteristics. Furthermore, velocity gain and joint protection capabilities due to the inherent

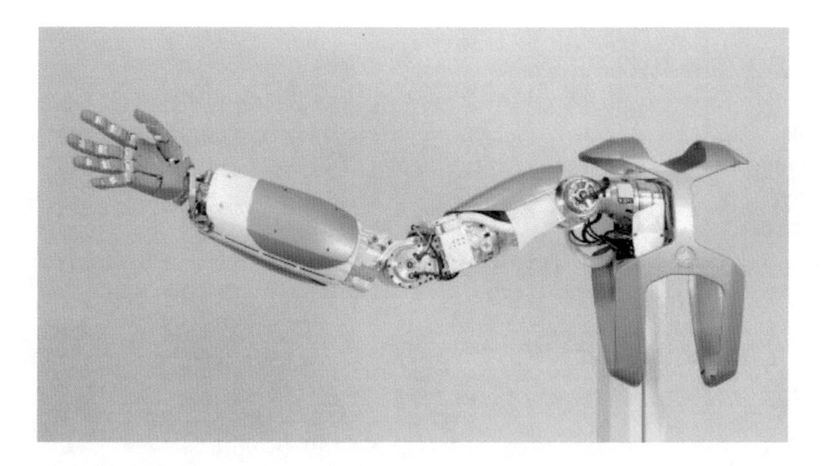

Fig. 10.1 The DLR hand-arm system

S. Haddadin: *Towards Safe Robots*,
Springer Tracts in Advanced Robotics 90,
DOI: 10.1007/978-3-642-40308-8_10, © Springer-Verlag Berlin Heidelberg 2014

elastic behavior of such mechanisms are investigated in detail, and the results are supported by numerous experiments. This is a continuation to the analysis in the previous chapter, extending the ideas significantly and contributing more theoretical background. Then, the effect joint stiffness has on safety properties of the robot is analyzed by means of intrinsic behavior and control schemes. Based on the results from Chapter 9, the theoretical insights are also verified by several simulations and experiments. It is shown that the discussion in literature in favor of intrinsically compliant actuation has left out important aspects, which to a certain extent contradict the paradigm of realizing safety by compliance. On the other hand, the circumstances under which this assumption actually holds are outlined and where a large benefit can be obtained.

The chapter is organized as follows. Section 10.1 gives an overview of existing intrinsically compliant joint designs. Section 10.2 introduces the design considerations for realizing these novel mechanisms. A novel joint design, its modeling, identification, and control are given in Sec. 10.3. Section 10.4 provides basic optimal control theory, Sec. 10.5 the corresponding numerical treatment of optimal control problems and Sec. 10.6 gives a more detailed description of the Nelder-Mead optimization algorithm. All of those previously mentioned sections are necessary to fully understand Section 10.7, where the achievable results in velocity increase are outlined in. Section 10.8 provides insights into the safety characteristics of intrinsically compliant joints. Various simulation and experimental results on the impact performance and characteristics of the prototypes are presented. Furthermore, the theoretical background for collision detection with such devices based on the results for flexible joint robots in Chapter 3.3 is introduced. Finally, the performance in joint protection during highly dynamic impacts is proven and the chapter concludes with Sec. 10.11.

10.1 Intrinsically Compliant Actuation

Since the early 1980's, different approaches were made to realize compliant joint coupling. The motivation originated mainly from using inherent elasticity to achieve stable behavior during hard contact, protecting the joints from impact shock, and storing elastic energy e.g. for energy efficient motions.

Table 10.1 Classification of intrinsically compliant joint architectures. *Distinct stiffness actuator* denotes whether one actuator is exclusively used for stiffness adjustment.

Example	SEA [33]	MIA [27]	MACCEPA [37], DLR VS-Joint[38]	McKibben [8]	GATECH [26]	VSA [4], VSA-II[35]	AMASC [22], DLR QA-Joint
Setup	Serial Spring	Serial tunable spring	Symmetric spring, progressive trans.	Antagonistic	Antagonistic	Antagonistic Push-Pull	Quasi-Antagonistic
Stiffness variation	Constant	Constant	Progressive	Progressive	Progressive	Progressive	Progressive
Adjustable stiffness	No	Yes	Yes	Yes	Yes	Yes	Yes
Stiffness characteristics variation type	No	Spring constant	Preload	Preload	Superposition	Superposition/ Double	Superposition
Distinct stiffness actuator	No	Yes	Yes	No	No	No	Yes

In an intrinsically compliant joint mechanism the relationship between the elastic force $F_E \in \mathbb{R}$, acting along the generalized displacement coordinate $x_E \in \mathbb{R}$ directly related to the axis of the compliant element, to the elastic joint torque $\tau_J \in \mathbb{R}$ is a possibly nonlinear transformation. In this chapter, only linear stiffness elements are treated, which are producing nonlinear output behavior via a nonlinear transmission. A selection of designs, known from literature and falling into this category, is shown in Tab. 10.1. Roughly, this wide array of different technical realization can be grouped into two main branches of development, namely

1. preload variable design, and
2. transmission variable design.

The *preload variable* branch evolved from constant stiffness towards symmetrically acting progressive stiffness assemblies as exemplified in Tab. 10.1. The *transmission variable* group showcases the development from human-like antagonistic actuation towards related actuation mechanisms that use superposition of torque/displacement characteristics for stiffness variation. Simultaneously, they intend to overcome the drawbacks of equally sized drives for opposing directions. On the one hand, the number of parts and expected complexity of this line of developments appears to be larger than for the preload variable type. On the other hand, the superposition of individual characteristics allows for new ways to influence the overall behavior of the mechanism.

10.2 Design Considerations

The human has the ability to co-contract his muscles for reacting with appropriate stiffness response to perturbations and to relax them almost instantaneously to become fully backdriveable. This is especially useful during high-performance tasks as throwing a ball or evading from external forces to prevent muscle damage due to overload. To mimic such capabilities, an electromechanical system requires series elastic coupling with variable impedance and high backdriveability. Current systems barely fulfill all requirements at the same time and are clearly outperformed by human actuation e.g. by means of load-to-weight ratio, payload, and speed capabilities. In this sense, the following general requirements are discussed, which are believed to be important in order to approach to human-like actuation performance. A joint design space is elaborated, taking into account external influences as well as internal relations of different design aspects.

The most characteristic properties of a robot joint are

- the maximum (stall) joint torque $\tau_{J,\max}$ and
- the maximum (transient) joint speed $\dot{\theta}_{\max}$.

In contrast to a stiff robot, additional design aspects have to be considered for intrinsically compliant mechanisms:

- Joint elastic deflection range $\varphi_{\max}$: possible range of elastic motion within mechanical limits
- Joint stiffness range K_J: appropriate shape and limits
- Energy storage capacity $E_{J,max}$: for energy absorption and dynamic tasks

For a passively elastic robot joint its characteristics can generally be visualized by two specific graphs. The torque-deflection (Fig. 10.2 (left)) and stiffness-torque (Fig. 10.2 (right)) plots are suitable for determining desired properties of a compliant mechanism.

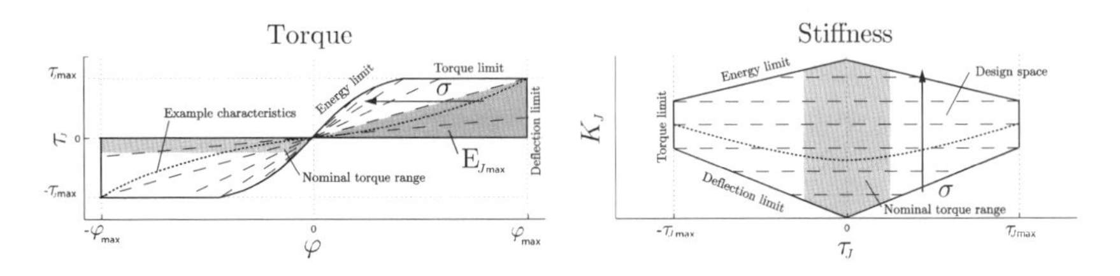

Fig. 10.2 Design space for torque displacement curves (left). Stiffness over torque (right).

In Figure 10.2 (left), limits due to maximum joint torque, maximum elastic deflection, and maximum potential energy span an elastic design space, in which the characteristics of the centering torque τ_J over passive deflection φ can be plotted, see Fig. 10.5. Stored potential energy through deflection is visualized as the area below a torque deflection curve. Consider the case of adjustable linear joint stiffness (dashed lines) with constant maximum deflection $\varphi_{\max}$ for all σ as the ideal joint. For realizing this, the aforementioned constraints limit the practically achievable design:

- Due to mechanical torque limits, there exists a maximum $\tau_{J,\max}$.
- Therefore, the maximum deflection $\varphi_{\max}$ is a function of σ.
- As a further consequence, the maximum joint torque is a function of σ.
- This in turn leads to higher strike-through risk for very low stiffness due to low energy storage and low maximum torque at the same time.
- In general, the *energy limit* is mainly caused by limited deflection $x_{E,\max}$ of physical springs.
- Consequently, the amount of energy required for stiffening the joint by internal tension is no longer accessible for further elastic deflection.

The second characteristic graph, depicted in Figure 10.2 (right), shows the stiffness characteristics K_J of the elastic element with respect to joint torque in the same design space limits. In particular, this plot visualizes the achievable stiffness under a given load. Again, it is considered desirable to achieve variable constant stiffness (dashed lines), especially in the so called *nominal torque range*.

In the following, some essential aspects regarding the above mentioned properties are addressed.

- *Joint torque*. It is desirable for an elastic joint mechanism to maintain the torque capacity for the entire stiffness preset range as steady as possible. Most elastic mechanisms show decreasing torque capacity in stiff operation mode due to internal spring preload.
- *Elastic motion range*. Since a robot has a limited range of motion, the maximum elastic deflection of the joints needs to be considered. The maximum joint torque required to prevent strike through must always be attainable before the mechanical limit, either by limiting θ to $q_{\max} - \varphi_{\max}$ and/or through active reaction schemes. In relation to the expected motion range and considering the range extension obtainable by reactive motion, a maximum elastic deflection of 15 ^{o} appears to be appropriate for humanoid arm joints.
- *Joint stiffness*. The predominant external load is expected to be within ≈ 25 % of the maximum joint torque $\tau_{J,\max}$, when assuming general manipulation tasks under gravity influence without major accelerations. In this nominal torque range, it is especially desired to be able to alter the joint stiffness over a wide range of values to cover differing stiffness demands. Since the external load may vary as a result of pose changes as well as due to reaction forces during contact, it is desirable to maintain constant stiffness behavior under varying load, easing manipulation tasks and simplifying control schemes.
- *Minimum stiffness*. In case of obtaining joint torque information by measuring deflection, zero joint stiffness is not considered as desirable, because torque information would be lost[1]. The same happens with controllability of the joint. In particular, it might not be restored quickly enough to ensure short reaction times.
- *Maximum stiffness*. Since one of the major purposes of elastic joints is robot and environment protection, limiting the maximum stiffness is an important issue. The maximum stiffness significantly defines the chance for reaction in case of an impact. Thus, it influences the available load capacity for heavy manipulation tasks, demanding safety reserves in deflection to sustain collisions. A relative collision, leading to $\dot{q}_c = \dot{\theta}_{\max}$, where $\dot{\theta}_{\max}$ is the maximum motor velocity, relates to the worst-case time t_{cr} required to react as

$$t_{cr} = t_{cd} + \frac{B\dot{\theta}_{\max}}{\tau_m}, \tag{10.1}$$

where t_{cr}, t_{cd} are the collision reaction and collision detection time, while B and τ_m are the motor inertia and motor torque. A purely geometric minimum deflection reserve is obtained as

$$\varphi_{\text{res}} = t_c \dot{\theta}_{\max}. \tag{10.2}$$

[1] Please note that this argument is purely motivated from a classical control point of view. In order to achieve highly dynamic/explosive motions, this presumably needs to be reconsidered. However, the usage of pure inertial "double pendulum" effects for these motions still needs to be investigated in more detail.

To be able to utilize most of the joint maximum torque (e.g. $1-c=95$ %), maximum stiffness has to be limited such that

$$\int_{\sigma-\varphi_{\text{res}}}^{\sigma} \mathrm{K}(\sigma,\varphi)\mathrm{d}\varphi \leq c\tau_{J,\max}. \tag{10.3}$$

These conditions determine the relationship between applicable load with safe speed and stiffness.

- *Energy storage*. The joint elasticity can be used for absorbing kinetic energy of an impact or during catching heavy objects. It can also be used for additional acceleration of the link [38, 19] by appropriate motion (see Sec. 10.7). However, the stored energy may also cause unwanted acceleration. This is the case when losing contact to an object or due to malfunction. Thus, the energy level should be kept moderate and the active reaction of the motors has to be fast enough to prevent severe damage in case of faults.

The properties described above influence the choice of the torque displacement characteristics significantly. However, they cannot all be maximized at the same time. In Table 10.2 the influence of selected torque/deflection characteristics is quantified, comparing rational, low progressive exponential, and quadratic torque displacement curves.

Table 10.2 Comparison of torque characteristics for different elastic characteristics

Characteristics	$\frac{1}{\varphi}$	e^{φ}	φ^2
Constant stiffness	$-$	$+$	$++$
Minimum stiffness	$--$	$-$	$+$
Maximum stiffness	$++$	$+$	$--$
Spring Energy	$-$	$+$	$++$
Joint protection	$\pm$	$+$	$-$

In the following, the resulting joint design and model of the designed joint prototype is outlined.

10.3 Joint Design, Modeling, Identification, and Control

For the technical realization of the joint it is important to achieve a compact design and lightweight structure for low inertia and thus high bandwidth of the robot. Furthermore, it is crucial for most control features developed at DLR to provide high quality torque feedback, which implies low friction and low hysteresis in the compliant mechanism.

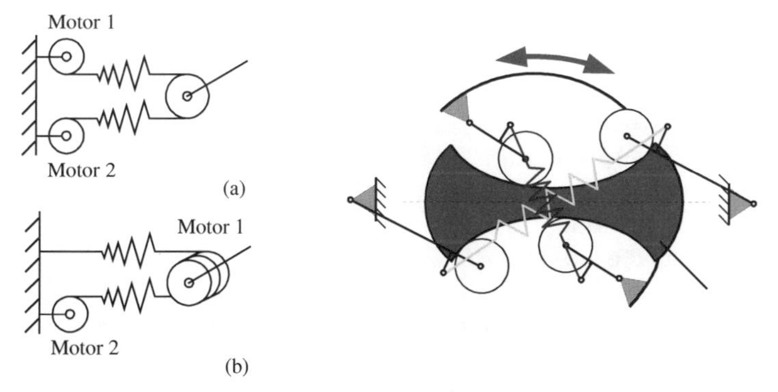

Fig. 10.3 Variable Stiffness Actuator with nonlinear progressive springs in antagonistic (a) and quasi-antagonistic (b) realization. Principle of the elastic mechanism (right).

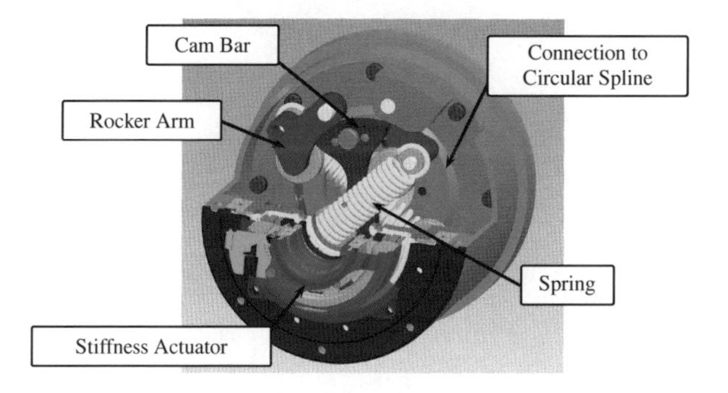

Fig. 10.4 Cross section of the Quasi-Antagonistic Joint design

10.3.1 Joint Design

The QA-Joint consists of a link positioning drive with Harmonic Drive gears and an elastic mechanism with the stiffness actuation drive. The main difference to a classical antagonistic joint (see Fig. 10.3 a) is that the two motors are not used in a symmetrical configuration as agonist and antagonist. Instead, one motor (the link drive) adjusts the link side position, while the second motor (the stiffness drive) operates stiffness adjustment, see Fig. 10.3 b. With this arrangement, the adjustment of position and stiffness is already decoupled to a large extent in hardware design. This special form of antagonistic actuation is advantageous for configurations with pronounced agonist actuation.

The compliance consists of two progressive elastic elements opposing each other with a variable offset that supports the link with variable range of elastic motion, see Fig. 10.4. The classically fixed Circular Spline of the Harmonic Drive gear for link positioning is held in a bearing and has a cam bar attached to it. Two pairs of rocker arms with cam rollers, each pair linked by a linear spring, act on different faces of this cam bar. External loads result in rotational displacement of the entire gear and force the rocker arms of the supporting direction to spread against the linear spring. This causes a progressive centering torque. The agonist rocker arms are fixed w.r.t. the housing. The opposing antagonist part is positioned with a rotational offset with respect to the stiffness actuator. This makes it possible to change stiffness independently from link speed in approximately 120 ms for the full stiffness range. In the QA-Joint the link position can be changed without moving the elasticity mechanism. This significantly reduces the inertia of the moving part of the joint.

The use of a cam-roller mechanism offers another advantage: The shape of the cam faces can be adapted to provide any progressive desired torque characteristic that stores the maximum potential energy in the linear spring. Thus, the design is well suited to realize different torque/displacement characteristics with little overhead. Overall, the superposition of agonist and antagonist action with different offsets results in the desired variable stiffness.

Table 10.3 Testbed properties

Property	Value
Torque capacity	$\tau_{J,\max} = 40$ Nm
Maximum positioning drive speed	$\dot{\theta}_{\max} = 3.8$ rad/s
Maximum elastic deflection	$\varphi_{\max} = 3 \ldots 15\,^{o}$
Maximum spring energy	$E_{\varphi,\max} = 2$ x 2.7 J
Stiffness range ($\tau_J = 0$)	20...750 Nm/rad
Maximum stiffness adjustment time	0.12 s
Mass	1.2 kg

Taking all the aforementioned design considerations into account, the key characteristics of the joint prototype were selected according to Tab. 10.3.

Next, the layout of the elastic torque characteristics is discussed to complete the design process.

10.3.2 Torque Characteristics Layout

For the shape of the torque/displacement curve an exponential characteristic is considered to be well suited, see Tab. 10.2. This is due to the fact that it results in a set of relatively constant stiffness curves over a wide load range, while providing large stiffness adjustment ranges at the same time. It allows moderate progression towards the elastic limits to protect the joint from strike through. The exponential stiffness has the general form

$$\tau_J = ae^{b(\varphi-\sigma)} = ae^{b((\theta-q)-\sigma)}, \tag{10.4}$$

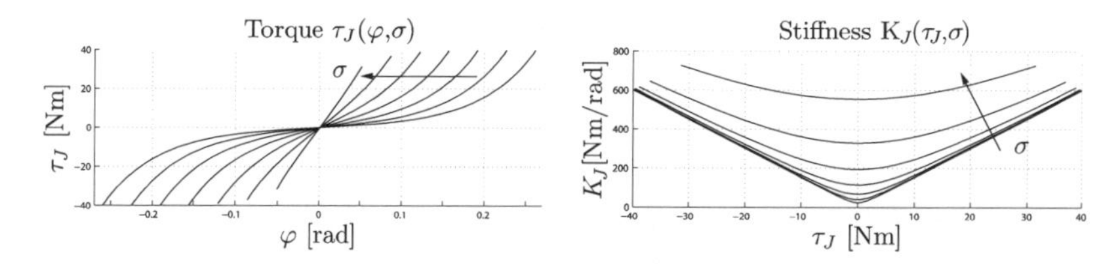

Fig. 10.5 Centering elastic joint torque over displacement curves for different stiffness presets (left). Stiffness values over elastic join torque (right).

where $\sigma \in \mathbb{R}$ denotes the displacement of the stiffness preset actuator. The corresponding stiffness is $K_J = \frac{\partial \tau_J}{\partial q}$. σ is also an upper limit for the elastic deflection φ which can be obtained for a given preset.

$$|\varphi| = |(\theta - q)| \leq \sigma \tag{10.5}$$

$a, b \in \mathbb{R}^+$ are design coefficients for setting the maximum torque and the elastic joint characteristic, which are chosen to be $a = 40.0$ Nm and $b = 15.0$ rad^{-1}. Therefore, the joint torque for the implemented design becomes

$$\tau_J = 40e^{15(\varphi-\sigma)}. \tag{10.6}$$

a denotes the torque at which the stored energy equals the maximum potential energy of the springs. For the full design of the hand-arm system (see Fig. 10.1) it is planned to use the even less progressive exponential characteristics $e^{12(\theta-q)}$ for each joint in the arm[2]. Thus, a is varied according to the desired maximum torque value and the available spring energy. The geometry of the joint, in particular of the cam-roller mechanism, is derived from this target torque curve. The superposition of the two opposing elastic elements for the complete joint model results in a centering torque

$$\tau_J = 40(e^{15(\varphi-\sigma)} - e^{15(-\varphi-\sigma)}), \tag{10.7}$$

leading to the torque/deflection curves shown in Fig. 10.5 (left). The corresponding stiffness adjustment range is shown in Fig. 10.5 (right). It is visible that changing τ_g or τ_{ext} (due to gravity or real external loads) results in only moderate change of stiffness until deflection comes close to the end of the elastic range. In the nominal torque area, stiffness can be varied from below 100 Nm/rad to more than 550 Nm/rad.

[2] The full arm design was shown at the most recent trade fair AUTOMATICA2010 in Munich. However, please note that the final joint design is a combination of the two prototypes used in this monograph. The first four joints of the arm are equipped with the so called FSJ mechanisms. Analogue to the VS-Joint, the torque is generated by a rotational cam disk and roller system. Furthermore, the FSJ is equipped with two opposing cam profiles, which originates from the QA-Joint. This leads to a similar torque-deflection profile as for the QA-Joint.

Given a current torque τ_J and a desired stiffness $K_{J,d}$, one can solve the system of equations for φ and σ.

$$\varphi = \frac{1}{b} \tanh^{-1} \left(\frac{K_{J,d}}{b\, \tau_J} \right) \tag{10.8}$$

$$\sigma = -\frac{1}{b} \ln \frac{\tau_J}{a(e^{b\varphi} - e^{-b\varphi})} \tag{10.9}$$

10.3.3 Model of the QA-Joint

The applied dynamics model of the QA-Joint incorporates the full motor dynamics, the elastic nonlinear joint torque as described in the previous subsection, and the link side inertia. Furthermore, the friction and gravity torques are taken into account. In [2] a generic model for variable stiffness joint based on Harmonic Drive gears was derived, which incorporates even the precise gear dynamics. However, in the present context such effects can be neglected, leading to a concise formulation.

$$B\ddot{\theta} = \tau_m - \tau_J \tag{10.10}$$

$$M\ddot{q} = \tau_J - \tau_F - \tau_g - \tau_{\text{ext}} \tag{10.11}$$

$B, M \in \mathbb{R}^+$ are the motor and link side inertia, respectively. $\theta, q \in \mathbb{R}$ are the motor and link side position, and $\tau_m, \tau_J, \tau_F, \tau_g, \tau_{\text{ext}} \in \mathbb{R}$ the motor, elastic, friction, gravity, and external torque, respectively. Please note the assumption that the stiffness actuator dynamics has no significant dynamic influence on the joint drive and the link.

In order to justify this assumption of symmetric torque deflection properties while keeping a desired stiffness value, the following measurements are shown to depict the compliance of the stiffness motor of the QA-Joint. Figure 10.6 shows the elastic properties of the stiffness adjuster during externally caused deflection φ of the link. Only little compliance is observed, which is several orders of magnitude smaller than the minimum compliance of the elastic mechanism.

As will be shown in Sec. 10.3.4, the dependencies on the friction torque are given by the elastic deflection, load, and stiffness preset of the joint (for the QA-Joint this is the internal tension). The position motor is PD position controlled for the identification phase, while motor torque saturation is taken into account.

$$\tau_m = \begin{cases} \tau_m^{\max} & \tau_d \geq \tau_m^{\max} \\ K_D(\dot{\theta}_d - \dot{\theta}) + K_P(\theta_d - \theta) & \tau_m^{\min} < \tau_d < \tau_m^{\max} \\ \tau_m^{\min} & \tau_d \leq \tau_m^{\min} \end{cases} \tag{10.12}$$

$K_P, K_D \in \mathbb{R}^+$ are the control gains, $\theta_d \in \mathbb{R}$ is the desired position of the positioning motor, $\tau_m^{\max}, \tau_m^{\min}$ are the maximum and minimum torque of the motor ($\tau_m^{\min} = -\tau_m^{\max}$), and $\tau_d = K_D(\dot{\theta}_d - \dot{\theta}) + K_P(\theta_d - \theta) \in \mathbb{R}$ the desired torque of the motor

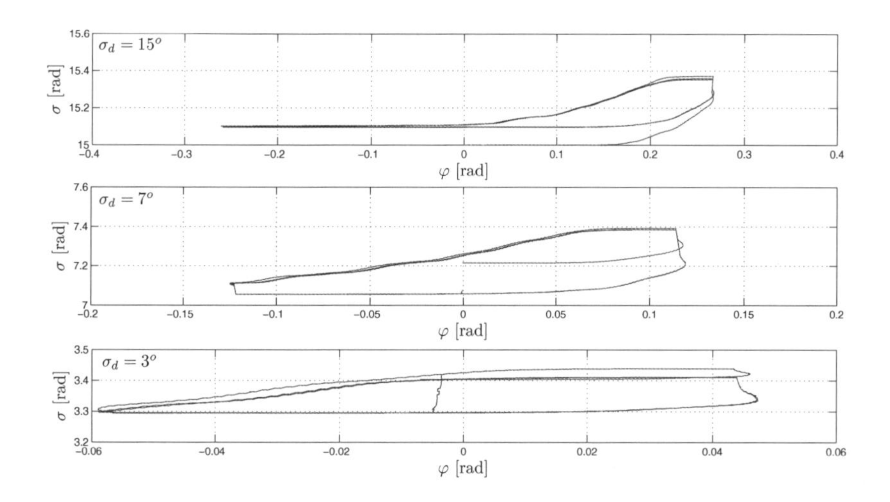

Fig. 10.6 Elastic behavior of the stiffness adjustment motor for different stiffness presets under loading conditions

controller. The structure of the nonlinear system is depicted in Fig. 10.7. Please note that the friction torque is modeled as pure nonlinear, parametric Coulomb friction, depending on $\text{sgn}(\dot{\varphi})$.

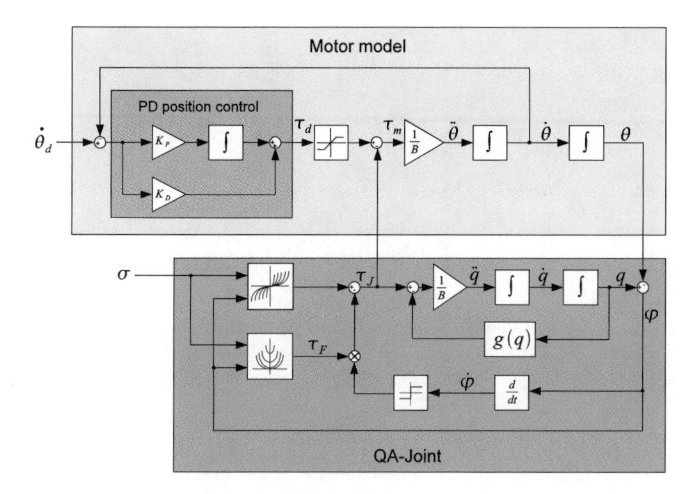

Fig. 10.7 Block diagram of the QA-Joint

10.3.4 Joint Identification

In order to generate the data for identifying the real elastic behavior and friction, the link is mechanically fixed (Fig. 10.8 (left)) and the position motor drives with different velocities at various stiffness presents within the elastic joint limits. The implemented sensors for identification are motor position sensors for θ, σ, q, and a link side joint torque sensor. The resulting measurements, together with the ideal model of the joint are given in Fig. 10.8 (right) and Fig. 10.9 for cyclic rectangular motions with $\theta_d \in \{30, 60, 90\}\ ^o/\mathrm{s}$. The real behavior is characterized by a hysteresis and significant deviation from the ideal one[3].

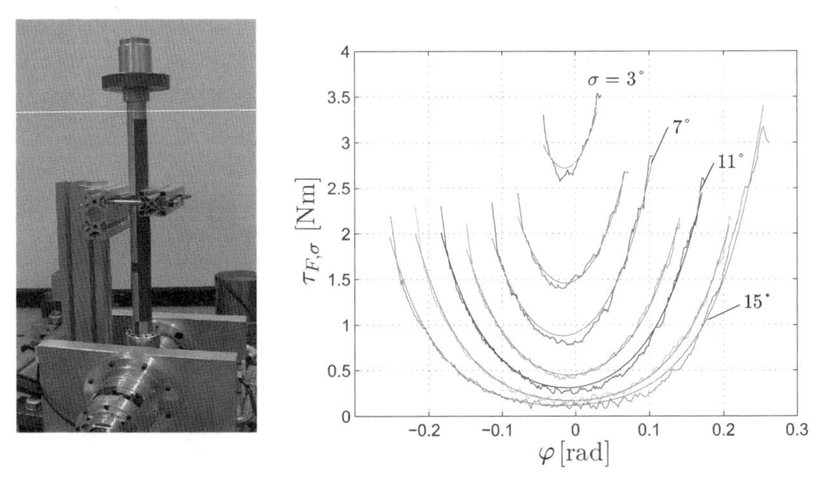

Fig. 10.8 QA-Joint with clamped link (left). Experimental friction torque over elastic deflection compared with the model friction (right).

The real torque characteristics $\tau_J^*(\theta, q) \in \mathbb{R}$ are estimated from the measurements. They are assumed to be the center lines of the hysteresis and are calculated as the arithmetic mean of the measured hysteresis. For the identification the following model is used, leaving the coefficients of the exponential function $a_S, b_S \in \mathbb{R}^+$ free for optimization.

$$\tau_J^* = a_S e^{(15(\varphi - \sigma))} - b_S e^{(15(-\varphi - \sigma))} \tag{10.13}$$

The compliance of the stiffness adjuster is directly taken into consideration by calculating φ, since its position is directly influencing this calculation. The parameter estimation of a_S, b_S is realized with least square error optimization.

[3] Please note that for all experiments except the control performance in Sec. 10.9.5 simple motor side PD control is used in order to fully exploit the intrinsic elastic capabilities of the joint.

$$\mathbf{y} = M\xi = M \begin{bmatrix} a_S \\ b_S \end{bmatrix}, \tag{10.14}$$

where $M \in \mathbb{R}^{N\times 2}$ is the data matrix consisting of the exponential parts, $\mathbf{y} \in \mathbb{R}^{N\times 1}$ the measurement vector containing joint torques, and $\xi \in \mathbb{R}^{2\times 1}$ the parameter vector. The calculated center line is denoted as τ_{mean}. Obtaining $\mathbf{p}$ is performed by calculating the pseudoinverse of the observation matrix.

$$\xi = (M^T M)^{-1} M^T \mathbf{y} \tag{10.15}$$

Table 10.4 Identified compliance a_S, b_s and friction coefficients a_F, b_F

σ	a_S [Nm]	b_S [Nm]	a_F [Nm]	b_F [Nm]
3°	26.2760	26.2221	2.5172	3.8480
5°	26.6049	27.1703	2.6125	3.2755
7°	26.8106	27.8989	2.9776	3.1582
9°	26.4714	28.4042	2.3106	3.0784
11°	26.3446	28.6174	2.8776	3.3393
13°	26.0417	29.1079	2.4160	3.3339
15°	22.7286	30.1821	3.3058	3.3004

The results of this calculation are given in Tab. 10.4 (first two columns). The real stiffness coefficients vary up to ≈ 35 % from the theoretical values. The asymmetry of the real values, which grows with increasing stiffness, can be explained by the slightly elastic behavior of the stiffness adjuster. Furthermore, the real friction torque τ_F depends on σ and φ. No relationship between velocity and friction could be observed, so viscous effects are neglected as already mentioned in Sec. 10.3.3, see Fig. 10.7.

Figure 10.8 depicts the friction torque for different values of σ. The results indicate an exponential relation between φ and τ_F. A closer look shows a linear relation between τ_J and τ_F. Therefore, is seems reasonable to model the friction as a sum of the torques resulting from the force input by each spring. This way the load free friction ($\varphi = 0$) can be established and explained by the internal tension of the joint, increasing with growing σ. This leads to the following friction model.

$$\tau_F(\varphi) = a_F e^{(15(\varphi-\sigma))} + b_F e^{(15(-\varphi-\sigma))} \tag{10.16}$$

The coefficient estimation is again obtained by least square error regression. The same structure as the one for the elastic joint torque can be obtained, except for the different sign for b_F. The obtained coefficients are given in Tab. 10.4 and a comparison of the measured and expanded model is shown in Fig. 10.9.

Next, the control scheme for providing active vibration damping of the intrinsically poorly damped joint design is outlined.

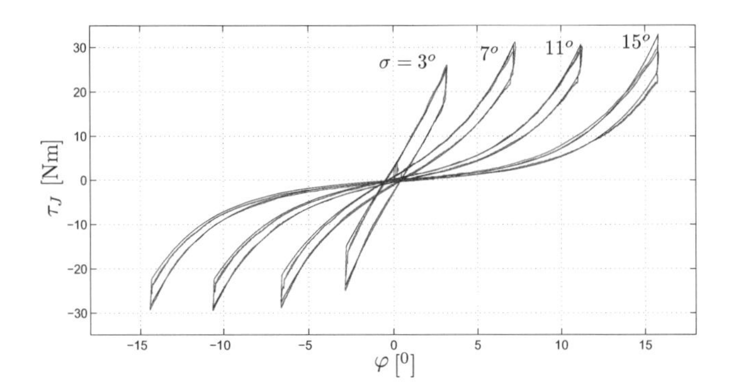

Fig. 10.9 Comparison of the measured and expanded simulated model

10.3.5 State Feedback Controller

Using a model based torque and stiffness estimation together with the known link side inertia, a full state feedback controller with gain scheduling is used. The method developed in [2] is utilized for motion control with active vibration damping of the intrinsically low damped joint. The stiffness actuator uses simple PD control. In the following, a brief overview of the full state feedback controller is given.

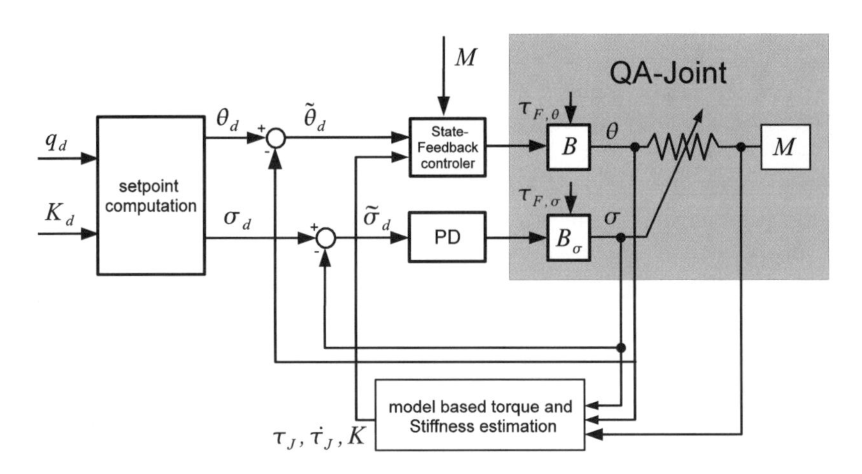

Fig. 10.10 Block diagram of the setpoint computation, control loop, dynamics model, and model-based torque and stiffness computation of the QA-Joint

Figure 10.10 depicts the control structure of the joint. In order to set a desired link position and nominal load stiffness a *setpoint computation* is carried out. The position motor controller consists of the state feedback loop

$$\tau_m = \tau_{m,ff} + k_P\tilde{\theta} + k_D\dot{\tilde{\theta}} + k_T\tilde{\tau}_J + k_S\dot{\tilde{\tau}}_J, \tag{10.17}$$

where k_P, k_D, k_T, k_S are gains depending on B, M, K_J, and the gravity potential $\frac{\partial g(q)}{\partial q}$. The gains are optimized such that critical damping for the linearized system is achieved and they stay within practically feasible bounds. The feed-forward term and the desired torque are given as

$$\tau_{m,ff} = B\ddot{\theta}_d + M\ddot{q}_d + g(q_d) \tag{10.18}$$

$$\tau_d = B\ddot{q}_d + g(q_d). \tag{10.19}$$

In order to provide theoretical background for the upcoming section, an overview of basic optimal control theory is given in the following.

10.4 Basic Optimal Control Theory

Some theoretical foundations of optimal control are summarized here that are being used in the following sections. The description is kept concise in order to only provide the relevant knowledge to understand the approach taken in this monograph.

10.4.1 Optimal Control of Dynamic Systems

Optimal control aims at finding a control input $\mathbf{u}$ for a dynamic system, which maximizes/minimizes an appropriately designed cost function within a certain time interval $t \in [0, t_f]$, often the initial and final state of the system are given[4]. The criterion can be a functional of the system state $\mathbf{x}(t)$, the control input $\mathbf{u}(t)$, and time t. $\mathbf{u}(t)$ and $\mathbf{x}(t)$ can be bounded. In the following, systems are assumed, which state space equations do not explicitly dependent on the time. Their mathematical description is a system of differential equations of first order.

$$\dot{\mathbf{x}}(t) = f(\mathbf{x}(t), \mathbf{u}(t), t) \tag{10.20}$$

A general optimality criterion is to be chosen such that the timely evolution of $\mathbf{x}(t)$ and $\mathbf{u}(t)$, as well as the final state of the system are weighted. Therefore, an integral cost functional is a reasonable choice, which weights the final state with the function h and the timely evolution of the state and control input with integrating the function g.

$$J = h(\mathbf{x}(t_f), t_f) + \int_0^{t_f} g(\mathbf{x}(t), \mathbf{u}(t), t)\, \mathrm{d}t \tag{10.21}$$

[4] In general, the theory permits also free final times.

The optimization consists of maximizing/minimizing J under the constraint of the state equations. This is a typical problem from Calculus Of Variations. For the optimization of dynamic systems the Hamiltonian method is a well known scheme. This formalism transforms a constrained optimization problem of the state equations into a problem without constraints. This is carried out by introducing Lagrange multipliers [36]. The cost function is extended and becomes

$$\tilde{J} = h(\mathbf{x}(t_f), t_f) + \int_0^{t_f} g(\mathbf{x}(t), \mathbf{u}(t), t) + \boldsymbol{\lambda}^T f(\mathbf{x}(t), \mathbf{u}(t), t)\, \mathrm{d}t. \tag{10.22}$$

The Hamiltonian is defined as follows.

$$H(\mathbf{x}(t), \boldsymbol{\lambda}(t), \mathbf{u}(t), t) = g(\mathbf{x}(t), \mathbf{u}(t), t) + \boldsymbol{\lambda}^T f(\mathbf{x}(t), \mathbf{u}(t), t) \tag{10.23}$$

The partial derivatives of the Hamiltonian with regards to the state and the co-states define a canonical system of differential equations.

$$\dot{\mathbf{x}} = \frac{\partial H}{\partial \boldsymbol{\lambda}} \tag{10.24}$$

$$\dot{\boldsymbol{\lambda}} = -\frac{\partial H}{\partial \mathbf{x}} \tag{10.25}$$

Thus, a canonical system of Hamiltonian differential equations of order $2n$ is created for a dynamic n-order system, if no further constraints are taken into consideration. The boundary values of the adjoint equations are obtained from the transversality condition for the final state.

$$\frac{\partial h(\mathbf{x}(t_f))}{\partial \mathbf{x}} - \boldsymbol{\lambda}(t_f) = \mathbf{0} \tag{10.26}$$

For general problems the minimization of the partial derivative of the Hamiltonian with respect to the control input (optimality or stationary condition)

$$\frac{\partial H}{\partial \mathbf{u}} = \mathbf{0} \tag{10.27}$$

yields an optimal control trajectory. With these control equations, which in general depend on the states and co-states, the canonical system of differential equations can be solved.

Finally, the Legendre-Clebsch condition, also known as convexity of the Hamiltonian, gives the confirmation of a local maximum of the optimality condition.

$$\frac{\partial^2 H}{\partial \mathbf{u}^2} \leq \mathbf{0} \tag{10.28}$$

10.4.2 Singular Control Problems

A distinctive feature of the problem that is investigated in this monograph comes with the linear dependency of the Hamiltonian on the control input. For this case the Hamiltonian can be split into two parts. The first one is independent from $\mathbf{u}(t)$ and the second one is linear in $\mathbf{u}(t)$.

$$H(\mathbf{x}(t),\boldsymbol{\lambda}(t),\mathbf{u}(t),t) = H_1(\mathbf{x}(t),\boldsymbol{\lambda}(t),t) + H_2^T(\mathbf{x}(t),\boldsymbol{\lambda}(t),t)\mathbf{u} \tag{10.29}$$

For this case (10.28) does not lead any optimal control trajectory since $\frac{\partial^2 H}{\partial \mathbf{u}^2}$ is not a function of $\mathbf{u}$. This is a so called singular control problem [30, 5]. However, by introducing a bounded control input, the maximum principle of Pontryagin, introduced in 1956, yields an optimal control trajectory.

10.4.3 The Maximum Principle of Pontryagin

The Pontryagin maximum principle states that the optimal control input is the one that maximizes the Hamiltonian at every time instant [23].

$$H(\mathbf{x}(t),\boldsymbol{\lambda}(t),\mathbf{u}(t),t) \leq H(\mathbf{x}(t),\boldsymbol{\lambda}(t),\mathbf{u}^*(t),t) \tag{10.30}$$

The stationary condition (10.27) has to be replaced by (10.30). The Pontryagin maximum principle is a necessary but no sufficient optimality condition. It is possible that several control trajectories exist, which are not optimal but satisfy the maximum principle. A mathematical proof for this statement is given in [32].

For singular control problems of the form (10.29) the maximum principle leads to bang-bang control, for which the control input $\mathbf{u}$ is assumed to be bounded.

$$u_{i,\min} \leq u_i \leq u_{i,\max} \tag{10.31}$$

Now, the maximization of the Hamiltonian only depends on the sign of H_2, which is labeled switching function. A maximum is given for positive switching function associated with maximum control input and negative switching function with smallest control input. For $H_2 = 0$ singular control inputs are obtained. Since the control input is bounded, the maximum principle is satisfied with following switching law [23].

$$u_i^*(t) = \begin{cases} u_{\mathrm{i,max}}, & H_{i,2} > 0 \\ u_{\mathrm{i,min}}, & H_{i,2} < 0 \\ \text{sing.}, & H_{i,2} = 0 \end{cases} \tag{10.32}$$

10.4.4 Bounded State Variables

For a given optimization it is often desired to satisfy further constraints apart from minimizing the cost function. For their solution this can yield to boundary control, boundary contact points, or singular control inputs. In the following, state and control bounds are defined as follows.

$$S := g(\mathbf{x}(t), \mathbf{u}(t), t) \leq 0 \tag{10.33}$$

The order of state bounds is defined as the q-th derivative of S

$$S^{(q)} := \frac{d^q S}{dt^q} = \frac{d^q g(\mathbf{x}(t), \mathbf{u}(t), t)}{dt^q} \leq 0 \tag{10.34}$$

that explicitly contains the control input $\mathbf{u}$ first. For the given optimization problem the compliance of bounds is necessary. This may lead to contact points, which can only occur for 2nd,4th,6th etc. order state bounds [7]. The state bounds are coupled to the Hamiltonian via a second set of Lagrange multipliers μ.

$$\begin{aligned} \tilde{H}(\mathbf{x}, \lambda, \mu, \mathbf{u}, t) &= -g(\mathbf{x}(t), \mathbf{u}(t), t) + \lambda^T f(\mathbf{x}(t), \mathbf{u}(t), t) \\ &\quad + \mu S^{(q)}(\mathbf{x}(t), \mathbf{u}(t), t) \end{aligned} \tag{10.35}$$

Equation (10.35) denotes the augmented Hamiltonian [5]. For μ following condition holds.

$$\mu \begin{cases} = 0, \; S < 0 \\ > 0, \; S = 0 \end{cases} \tag{10.36}$$

The second set of Lagrange multipliers is zero if (10.34) is fulfilled and larger zero if the bound is active. The derivation of the canonical system of differential equations from (10.35) leads to jumps in the adjoint equations. According to [5] following conditions hold for occurring contact points.

$$\lambda^T(t_b^+) = \lambda^T(t_b^-) - \mu_0 \frac{\partial S^{(0)}}{\partial \mathbf{x}} \tag{10.37}$$

$$H(t_b^+) = H(t_b^-) + \mu_0 \frac{dS^{(0)}}{dt} \tag{10.38}$$

For contact points, which show extremal behavior at contact time t_b, the tangency condition

$$N^{q-1}(\mathbf{x}, t_b) = \begin{bmatrix} S^{(1)}(\mathbf{x}, t_b) \\ \vdots \\ S^{(q-1)}(\mathbf{x}, t_b) \end{bmatrix} = 0 \tag{10.39}$$

has to be fulfilled. For the problem with $q-1=1$

$$N^1(\mathbf{x},t_b) = \begin{bmatrix} S^{(0)}(\mathbf{x},t_b) \\ S^{(1)}(\mathbf{x},t_b) \end{bmatrix} = \begin{bmatrix} 0 \\ 0 \end{bmatrix} \tag{10.40}$$

is obtained. After the discussion of the effects of state constraints on the optimal solution, a short introduction to solving nonlinear optimal control problems without state constraints is given. This generally leads to two-point Boundary Value Problems (BVP), which can only be solved analytically in simple cases [30]. For their numerical treatment methods were developed that base on powerful methods for solving initial value problems. A well known class of such schemes are shooting methods.

10.5 Shooting Methods for Solving MPBVPs

In this section, two of the main methods for solving Multi-Point Boundary-Value Problems (MPBVP) are discussed: **single-shooting** and **multiple-shooting** methods.

10.5.1 Single-Shooting

According to [21] MPBVPs can be formulated as follows.

$$\dot{\mathbf{x}} = f(\mathbf{x}(t),\mathbf{u}(t),t) \quad \text{with } t \in [0,t_f] \tag{10.41}$$

$$g(\mathbf{x}(0),\mathbf{x}(t_f)) = \mathbf{0} \tag{10.42}$$

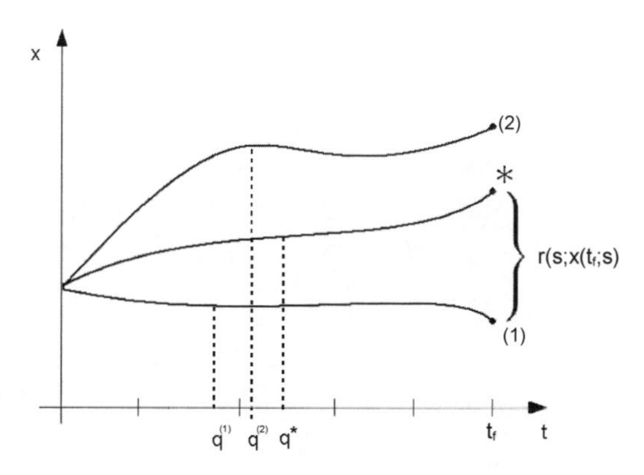

Fig. 10.11 Single-shooting method

The state equations (10.41) have to be solved while fulfilling the constraints (10.42). If for each differential equation only an initial **or** a final value are given, the boundary values are called entirely separated. The basic idea of the single-shooting method is to parameterize the Initial Value Problem (IVP) appropriately. An estimated parameter vector $\mathbf{s}_0 \in \mathbb{R}^n$, which is called shot parameter contains the initial values of the IVP. For switching operations as e.g. for bang-bang control, the switching times $\mathbf{q}$ are to be estimated as well. With $\mathbf{s} \in \mathbb{R}^n$ the initial value problem is solved by numerical integration forward in time. The solution usually deviates from the exact solution, producing an accordingly designed residual $\mathbf{r}$ that vanishes for the exact solution, see Fig. 10.11. The residuum is formulated as a functional relationship between $\mathbf{s}$ and the solution for final time $\mathbf{x}(t_f)$.

$$F(\mathbf{s}) = \mathbf{r}(\mathbf{s}, \mathbf{x}(t_f, \mathbf{s})) = \mathbf{0} \tag{10.43}$$

Now, the problem is reduced to finding the root of $\mathbf{r}$, which can be solved with a (modified) Newton or bisection method [21]. After each iteration, the initial value problem is solved with new initial values and the residuum is calculated and analyzed. For nonlinear differential equations there exists usually no unique solution of the MPBVP. Therefore, it must be ensured that the initial estimation of the start parameters are close enough to the correct solution for ensuring convergence.

Next, the concept of multiple-shooting is shortly discussed.

10.5.2 Multiple-Shooting

In [6] it was shown that the deviation in the start parameters influences the error of the solution exponentially with growing final time. Due to this fact the multiple-shooting method was developed, which divides the entire considered time interval into N partial ones.

$$t_0 < t_1 < t_2 < \ldots < t_{N-1} < t_f \tag{10.44}$$

For each of these intervals the single-shooting method is applied and thus for every partial interval estimations of the partial initial conditions are needed, see Fig. 10.12. This can be difficult for differential equations of higher order and many nodes. In other words a good intuition about the solution is needed in advance. As described in [7] other optimization techniques as collocation or homotopy are used for this purpose. Inner point conditions, as they occur due to state and control input bounds, can be incorporated with additional nodes and their respective boundary conditions.

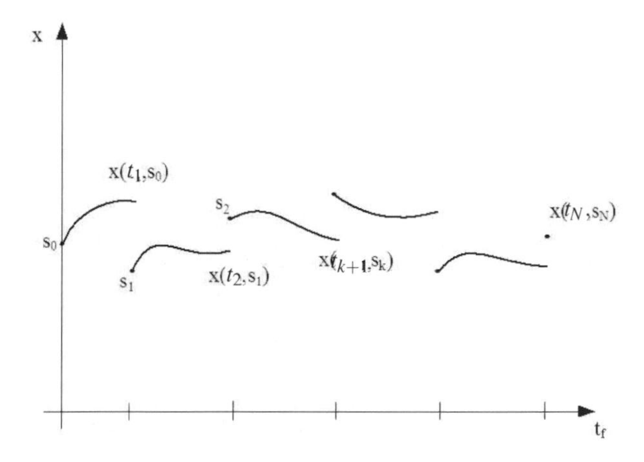

Fig. 10.12 General idea of the multiple-shooting approach

Analogous to the single-shooting method the solution is obtained via a root-finding problem, which can be approximated with one of the already mentioned iterative methods.

A powerful implementation of the multiple-shooting method is the program packet BNDSCO [29]. The sourcecode written in FORTRAN77 is freely available and was used in this monograph for solving some of the occurring problems.

For the QA-Joint, the insights from optimal control theory were used to formulate the optimization of the control trajectory as a simple parameter optimization problem, see later. These can be solved with classical optimization techniques. As the Nelder-Mead Simplex-Downhill algorithm was used for solving the QA-Joint optimization problem, this is shortly described next.

10.6 The Nelder-Mead Simplex-Downhill Algorithm

The Nelder-Mead Simplex-Downhill algorithm is due to its high robustness a widely spread optimization algorithm for solving multi-dimensional nonlinear and unbounded optimization problems. It does not use gradients and is classified as a direct search algorithm [24]. It seeks for a local minimum of a scalar quality function J, which can depend on several parameters $\mathbf{p}$.

$$\min_{\mathbf{p}}\{J(\mathbf{p})\} \tag{10.45}$$

In a first step $n+1$ points are generated for n parameters around the given initial value and the cost function is evaluated for all points. According to the result the points are indexed, and then varied by four mappings according to Fig. 10.6.

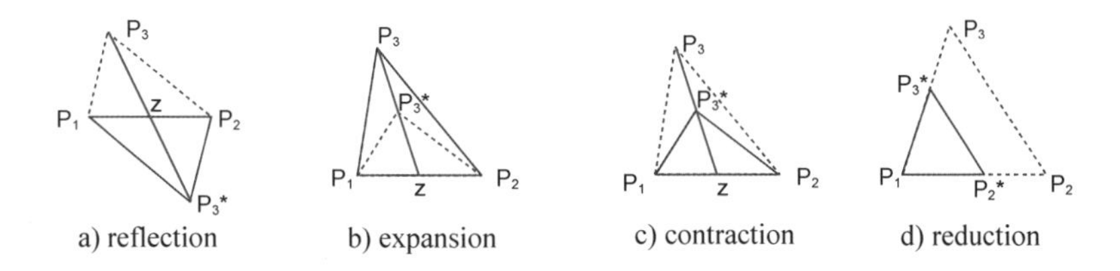

Fig. 10.13 The four simplex mappings of the Simplex-Downhill algorithm

The algorithm consists of following steps, iteratively applied until a convergence criterion is satisfied:

1. Calculate the center of the best n points of the simplex as the averaged mean

$$\mathbf{z} = \frac{1}{n} \sum \mathbf{p}_n \tag{10.46}$$

2. The point with lowest quality is mirrored along this center $\mathbf{z}$ (reflexion). If the mirrored quality value is larger, the old one is depleted.
3. If the mirrored point has the highest quality of the ensemble, the simplex is stretched along the line $(\mathbf{p}_i - \mathbf{z})$ by a constant factor (expansion). If the quality is higher, the old point is depleted, otherwise the originally mirrored point is kept.
4. If the mirrored point has the lowest quality a contracted point is calculated. In case this leads to an increase in quality, the contracted point is kept and the originally mirrored one is depleted.
5. If none of the operations depicted in Fig. 10.6 (a.-c.) leads to an increase in quality, the simplex is reduced in size while keeping the point with the best quality result (Fig. 10.6 d.) and the algorithm is repeated.

The convergence rate of the method is comparably slower to gradient-based techniques. More detailed discussions are given in [24].

A possibility to use the Simplex-Downhill method for optimization problems with bounds is to extend the quality function with a penalty term

$$\min_{\mathbf{p}} \left\{ J(\mathbf{p}) + J_{pen} \right\}, \tag{10.47}$$

increasing the cost if boundaries are violated. This reduces a bounded problem to an unbounded one.

In the next section, the theoretical foundations to optimally excite a VSA joint to achieve maximum link side velocity are systematically developed. For maximizing the link side velocity, vibration damping is again switched to a simple motor side PD control in order to utilize the eigenvibrations of the system, which would otherwise be damped out.

10.7 Performance Increase through Joint Compliance

In this section, the theory to maximize the link side velocity of a variable impedance joint is developed and the results are experimentally verified. For solving this problem, the aforementioned methods from optimal control theory are used. In order to systematically analyze the different effects and constraints, the complexity of the used models is increased and analytical solutions are derived if possible. Table 10.5 depicts the consecutive steps made and points out whether analytical or numerical solutions were obtained. First, the constant stiffness case (case A) is solved with different motor models (case B+C+D). Then, the presence of bounds on the state variables (case A'+E) is incorporated, the influence of adjusting the stiffness (case F+G) is analyzed, and finally experimental results on the DLR QA-Joint (case H) are discussed. Each step contributes particular insights, as e.g. the influence of constrained motor dynamics, constraints on the elastic deflection, or stiffness adjustment, which makes it possible to formulate a full view on the problem. As mechanical damping is usually unwanted due to energetic arguments, most VIA implementations realize damping via active control and not through a mechanically complex solution. Therefore, damping is not considered, i.e. $D_J = 0$. Furthermore, $K_J = \sigma_1$ is assumed for the theoretical analysis in order to keep it clear for the reader. Therefore, only σ is used to denote the stiffness actuation variable from now on. In general, the present section extends the initial results from Chapter 9.

Table 10.5 Analyzed models (SEA= Series Elastic Actuation, JTF = joint torque feedback, CD = constrained deflection, VS = variable stiffness, CMT = constrained motor torque)

case	model	solution	achieved insights
A	Velocity source + SEA	analytical	principal effect of significant joint elasticity
A'	Velocity source + SEA + CD	analytical	influence of constrained deflection
B	PT1 + SEA	analytical	influence of constrained motor dynamics, 1st order
C	PT2 + SEA	analytical	influence of constrained motor dynamics, 2nd order
D	PT2 + SEA + JTF	numerical	influence of joint torque feedback on motor inertia
E	PT2 + SEA + JTF + CD	numerical	influence of deflection constraints
F	Velocity source + VS	analytical	principle effect of stiffness adjustment
G	Velocity source + VS + CD	numerical	influence of stiffness adjustment and constrained deflection
H	PT2 + VS + CMT	numerical	real VIA design behavior and constrained motor torque

10.7.1 Maximization of Link Velocity

As systems whose state space equations do not explicitly depend on time are assumed, the description of their dynamics is a system of differential equations of first order.

$$\dot{\mathbf{x}}(t) = f(\mathbf{x}(t), \mathbf{u}(t)), \tag{10.48}$$

with $\mathbf{x}$ and $\mathbf{u}$ being the state vector and control input, respectively. For achieving an optimal control input, a general optimality criterion is usually to be chosen such that the timely evolution of $\mathbf{x}(t)$ and $\mathbf{u}(t)$, as well as the final state of the system $\mathbf{x}(t_f)$ are weighted with respect to each other, see Section 10.4.1.

$$J = h(\mathbf{x}(t_f), t_f) + \int_0^{t_f} g(\mathbf{x}(t), \mathbf{u}(t), t)\, \mathrm{d}t \tag{10.49}$$

Together with the Hamiltonian

$$H(\mathbf{x}(t), \boldsymbol{\lambda}(t), \mathbf{u}(t), t) = g(\mathbf{x}(t), \mathbf{u}(t), t) + \boldsymbol{\lambda}^T f(\mathbf{x}(t), \mathbf{u}(t), t) \tag{10.50}$$

the constrained optimization problem is transformed into a problem without constraints. However, in order to maximize[5] the link side velocity at a certain time instant t_f only, (10.49) reduces to:

$$J = h(\mathbf{x}(t_f), t_f)) = \dot{q}(t_f) \tag{10.51}$$

Since no other constraints are taken into consideration (10.50) reduces to

$$H(\mathbf{x}, \boldsymbol{\lambda}, \mathbf{u}, t) = \boldsymbol{\lambda}^T f(\mathbf{x}(t), \mathbf{u}(t), t). \tag{10.52}$$

For the optimization of the final state the boundary conditions of the adjoint equations result from the transversality condition

$$\boldsymbol{\lambda}(t_f) = \frac{\partial h(t_f)}{\partial \mathbf{x}}. \tag{10.53}$$

Together with the initial boundary conditions of the state space equation and the final boundary conditions of the adjoint equations, this leads to a two-point boundary value problem. The partial derivatives of the Hamiltonian with regard to the state and co-states define a canonical system of differential equations that needs to be solved:

[5] Later on, the problem is treated as an equivalent minimization problem and therefore the sign of the cost function changes.

$$\dot{\mathbf{x}} = \frac{\partial H}{\partial \lambda} \tag{10.54}$$

$$\dot{\lambda} = -\frac{\partial H}{\partial \mathbf{x}} \tag{10.55}$$

Next, models of increasing complexity are analyzed in order to elaborate the fundamental aspects about optimizing the link side velocity at a certain time instant t_f.

10.7.2 *Optimal Control for Linear Cases*

In this section, the constant elasticity case (K_J = const.) is treated. Constraints, stiffness adjustment and other nonlinear effects are discussed in Sec. 10.7.3 and Sec. 10.7.4. For the first model the motor behaves as a velocity source, which gives insight into the principles of utilizing joint elasticity. In order to investigate the influence of motor dynamics on the switching trajectory, the motor is considered to be position controlled. Both PT1 and PT2 behavior are investigated for the controlled

Table 10.6 Summary of the investigated linear optimal control problems

	Vel. source (A)	PT1 (B)	PT2 (C)	PT2+τ_J (D)
1	$\theta = \int_0^{t_f} \dot{\theta}_d \, \mathrm{d}t$ $M\ddot{q} = K_J(\theta - q)$	$\tau_m = K_P(\theta_d - \theta)$ $\tau_m = B\ddot{\theta}$ $M\ddot{q} = K_J(\theta - q)$	$\tau_m = K_D(\dot{\theta}_d - \dot{\theta}) + K_P(\theta_d - \theta)$ $\tau_m = B\ddot{\theta}$ $M\ddot{q} = K_J(\theta - q)$	$\tau_m = K_D(\dot{\theta}_d - \dot{\theta}) + K_P(\theta_d - \theta)$ $\tau_m = B\ddot{\theta} + K_J(\theta - q)$ $M\ddot{q} = K_J(\theta - q)$
2	$\mathbf{x}^T = [\theta \; q \; \dot{q}]$ $u = \dot{\theta}_d$	$\mathbf{x}^T = [\theta \; \dot{\theta} \; q \; \dot{q}]$ $u = \dot{\theta}_d$	$\mathbf{x}^T = [\theta_d \; \theta \; \dot{\theta} \; q \; \dot{q}]$ $u = \dot{\theta}_d$	$\mathbf{x}^T = [\theta_d \; \theta \; \dot{\theta} \; q \; \dot{q}]$ $u = \dot{\theta}_d$
3	$\dot{x}_1 = u$ $\dot{x}_2 = x_3$ $\dot{x}_3 = \omega^2(x_1 - x_2)$	$\dot{x}_1 = x_2$ $\dot{x}_2 = \frac{K_P}{B}(u - x_2)$ $\dot{x}_3 = x_4$ $\dot{x}_4 = \omega^2(x_1 - x_3)$	$\dot{x}_1 = u$ $\dot{x}_2 = x_3$ $\dot{x}_3 = \frac{1}{B}(K_D(u - x_3) +$ $+ K_P(x_1 - x_2))$ $\dot{x}_4 = x_5$ $\dot{x}_5 = \omega^2(x_2 - x_4)$	$\dot{x}_1 = u$ $\dot{x}_2 = x_3$ $\dot{x}_3 = \frac{1}{B}(K_D(u - x_3) +$ $+ K_P(x_1 - x_2) - K_J(x_2 - x_4))$ $\dot{x}_4 = x_5$ $\dot{x}_5 = \omega^2(x_2 - x_4)$
4	$H(\mathbf{x}(t), \lambda(t), u(t), t) =$ $\lambda_1 u + \lambda_2 x_3 + \lambda_3 \omega^2(x_1 - x_2)$	$H(\mathbf{x}(t), \lambda(t), u(t), t) = \lambda_1 x_2 +$ $+ \lambda_2 \frac{K_P}{B}(u - x_2) + \lambda_3 x_4 + \lambda_4 \omega^2(x_1 - x_3)$	$H(\mathbf{x}(t), \lambda(t), u(t)) = \lambda_1 u + \lambda_2 x_3$ $+ \lambda_3 \frac{1}{B}(K_D(u - x_3) + K_P(x_1 - x_2)) +$ $+ \lambda_4 x_5 + \lambda_5 \omega^2(x_2 - x_4)$	$H(\mathbf{x}(t), \lambda(t), u(t)) = \lambda_1 u + \lambda_2 x_3$ $+ \lambda_3 \frac{1}{B}(K_D(u - x_3) +$ $+ K_P(x_1 - x_2) - K_J(x_2 - x_4)) +$ $+ \lambda_4 x_5 + \lambda_5 \omega^2(x_2 - x_4)$
5	$\dot{\lambda}_1 = -\lambda_3 \omega^2$ $\dot{\lambda}_2 = \lambda_3 \omega^2$ $\dot{\lambda}_3 = -\lambda_2$	$\dot{\lambda}_1 = -\lambda_4 \omega^2$ $\dot{\lambda}_2 = -\lambda_1 + \frac{K_P}{B}\lambda_2$ $\dot{\lambda}_3 = \lambda_4 \omega^2$ $\dot{\lambda}_4 = -\lambda_3$	$\dot{\lambda}_1 = -\lambda_3 \frac{K_P}{B}$ $\dot{\lambda}_2 = \lambda_3 \frac{K_P}{B} - \lambda_5 \omega^2$ $\dot{\lambda}_3 = -\lambda_2 + \lambda_3 \frac{K_D}{B}$ $\dot{\lambda}_4 = \lambda_5 \omega^2$ $\dot{\lambda}_5 = -\lambda_4$	$\dot{\lambda}_1 = -\lambda_3 \frac{K_P}{B}$ $\dot{\lambda}_2 = \lambda_3 \left(\frac{K_P}{B} + \frac{K_J}{B}\right) - \lambda_5 \omega^2$ $\dot{\lambda}_3 = -\lambda_2 + \lambda_3 \frac{K_D}{B}$ $\dot{\lambda}_4 = -\lambda_3 \frac{K_J}{B} + \lambda_5 \omega^2$ $\dot{\lambda}_5 = -\lambda_4$
6	$\lambda^T(t_f) = [0\,0\,1]$ $\mathbf{x}^T(0) = [0\,0\,0]$	$\lambda^T(t_f) = [0\,0\,0\,1]$ $\mathbf{x}^T(0) = [0\,0\,0\,0]$	$\lambda^T(t_f) = [0\,0\,0\,0\,1]$ $\mathbf{x}^T(0) = [0\,0\,0\,0\,0]$	$\lambda^T(t_f) = [0\,0\,0\,0\,1]$ $\mathbf{x}^T(0) = [0\,0\,0\,0\,0]$
7	$\dot{\theta}_d^* = \begin{cases} \dot{\theta}_{max}, & \lambda_1 > 0 \\ \dot{\theta}_{min}, & \lambda_1 < 0 \\ \text{singular}, & \lambda_1 = 0 \end{cases}$	$\dot{\theta}_d^* = \begin{cases} \dot{\theta}_{max}, & \lambda_2 > 0 \\ \dot{\theta}_{min}, & \lambda_2 < 0 \\ \text{singular}, & \lambda_2 = 0 \end{cases}$	$\dot{\theta}_d^* = \begin{cases} \dot{\theta}_{max}, & \lambda_1 + \frac{K_D}{B}\lambda_3 > 0 \\ \dot{\theta}_{min}, & \lambda_1 + \frac{K_D}{B}\lambda_3 < 0 \\ \text{singular}, & \lambda_1 + \frac{K_D}{B}\lambda_3 = 0 \end{cases}$	$\dot{\theta}_d^* = \begin{cases} \dot{\theta}_{max}, & \lambda_1 + \frac{K_D}{B}\lambda_3 > 0 \\ \dot{\theta}_{min}, & \lambda_1 + \frac{K_D}{B}\lambda_3 < 0 \\ \text{singular}, & \lambda_1 + \frac{K_D}{B}\lambda_3 = 0 \end{cases}$

motor. In a first step the influence of the elastic joint torque feedback on the motor inertia is neglected as this allows to find a closed solution[6]. Finally, the feedback of the elastic joint torque is also considered. The actuating variable u is chosen to be the desired motor speed $\dot{\theta}_d$. The proportional and damping gain values for the motor controller are denoted as K_P and K_D, respectively.

As the principal approach is the same throughout this chapter, the relevant equations and conditions are summarized for the interested reader in Tab. 10.6 and focus is only laid on the most significant general insights in the following description. Table 10.6 lists the system dynamics (1), the state and input vector (2), the state space equations (3), the Hamiltonian (4), the adjoint system (5), the boundary conditions (6), and the solution of the switching system (7). The eigenfrequency is denoted as $\omega = \sqrt{K_J/M}$.

Since all system equations (row 3) are linear in u, Pontryagin's maximum principle leads to bang-bang control if no singular arcs occur. The optimal switching functions are the terms of the particular Hamiltonian (row 4) that linearly depend on u. Together with its final conditions (row 6) the adjoint equation system (row 5) forms a final value problem.

For case A following solution is obtained for the relevant adjoint λ_1.

$$\lambda_1 = \omega \sin(\omega(t - t_f)) \tag{10.56}$$

Since this function never remains at zero, no singular arcs occur. The switching law is therefore

$$\dot{\theta}_d^* = \dot{\theta}_{\max} \operatorname{sgn}(\sin(\omega(t - t_f))). \tag{10.57}$$

This rectangular function, whose frequency is the resonance frequency of the joint has a phase shift that depends on t_f in order to maximize the link side velocity at this particular time instant. Figure 10.14 depicts an example for the solution of the adjoint and system equation as well as the input. This result leads to the conclusion that with half period $t = \omega/(4\pi)$ the link side velocity is doubled.

As for case A, the optimal control trajectory of case B is also derived from Pontryagin's maximum principle. The solution is again linear in u and of bang-bang type because the switching function never stays at zero for a nonempty time interval. The switching times depend for case B on sign(λ_2), which is found to be

$$\lambda_2(t) = \left(B^2 K_J \mathrm{e}^{\frac{K_P(t-t_f)}{B}} - B^2 K_J \cos\left(\omega\left(t - t_f\right)\right) - B K_P \sin\left(\omega\left(t - t_f\right)\right) \sqrt{K_J M} \right) \left({K_P}^2 M + K_J B^2\right)^{-1}. \tag{10.58}$$

[6] Please note that the stiffness of the motor PD controller is three order of magnitudes larger than the joint stiffness. Therefore, the effect of the elastic torque is expected to be reasonably small to neglect this effect. Later on this will be confirmed with realistic simulation parameters. Furthermore, the result for the optimal control basically leads to switching the motor velocity sign when the elastic joint torque is zero, i.e. it does not significantly affect the motor velocity during the switching instance.

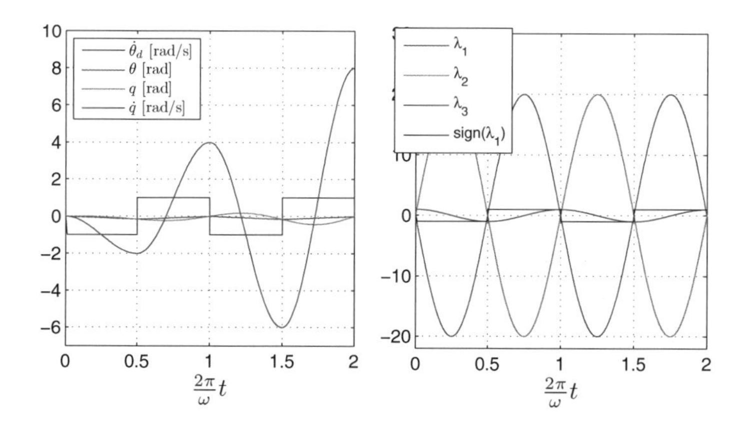

Fig. 10.14 Solution of the adjoint and system equations

Compared to case A the switching condition consists of an additional trigonometric and exponential lag term. However, the principal structure remains the same.

For case C the solution is also similar to the previous ones, except for some additional trigonometric and exponential terms. Again, they do not alter the principal switching structure. The switching condition is

$$\begin{aligned}\lambda_1 + \frac{K_D}{B}\lambda_3 = &-\frac{K_D K_J^2 B}{X1}\cos\left(\omega\left(t-t_f\right)\right)\\ &+\frac{\left(K_J B K_P - K_J K_D{}^2 - K_P{}^2 M\right)\sqrt{K_J M}}{X1}\sin\left(\omega\left(t-t_f\right)\right)\\ &+\frac{X4}{X1X2}\mathrm{e}^{\frac{\left(t-t_f\right)\left(X2+K_D\right)}{2B}}+\frac{X3}{X1X2}\mathrm{e}^{\frac{\left(t-t_f\right)\left(-X2+K_D\right)}{2B}}\end{aligned} \tag{10.59}$$

with

$$\begin{aligned}&\mathrm{X1} = K_J K_D{}^2 M + K_J{}^2 B^2 - 2K_J B K_P M + K_P{}^2 M^2\\ &\mathrm{X2} = \sqrt{K_D{}^2 - 4K_P B}\\ &\mathrm{X3} = 1/2K_J B\left(-2K_J B K_P + 2K_P{}^2 M + K_J K_D\sqrt{K_D{}^2 - 4K_P B} + K_J K_D{}^2\right)\\ &\mathrm{X4} = 1/2K_J B\left(2K_J B K_P - 2K_P{}^2 M + K_J K_D\sqrt{K_D{}^2 - 4K_P B} - K_J K_D{}^2\right).\end{aligned} \tag{10.60}$$

In order to complete the motor model, the feedback of the elastic joint torque shall be considered now (case D). Table 10.6 lists all relevant equations and also the switching law. However, an analytical solution could not be found for this system. Therefore, numerical methods have to be applied. Since the adjoints are not coupled with the system's differential equations they can be solved via numerical integration (e.g. with Runge-Kutta methods).

A comparison of the different motor models is depicted in Fig. 10.15, showing the dynamic response of $\dot{\theta}$ for $\dot{\theta}_d$, being the step function. Two main observations

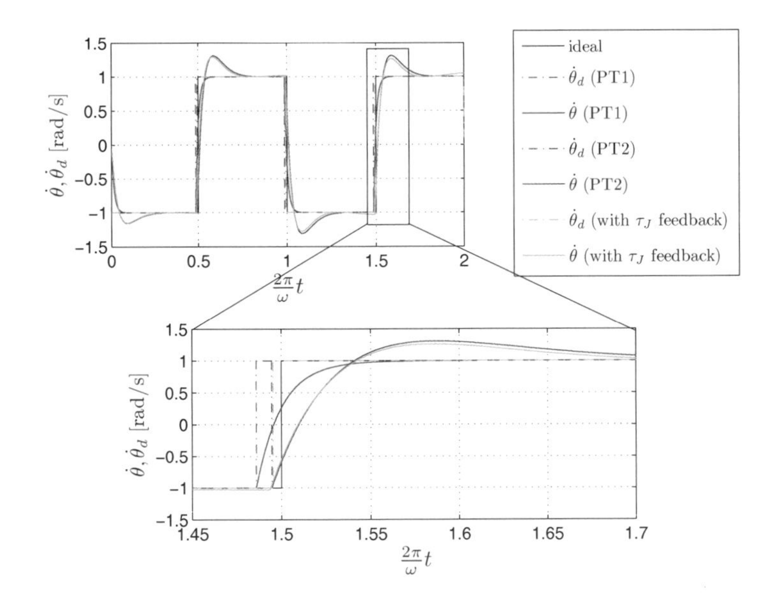

Fig. 10.15 Comparison of the different models

can be made: The significant switching time between PT1 and PT2 and the negligible influence of the elastic joint torque τ_J on the motor response of the PT2 model.

The main conclusions up to now are

- Motor dynamics do not influence the principal switching structure.
- Every delay element leads to a phase shift of the switching times.
- No analytical solution was found[7], when adding the influence of the elastic joint torque τ_J.
- Insufficient motor dynamics lead to a saturation of the characteristic velocity increase curve (not described for brevity).

In the next section the influence of an important real-world constraint of VIA joints is discussed: the elastic deflection limit $\varphi_{\max}$.

10.7.3 Constrained Deflection

Now, an optimal strategy is derived which maximizes the end link velocity of the system A'. The states will be redefined here as $\mathbf{x}^T = (\theta - q, \dot{q}) = (\varphi, \dot{q})$ for convenience and also to be able to visualize the solution in a 2 dimensional plot. The motor velocity $u_1 = \dot{\theta}$ is directy controlled, hence the system dynamics are

[7] Although this problem is theoretically solvable because of its linearity, state of the art symbolic solvers are not capable to find the solution due to its size.

$$\dot{\mathbf{x}} = f(\mathbf{x}, u) = \begin{pmatrix} u_1 - x_2 \\ \omega^2 x_1 \end{pmatrix}. \tag{10.61}$$

Both the motor velocity and the angular deflection φ are now constrained:

$$\forall t \in [t_0, t_f]: \qquad |u_1(t)| \leq u_{1max} \wedge |\varphi(t)| \leq \varphi_{\max} \tag{10.62}$$

Without loss of generality, the initial time t_0 is chosen to be equal 0. Note that the constraints in (10.62) uniquely determine the maximum end link velocity, which the link can move at. This velocity consists of the maximum motor velocity and an additional term $\Delta\dot{q}$, which is related to the potential energy that can maximally be stored in the spring.

Note also that the following condition holds whenever the maximum spring deflection $\varphi_{\max}$ is reached:

$$K_J \varphi_{\max} = M\ddot{q}_{max}, \tag{10.63}$$

where $\ddot{q}_{max}$ is the maximum acceleration of the link. The additional gain $\Delta\dot{q}$ will thus depend on the maximum deflection according to (10.63). The maximum potential energy $E_{pot_{max}} = \frac{1}{2}K_J\varphi_{\max}^2$ may be used to obtain this additional velocity gain:

$$\frac{1}{2}M\dot{q}_{max}^2 = \frac{1}{2}M\dot{\theta}_{max}^2 + E_{pot_{max}} \tag{10.64}$$

$$\dot{q}_{max} = \dot{\theta}_{max} + \omega\varphi_{\max} \tag{10.65}$$

$$\Rightarrow \Delta\dot{q} = \omega\varphi_{\max}, \tag{10.66}$$

where the first two equations can be obtained from energetic and velocity considerations, and $\omega = \sqrt{\frac{K_J}{M}}$ is again the eigenfrequency of the mass-spring system.

In principle, $\dot{q}_{max}$ can be obtained with various control trajectories, all of which make use of the maximal potential energy in the springs. The aim is here, however, to exploit the capabilities of the joint as fast as possible. Therefore, the minimum time to gain the maximum link velocity is seeked for, under the constraints (10.62). This means that $x_2(t_f) = \dot{q}(t_f) = \dot{q}_{max}$ and unlike the optimal control problems solved so far, the final state is not free.

The problem mentioned above is a minimum-time optimal control problem with the cost functional:

$$J(u) = \int_0^{t_f} 1dt, \tag{10.67}$$

and the *free* end time t_f.

Before formally applying the optimal control formalism, i.e. Hamiltonian, costate dynamics etc., some physical reasoning should be started with. This will simplify the problem by dividing it into two subproblems that can be solved separately. Going backwards in time, the problem of reaching a charged state (fully deflected joint spring) from the final state (maximum link velocity) is addressed first, and then how to hit the initial state (resting position) from the charged state.

Reaching Charged State from Final State

The maximum energetic state the joint can occupy is $x(t_{ch}) = \left(\varphi_{\max}\ u_{1max}\right)^T$. To reach $\dot{q}_{max}$ the potential energy $E_{pot_{max}}$ must be completely transformed into kinetic link energy, see Fig. 10.16. The state at final time is therefore $x(t_f) = \left(0\ \dot{q}_{max}\right)^T$. To reach $x(t_f)$ from $x(t_{ch})$ one must simply apply the maximum motor velocity. From an energetic point of view this is the only way to achieve the theoretical maximum link velocity and therefore also the time-optimal solution.

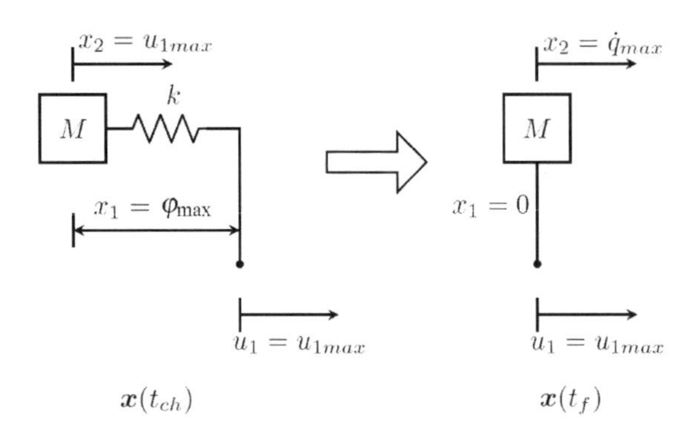

Fig. 10.16 Releasing the spring from $x(t_{ch})$ to reach $x_2^*(t_f) = \dot{q}_{max}$

In general, solving the system dynamics for an arbitrary initial condition $x(0) = \left(x_{10}\ x_{20}\right)^T$ and for a constant input u_1 yields:

$$\frac{x_1^2}{a^2} + \frac{(x_2 - u_1)^2}{b^2} = 1, \tag{10.68}$$

where $a^2 = \frac{1}{\omega^2}\left((u_1 - x_{20})^2 + (\omega x_{10})^2\right)$ and $b^2 = (u_1 - x_{20})^2 + (\omega x_{10})^2$. If a phase plot is used to display the timely evolution of the states, (10.68) indicates that they follow an ellipse. For $u_1 = u_{1max}$ the center of this ellipse is $\left(0\ u_{1max}\right)^T$, while for $u_1 = -u_{1max}$ its center is $\left(0\ -u_{1max}\right)^T$. The size of the ellipse is determined by the semi-minor axis a and the semi-major axis b. Both depend on the deflection $\varphi(0)$ and its time derivative $\dot{\varphi}(0)$ at the initial time.

Tracking the ellipse starting from $x(t_f)$ with $u_1 = u_{1\max}$ backwards in time, we reach $x(t_{ch})$ after a quarter of the periodic time of the spring-mass system, i.e. $t_f - t_{ch} = \frac{\pi}{2\omega}$, see Fig. 10.17. At this instant the angular deflection constraint is active, $\varphi(t_f) = \varphi_{\max}$. If $u_1 = u_{1max}$ is further applied, no greater elastic deflection than the maximum deflection can be reached, because $x(t_{ch})$ lies on the semi-minor axis of the ellipse. Please note that if $u_1 = -u_{1max}$ is applied backwards in time after $x(t_{ch})$, $\varphi_{\max}$ will be exceeded, which is therefore not a valid solution.

Reaching Resting Position from Charged State

The remaining problem is to hit $x(0) = \left(0\ 0\right)^T$ starting from $x(t_{ch})$ in minimum time. For the considered dynamics the time-optimal solution for reaching $x(0)$ from an arbitrary initial state *without* state constraints can e.g. be found in [30]. Here, the inverse problem is considered, which is approached similarly.

For system A', the Hamiltonian yields $H = \lambda_1(u_1 - x_2) + \lambda_2\omega^2 x_1 + 1$. The control input enters the Hamiltonian linearly and the optimal switching function is therefore $\sigma^* = \frac{\partial H}{\partial u_1} = \lambda_1^*$. According to Minimum Principle the optimal control law is thus:

$$u_1^* = \begin{cases} -u_{1max}\text{sign}(\lambda_1^*), & \lambda_1^* \neq 0, \\ \text{singular}, & \lambda_1^* = 0. \end{cases} \tag{10.69}$$

The costate dynamics are

$$\dot{\lambda} = -\frac{\partial H}{\partial x} = \begin{pmatrix} -\omega^2\lambda_2 \\ \lambda_1 \end{pmatrix}. \tag{10.70}$$

This means that the costates of this optimal control problem describe an undamped harmonic oscillator and we need boundary conditions in order to uniquely determine the optimal costates λ^* and the corresponding switching function σ^*. Notice that for this optimal control problem (10.26) can not be used to determine the boundary conditions of the costates as done in previous sections, because the final state $x(t_{ch})$ is fixed. Nevertheless, the final time is free and consequently, the Hamiltonian equals to zero and stays constant along the optimal trajectory $H^* \equiv 0$, assuming the deflection constraint is never active [25]. Using this and the additional condition for free end time [5]

$$\underbrace{\left.\frac{\partial h}{\partial t}\right|_{t_f}}_{=0} + H^* = 0, \tag{10.71}$$

it can be concluded:

$$\lambda_1^*(0)u_1^*(0) + 1 = \lambda_2^*(t_{ch})\omega^2\varphi_{\max} + 1 = 0. \tag{10.72}$$

The left hand side of (10.72) shows that $\sigma^* = \lambda_1^*$ can never stay identically at zero, since $\lambda_1^*(0) = -\frac{1}{u_1^*(0)} \neq 0$. Consequently, the optimal control will be bang-bang with the switching function σ^*:

$$\sigma^* = -\omega\lambda_2^*(0)\sin(\omega t) + \lambda_1^*(0)\cos(\omega t). \tag{10.73}$$

Notice that σ^* decribes a harmonic oscillation, which is 2π-periodic and u_1 switches sign at the latest after a half period π, i.e. one half ellipse in a phase plot. This insight can now be used to construct a switching curve $\mathscr{S}$, see Fig. 10.17. The origin $x(0) = \left(0\ 0\right)^T$ can only be hit if the state trajectory approaches one of the following half ellipses:

$$(\omega x_1)^2 + (x_2 - u_{1max})^2 = u_{1max}^2, x_1 \geq 0 \tag{10.74}$$
$$(\omega x_1)^2 + (x_2 + u_{1max})^2 = u_{1max}^2, x_1 \leq 0 \tag{10.75}$$

The adjacent ellipses of $\mathscr{S}$ be can constructed by drawing a half ellipse from every point of (10.74) or (10.75). This procedure can be repeated in order to get the other ellipses of the switching curve. To approach $x(0) = \begin{pmatrix} 0 & 0 \end{pmatrix}^T$, one must apply $u_1 = u_{1max}$ if the current system state is above $\mathscr{S}$. Below the switching curve $u_1 = -u_{1max}$ must be chosen, respectively. Otherwise, the solution diverges. Please note that the switching curve does not hold for the first part of the problem, i.e. reaching $x(t_{ch})$ from the final state $x(t_f)$. This is because the state inequality constraint $|x_1| \leq \varphi_{\max}$ is active at t_{ch}, which means that σ^* has a discontinuity at this time instant [30]. Furthermore, if the maximum deflection of the spring is not sufficiently large, the state contraint can become active, before $x(t_{ch})$ is reached (see Fig. 10.17, right row). In this case, the analysis loses its validity as well and results in a singular problem. The next part discusses the optimal strategy for this case.

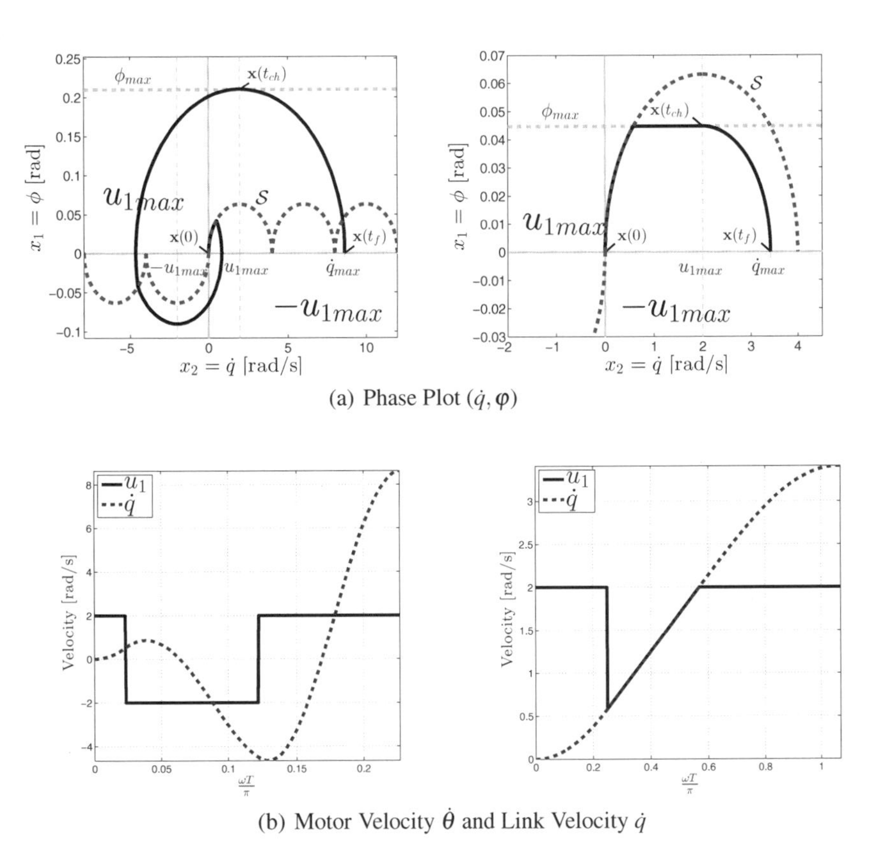

Fig. 10.17 Time-optimal control strategy (M = 0.1 $[kgm^2]$, $K_J = 100$ $[\frac{Nm}{rad}]$, u_{1max} = 2 $[\frac{rad}{s}]$)

Dependency of Solution on the Energy Ratio e_{SL}

The number of control switchings to reach $x(0)$ depends on the spring energy and motor velocity. If the link travels at maximum motor velocity, its kinetic energy is $E_{kin} = \frac{1}{2}M\dot{\theta}_{max}^2$. Next, the ratio of the maximum potential energy in the spring to this kinetic energy is introduced:

$$e_{SL} = \frac{E_{pot_{max}}}{E_{kin}} = \left(\frac{\omega\varphi_{\max}}{\dot{\theta}_{max}}\right)^2. \tag{10.76}$$

In order to express the benefit of the spring on the maximum link velocity in terms of energies, (10.65) and (10.66) can be reformulated as

$$\dot{q}_{max} = \dot{\theta}_{max}(1+\sqrt{e_{SL}}). \tag{10.77}$$

The term $\varepsilon := \dot{q}_{max}/\dot{\theta}_{max} = 1+\sqrt{e_{SL}}$ may be denoted as the joint speed gain.

The switching curve indicates that the motor needs to reverse its direction of speed each time the link speed grows more than two times the motor speed, i.e.

$$n_c = \left\lceil \frac{\dot{q}_{\max}}{2\,\dot{\theta}_{\max}} \right\rceil = \left\lceil \frac{1+\sqrt{e_{SL}}}{2} \right\rceil, \tag{10.78}$$

where n_c is the number of motor cycles. For an odd number of intervals the alternating switching sequence starts with $u_1(0) = u_{1max}$, the first control for an even number of n is $u_1(0) = -u_{1max}$. In Fig. 10.17 (upper left) $e_{SL} = 11$ is chosen, which means that there are three intervals starting with u_{1max}.

As mentioned above, if the maximum deflection is lowered it may intersect the switching curve. Fig. 10.17 (upper right) visualizes this case in a phase plot, which is obtained by the numerical software GPOPS [34], which uses the *Gauss Pseudospectral Method* to solve optimal control problems [3, 12]. The case occurs only if $\varphi_{\max} < \frac{u_{1max}}{\omega}$, respectively $e_{SL} < 1$. Keeping $u_1 = u_{1max}$, once $\varphi_{\max}$ is reached would lead to a violation of the constraint, so that the motor velocity is decreased as soon as $x_1 = \varphi_{\max}$. Although the motor velocity is being lowered, maintaining $x_1 = \varphi_{\max}$ means that the maximum possible elastic torque, $\tau_{J_{max}} = K_J\,\varphi_{\max}$ is used to accelerate the link. It is therefore optimal to follow the constraint in this singular case.

General Treatment of Constraints for System D

Now, the formulation of the optimal control problem with constraints is based on the model of case D. As seen, $\varphi_{\max}$ can be expressed as an inequality constraint on the difference of motor and link side position. Its second derivative incorporates the control variable. Thus, the order of the constraint is $q = 2$ and one contact point exists[8].

---

[8] A single contact point is assumed in order to formulate an unilateral deflection constraint only.

$$S^{(0)} := (\theta - q) - \varphi_{\max} \leq 0 \tag{10.79}$$

$$S^{(1)} := (\dot{\theta} - \dot{q}) \leq 0 \tag{10.80}$$

$$S^{(2)} := (\ddot{\theta} - \ddot{q}) \leq 0 \tag{10.81}$$

The Hamiltonian is extended by a term that incorporates new Lagrange multipliers μ. In total one obtains an 11-th order canonical system of differential equations with side constraints. For contact points the conditions given in [5] count. This leads to a jump in the adjoint variables for the contact time t_b. Because $\frac{\partial S^{(2)}}{\partial x_i} = 0$ and for choosing $\mu_1 = 0$ the jump conditions may be written as

$$\lambda_2(t_b^+) = \lambda_2(t_b^-) + \mu_0 \frac{dS^{(0)}}{dx_2} \tag{10.82}$$

$$\lambda_4(t_b^+) = \lambda_4(t_b^-) + \mu_0 \frac{dS^{(0)}}{dx_4}, \tag{10.83}$$

leading to

$$\begin{aligned} \lambda_2(t_b^+) &= \lambda_2(t_b^-) + \mu_0 \\ \lambda_4(t_b^+) &= \lambda_2(t_b^-) - \mu_0. \end{aligned} \tag{10.84}$$

The additional trivial differential equation is

$$\dot{\mu}_0 = 0. \tag{10.85}$$

The full system of equations can be solved with a numerical multiple-shooting method as e.g. described in [6, 7] or Sec. 10.5.

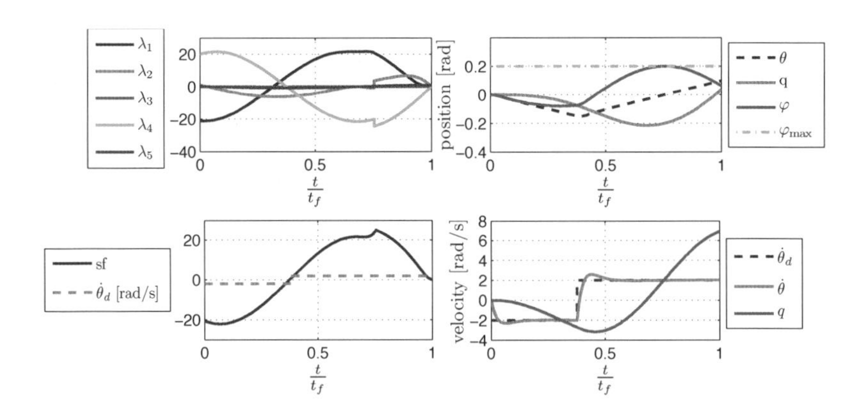

Fig. 10.18 Optimization with limited elastic deflection. sf denotes the switching function.

Figure 10.18 depicts such a numerical solution of the MPBVP obtained with the multiple-shooting method implemented in the program BNDSCO [29]. For the constrained deflection case the optimization aims at the maximal elastic deflection (upper right). The optimal switching time is defined by keeping the constraints, rather than resonant excitation.

In the next section it is discussed to what extent the stiffness adjustment during motion contributes to an increase in maximum link side velocity.

10.7.4 Stiffness Adjustment

First, the influence of stiffness adjustment without a deflection constraint is taken into consideration and then the effect such limits have is analyzed.

10.7.4.1 Unconstrained Deflection

In order to elaborate the effect of stiffness adjustment, the underlying model for this analysis is chosen to be the one of case A. The joint stiffness is now considered as an additional control input. Overall, the system equations are

$$\theta = \int \dot{\theta}_d \, dt \qquad \text{with } \dot{\theta}_{\min} \leq \dot{\theta} \leq \dot{\theta}_{\max} \tag{10.86}$$

$$M\ddot{q} = K_J(t)(\theta - q) \quad \text{with } K_{J,\min} \leq K_J(t) \leq K_{J,\max}, \tag{10.87}$$

with $\mathbf{x} = [\theta, q, \dot{q}]^T$ being the state vector and $\mathbf{u} = [\dot{\theta}_d(t), K_J(t)]^T$ the control input vector. The canonical system of differential equations is

$$\dot{x}_1 = u_1 \tag{10.88}$$

$$\dot{x}_2 = x_3 \tag{10.89}$$

$$\dot{x}_3 = \frac{u_2}{M}(x_1 - x_2) \tag{10.90}$$

$$\dot{\lambda}_1 = -\lambda_3 \omega^2 \tag{10.91}$$

$$\dot{\lambda}_2 = \lambda_3 \omega^2 \tag{10.92}$$

$$\dot{\lambda}_3 = -\lambda_2. \tag{10.93}$$

The corresponding Hamiltonian can be derived as

$$H(\mathbf{x}(t), \lambda(t), \mathbf{u}(t), t) = \lambda_1 u_1 + \lambda_2 x_3 + \lambda_3 \tfrac{u_2}{M}(x_1 - x_2). \tag{10.94}$$

The Hamiltonian is linear in u_1 and u_2, leading directly to following switching laws.

$$\dot{\theta}_d^* = \begin{cases} \dot{\theta}_{\max}, & \lambda_1 > 0 \\ \dot{\theta}_{\min}, & \lambda_1 < 0 \\ \text{singular}, & \lambda_1 = 0 \end{cases} \tag{10.95}$$

$$K_{J,d}^{*} = \begin{cases} K_{J,\max}, & \lambda_3 \frac{x_1 - x_2}{M} > 0 \\ K_{J,\min}, & \lambda_3 \frac{x_1 - x_2}{M} < 0 \\ \text{singular}, & \lambda_3 \frac{x_1 - x_2}{M} = 0 \end{cases} \tag{10.96}$$

Due to the bang-bang structure of the desired stiffness the solution of the adjoints is similar to (10.56):

$$\lambda_1 = \sqrt{\frac{u_{2,\max}}{M}} \sin\left(\sqrt{\frac{u_{2,\max}}{M}}(t - t_f)\right) \tag{10.97}$$

$$\lambda_3 = \cos\left(\sqrt{\frac{u_{2,\max}}{M}}(t - t_f)\right) \tag{10.98}$$

This solution is derived for $t > t_f - \frac{T_{\max}}{4} = t_f - \frac{\pi}{2}\sqrt{\frac{M}{K_{J,\max}}}$. It is not straightforward to obtain the full solution for the system in case $t_f > \frac{T_{\max}}{4}$. For this, all switching times have to be found in order to identify the initial conditions for each cycle. For the present case two adjoints influence the switching condition. λ_1 determines the excitation of the system with $\dot{\theta}_d$. The stiffness switching function is characterized by two terms. First, the sign of the elastic deflection $\text{sgn}(x_1 - x_2)$ and secondly, the switching function λ_3.

10.7.4.2 Constrained Deflection

Based on Sec. 10.7.3 it is clear that the stiffness adjustment between maximal elastic deflection (maximum potential energy stored) and the time instant of maximal velocity (moment of launch) is critical. Therefore, the maximization of the Hamiltonian (10.94) during this particular time interval is investigated. The term containing the stiffness u_2 and the elastic deflection $(x_1 - x_2) = (\theta - q)$ is to be maximized.

$$\max\left\{\lambda_3 \frac{u_2}{M}(x_1 - x_2)\right\}. \tag{10.99}$$

$(x_1 - x_2)$ is always larger than zero between the moment of its maximal value and launch. The maximal value will be achieved the earliest at $t_f - \frac{1}{2\pi\omega}$. Due to the transversality condition $\partial h(\mathbf{x}(t_f))/\partial x_3 = \partial \dot{q}(t_f)/\partial \dot{q} = 1$ the last adjoint λ_3 reaches its maximal value $\lambda_3 = 1$ at t_f (see (10.98)). Furthermore, it changes its sign also at a quarter of the periodicity before the launch time. The switching function λ_3 is consequently positive in the considered time interval. This leads, according to the maximum principle, to maximizing the stiffness (see (10.96)) towards the moment of launch.

$$K_J^{*} = K_{J,\max} \qquad t_b \leq t \leq t_f \tag{10.100}$$

Up to now, it was assumed that the stiffness trajectory before the boundary point does not influence the end velocity. Therefore, it would be reasonable to set the stiffness to its maximum value during the throwing trajectory without additionally adjusting the stiffness. However, from a practical point of view it may be necessary to start the motion at low stiffness adjustment and enlarge it towards the launch time. This can have three main causes:

- The motor dynamics is not sufficient to excite the joint at maximum stiffness at the corresponding eigenfrequency.
- The motor power is not sufficient to deflect the joint with an adequately low number of switching cycles.
- Limits on the elastic deflection can lead to higher energy storage for lower stiffness ranges due to higher possible deflection than for higher stiffness presets.

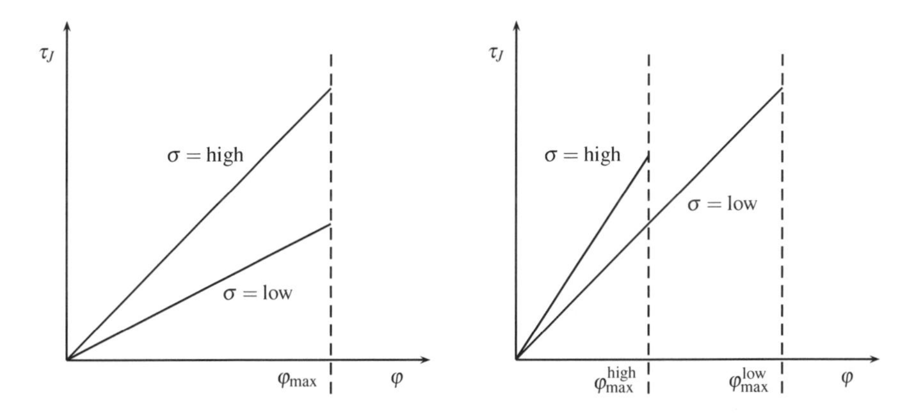

Fig. 10.19 Deflection limits φ_{max} for different stiffness presets σ. The left figure shows a design, where φ_{max} is constant for every σ and the right one depicts a functional relationship between φ_{max} and σ.

The last aspect can be explained with Fig. 10.19 and is caused by the implemented working principle of the VIA mechanism. The left figure shows two different linear stiffness curves for which the maximum deflection is constant for all presets. On the right one a characteristic is depicted, where a functional relationship between maximum deflection and stiffness preset exists.

First, the former is discussed. According to the maximum principle the Hamiltonian is maximized through the entire motion process and therefore the joint stiffness as well. Consequently, the potential energy stored in the joint elasticity is maximized for every deflection. This is not optimal for changing the stiffness, on the contrary, it reduces the achievable link velocity.

For the latter characteristic the maximum elastic energy that can be stored depends on the deflection. For large deflection a soft preset and for small deflection a stiff one are preferred. Maximization of joint torque is therefore directly coupled with adjusting stiffness along the admissible deflection.

All the theoretical results summarized so far, imply that using elastic joints can significantly increase the performance of a robot joint in terms of achieving high link velocities compared to a rigid robot. Next, the analysis for a concrete joint design is discussed and various experimental results are presented.

10.7.5 Performance Analysis for the QA-Joint

In this section, the elaborated insights are applied to the DLR QA-Joint. The motor in the joint will be modeled as a PT2 actuator and damping effects will be ignored. The maximum velocity will be numerically investigated, which can be obtained by one switching of the desired motor velocity, and the results will be verified experimentally. In order to find boundaries for this velocity, the theoretical results from the previous sections are used and extended by including a nonlinear spring into the undamped joint model.

10.7.5.1 Without Stiffness Adjustment

From an energy point of view, the maximum kinetic energy of the joint $E_{kin_{max}}$ will always depend on the maximum potential energy stored in the nonlinear spring and the maximum motor velocity. In other words, (10.64) holds regardless of the spring characteristics and (10.77) is still valid with e_{sl} now being defined as:

$$e_{sl} = \frac{E_{pot_{max}}}{E_{kin}} = \frac{2\int_0^{\varphi_{\max}} \tau_J(\varphi)d\varphi}{M\dot{\theta}_{max}^2} = \left(\frac{\dot{\varphi}_{\max}}{\dot{\theta}_{max}}\right)^2$$

The elastic joint torque is again denoted by τ_J and $\dot{\varphi}_{\max}$ stands for the maximum velocity of the joint relative to the motor, which is obtained when all the potential energy is transformed to the kinetic energy. Similar to the previous section, we can numerically compute time-optimal trajectories, which result in this maximum link velocity $\dot{q}_{max} = \dot{\varphi}_{\max} + \dot{\theta}_{max}$ in minimum time. Figure 10.20 shows phase plots of these trajectories for different e_{sl} values. The simulations were done again with GPOPS and e_{sl} was increased by decreasing $\dot{\theta}_{max}$ while keeping $E_{pot_{max}}$ and thus $\varphi_{\max}$, $\dot{\varphi}_{\max}$ constant. The simulated joint has the same nonlinear spring chracteristics as the QA-Joint and its elastic joint torque τ_J can be computed from:

$$\tau_J = \frac{1}{15}\left(a_s e^{15(\varphi-\sigma)} - b_s e^{15(-\varphi-\sigma)}\right),$$

where a_s and b_s depend on the stiffness actuator position σ. The equation above corresponds to eq. (10.7) for $a_s = b_s = 600$.

According to the illustrated phase plots, the energy ratio e_{sl} can still be used as an indicator for the singularity as well as the number of the switches for the time-optimal trajectory. As observed from Figure 10.20 if $e_{sl} < 1$, the angular deflection constraint becomes active before $\dot{q}_{max}$ reaches $\dot{\theta}_{max}$ and yields a singular solution along the time-optimal trajectory. Furthermore, it can be shown that for a fixed

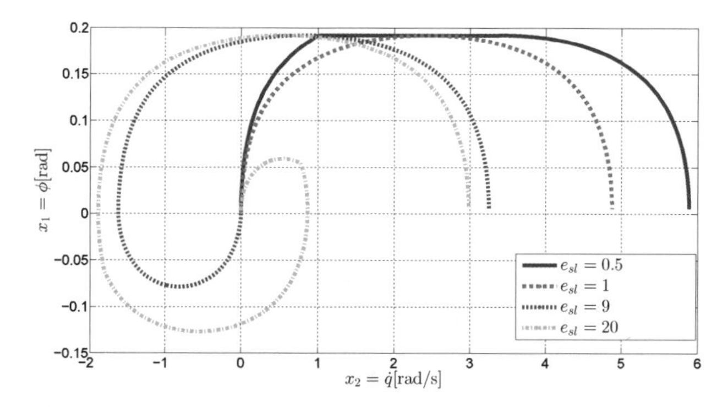

Fig. 10.20 Time-Optimal Control Strategy for the QA-Joint ($u_1 = \dot{\theta}, a_s \approx 26.34, b_s \approx 28.62, \varphi_{\max} = \sigma = 11^o, \dot{\varphi}_{\max} \approx 2.44\frac{rad}{s}$)

number of switches n_c, the motor velocity must always change at $\varphi = 0$ to obtain the maximum values for $\dot{\varphi}$ and $\dot{q}$ after the switches. Since $|\dot{\varphi}(0)| = u_{1max}$ and switching of u_1 changes $\dot{\varphi}$ by $2u_{1max}$, the relative velocity $\dot{\varphi}_{nc}$, which denotes $\dot{\varphi}$ obtained with n_c motor cycles is bounded above[9]:

$$\dot{\varphi}_{n_c} \leq (2n_c - 1)\dot{\theta}_{max}. \tag{10.101}$$

Consequently, the link velocity after n_c cycles satisfies $\dot{q}_{n_c} \leq 2n_c\dot{\theta}_{max}$ yielding a similar relation for n_c as in (10.78). For ex. with switching of the motor ($n_c = 2$), $\dot{q}_{n_c}$ can theoretically be maximum $4\dot{\theta}_{max}$ according to this relation. Figure 10.20 shows that this velocity is indeed obtained in a time-optimal way by one switching of the motor at $\varphi = 0$, when $e_{sl} = 9$ or equivalently $\dot{q}_{max} = 4\dot{\theta}_{max}$. Note that for $e_{sl} > 9$, the maximum velocity $\dot{q}_{max}$ can not be obtained with one switching anymore.

It could be seen how the maximal velocity of the constrained joint is bounded by the number of the switches n_c, the maximum motor velocity $\dot{\theta}_{max}$ and of course the maximum potential energy in the spring $E_{pot_{max}}$.

In the experiments, the motor velocity cannot be controlled directly but the motor torque can. For that reason, in the search for the optimal control strategy, the motor will be modelled as a PT2 actuator. The state vector $x = \left(\theta_d\ \theta\ \dot{\theta}\ q\ \dot{q}\right)^T$, the control input $u_1 = \dot{\theta}_d$, and the initial conditions $x(0) = 0$ will be used to obtain the following system of differential equations that describe the system when taking account for the elastic torque feedback to the motor with a PT2 behaviour:

$$\dot{x} = \begin{pmatrix} u_1 \\ x_3 \\ \frac{1}{B}(\tilde{\tau}_m - \tau_J) \\ x_5 \\ \frac{\tau_J}{M} \end{pmatrix},$$

[9] See Appendix B for a detailed analysis.

with $\tilde{\tau}_m$ denoting the bounded motor torque

$$\tilde{\tau}_m = \begin{cases} \tau_{m,\max} & \tau_{m,d} \geq \tau_m^{\max} \\ \tau_{m,d} & \tau_m^{\min} < \tau_{m,d} < \tau_m^{\max} \\ \tau_{m,\min} & \tau_{m,d} \leq \tau_{m,d}^{\min}. \end{cases} \quad (10.102)$$

$\tau_{m,d} = K_D(u - x_3) + K_P(x_1 - x_2)$ is the desired motor torque from the PD controller. The Hamiltonian for the optimal control problem of maximizing the end link velocity, i.e. for a cost function $J = -\dot{q}(t_f)$ is then:

$$H(.) = \lambda_1 u + \lambda_2 x_3 + \lambda_3 \frac{1}{B}(\tilde{\tau}_m - \tau_J(\sigma)) + \lambda_4 x_5 + \lambda_5 \frac{1}{M} \tau_J(\sigma). \quad (10.103)$$

The optimal control problem to be solved consists of a system of differential equation of 11-th order (adjoint and system equations), including the additional trivial differential equation if taking into account the elastic deflection limit with one boundary point, see Sec. 10.7.3. The nonlinearity causes a coupling of the adjoint and state equations, leading to a MPBVP with separate initial and end conditions for the canonical system of differential equations[10]. The limits of motor torque eventually lead to a necessary formulation of boundary control. Solving this problem with multi-shooting methods did not converge. This is because on the one side $5n$ starting conditions need to be estimated for n nodes and their deviation from the solution is highly influencing the convergence of the method. Furthermore, a physical interpretation of the adjoint variables is also not given. Thus, the estimation of their start values would lead to a solution that is not straightforward.

A possibility to solve this optimization problem is a parameter estimation method by utilizing the information that the optimal control trajectory shows bang-bang behavior (which comes from the linear occurrence of the input into the state equation). This is also independent from the limit in motor torque $\tilde{\tau}_m$ (see (10.102)), as the principal structure of the Hamiltonian remains the same regardless of the saturation[11]:

$$\widetilde{H}(\lambda(t), u(t)) = \left(\lambda_1 + \lambda_3 \tfrac{K_D}{B}\right) u, \quad \tau_{m,\min} < \tau_{m,d} < \tau_{m,\max} \quad (10.104)$$

$$\widetilde{H}(\lambda(t), u(t)) = \lambda_1 u, \quad (\tau_{m,d} < \tau_{m,\min}) \vee (\tau_{m,d} > \tau_{m,\max}) \quad (10.105)$$

The parameter to be estimated is the switching time. The optimization is carried out by multiple solving of the system equations with the jumping times in the control variable being timely varied via appropriate optimization. The algorithm applied is the Nelder-Mead Simplex-Downhill method with the following optimization criterion.

$$J = -\dot{q}(t_f) + J_p \quad (10.106)$$

[10] The adjoint system is given in Sec. 10.7.5.2.

[11] Please note that only the relevant term of the Hamiltonian is shown, which linearly depends on u.

$$J_p = \begin{cases} 0 & \varphi_{\min} \leq \varphi \leq \varphi_{\max} \\ \exp(|\varphi| - \varphi_{\max}) & |\varphi| > \varphi_{\max} \end{cases} \quad (10.107)$$

Complying with the constraints is ensured with penalty term J_p.

Similar to the results regarding the simpler joint model, under the premise of achieving maximal deflection with one switching cycle (throwing with striking out once), a limited velocity range for the position motor complies in this case, as well. On one hand, a minimum motor velocity $\dot{\theta}_{max}$ for achieving the maximal deflection ϕ_{max} is needed. Decreasing this motor velocity will increase e_{sl} and one switching will not be sufficient to make use of the maximum potential energy. On the other hand, there exists a maximum velocity at which the total potential energy can be used without staying on the singular trajectory. Simulation results are depicted in Figures 10.21 and 10.22 with blue curves, whereas the experimental results are indicated with green crosses. Note that in our new model, the minimum boundary for $\dot{\theta}_d$ correspond to a e_{sl} value lower than 9, which is due to the change in our control model and the additional motor torque limitation.

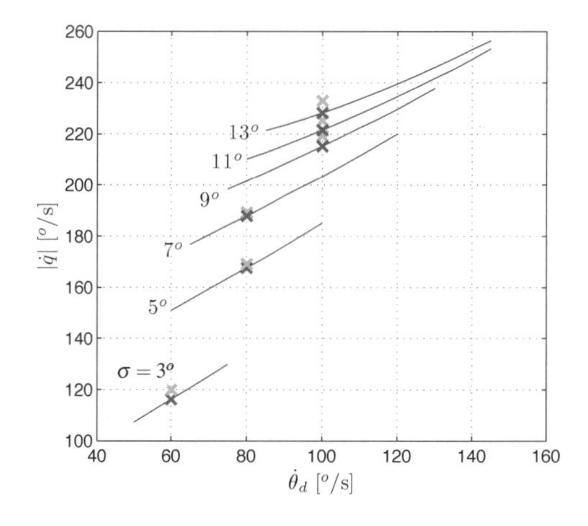

Fig. 10.21 Final link velocity $\dot{q}(t_f)$ as a function of the desired motor velocity $\dot{\theta}_d$

Fig. 10.21 shows the absolute achievable final velocity, as a function of commanded motor velocity characterized by the almost linear relationship. This induces a continuous velocity increase with stored potential energy as expected from (10.77). Furthermore, it becomes clear that too low elasticity leads to a degradation of achievable link velocity. The relative velocity increase with respect to the motor velocity at the final time, $\varepsilon = \frac{\dot{q}_{max}}{\dot{\theta}_{max}}$, is depicted in Fig. 10.22. It can be stated that ε degrades with increasing motor velocity and increasing stiffness. As already explained, it is necessary to drive with higher motor velocities to achieve the maximum deflection for low stiffness. For the QA-Joint the largest speed gain can be obtained at $\dot{\theta}_d = 65^o$/s and moderate stiffness. This is equivalent to $\varepsilon = 2.7$.

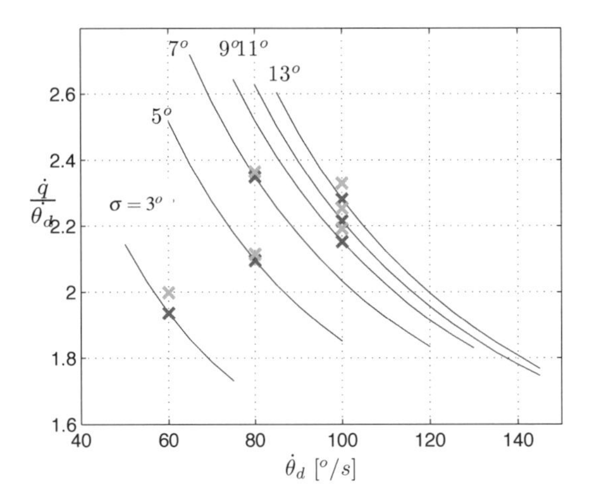

Fig. 10.22 Relative final link side velocity as a function of motor velocity

In Fig. 10.23, the time courses of measurements and simulations for high and low stiffness presets are shown. The upper row shows the motion for $\theta_d = 60^o$/s, $\sigma = 3^o$ and the lower row depicts the results for $\dot{\theta}_d = 100^o/$s, $\sigma = 11^o$. The subscript *mdl* indicates the simulation results and *msr* the measurements. The relevant variables are the link side velocity, deflection, and the elastic joint torque:

- **Link Velocity (left)**
  The trajectory of the link velocity shows very good consistency with the simulation. At final time the velocity is approximately twice the motor velocity. The deviation in joint torque are almost not reflected in the velocity profile.
- **Deflection (middle)**
  In contrast to the simulation a slight exceedance of the deflection constraints can be observed in the lower row. This is mainly due to the variance in the identified stiffness and friction parameters, calibration errors, and simplified assumptions for the friction model.
- **Joint Torque (right)**
  The principal time course of the joint torque confirms the joint model with respect to the identification of stiffness and friction. The discontinuities in the simulation are caused by a Coulomb friction model during direction changes (See [20] for details).

To sum up, the experiments verified the use of the maximal angular deflection ϕ_{max} in springs to maximize the end link velocity $\dot{q}(t_f)$. Also the energy ratio e_{sl} is shown to be suitable in analysing maximum velocities obtained by a fixed number of motor cycles.

Next, elastic joints with variable stiffness actuators are discussed, where the stiffness can explicitly be regulated, i.e. the torque-deflection relation is assumed to be adjustable but linear. The focus is on maximizing the angular deflection and based

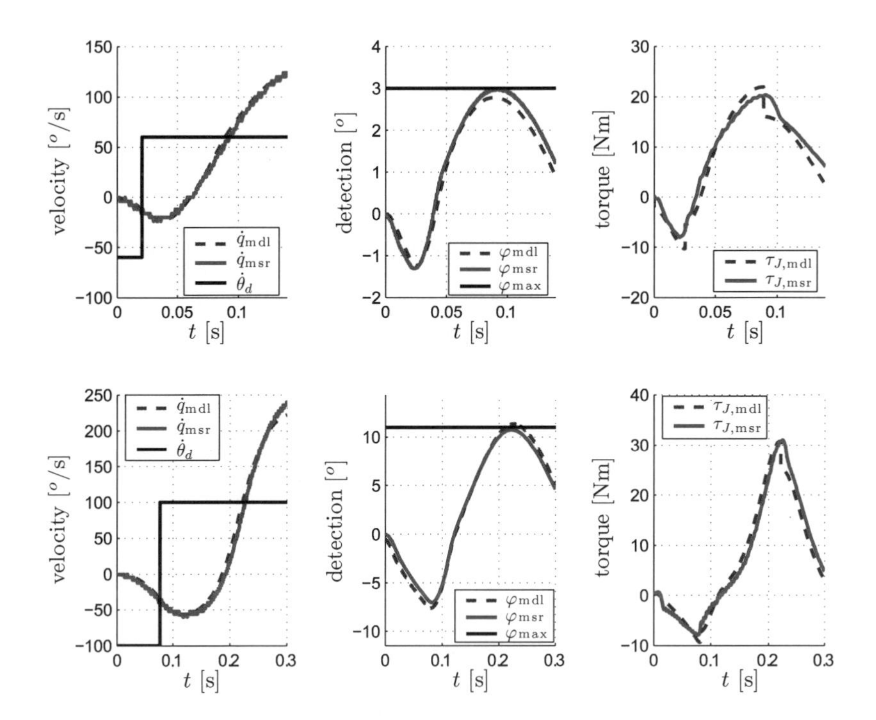

Fig. 10.23 Comparison of simulation and measurements for different stiffness presets. The upper row shows the motion for $\dot{\theta}_d = 60\ ^o$/s and $\sigma = 3\ ^o$. The lower row depicts the results for $\dot{\theta}_d = 100\ ^o$/s, $\sigma = 11\ ^o$.

on this investigation the basic problem of how to optimally store potential energy into a variable stiffness joint is analyzed. The obtained strategy will turn out to be strongly related to the maximization of its kinetic energy and thus its end link velocity.

10.7.5.2 Stiffness Adjustment

For the stiffness adjustment during the motion there are also some conclusions to be drawn. For the linear joint stiffness it was shown that the relation between stiffness and deflection is critical, see Sec. 10.7.4.1. For the QA-Joint this constraint is formally defined as

$$\sigma \geq \varphi \qquad \sigma \in [3^o, 15^o]. \tag{10.108}$$

For maximizing the Hamiltonian (10.103), following term is considered, which explicitly depends on the stiffness adjustment σ.

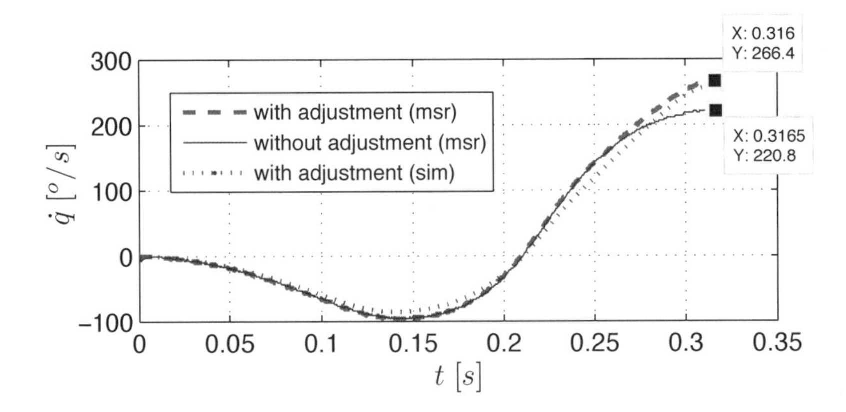

Fig. 10.24 Link side velocity with stiffness adjustment

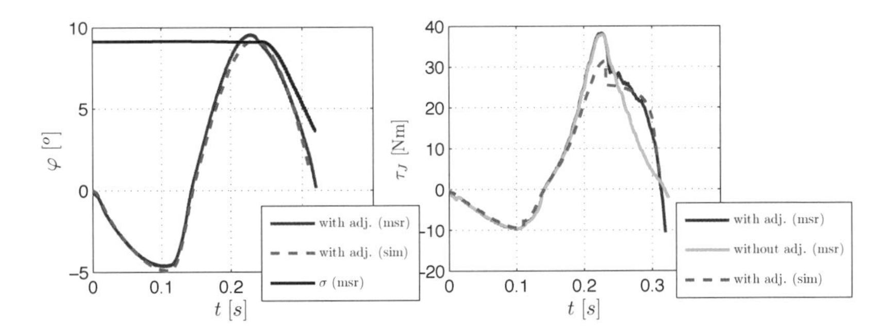

Fig. 10.25 Deflection (left) and joint torque (right) with stiffness adjustment

$$\widetilde{H}(\lambda(t), \mathbf{x}(t), \sigma(t)) = \underbrace{\left(\lambda_5 \frac{1}{M} - \lambda_3 \frac{1}{B}\right)}_{\lambda^*} \tau_J(\sigma) \tag{10.109}$$

As assumed in Sec. 10.7.4.1 only a stiffening during the relaxation phase is essential. Thus, the sign of $\dot{\varphi}$ does not change. In order to confirm the assumption $\lambda^* \geq 0$ for the experiment, the adjoint equations have to be solved for the time interval of stiffness adjustment. Since they do not show discontinuities they can be solved numerically as a final value problem by utilizing the already optimized solution of the state equations.

The solution of the adjoint equation systems in the time interval $[t_b, t_f]$ gives the confirmation that the stiffness adjustment presented in Sec. 10.7.3 is indeed satisfying the necessary conditions of optimal control theory. For this, the switching

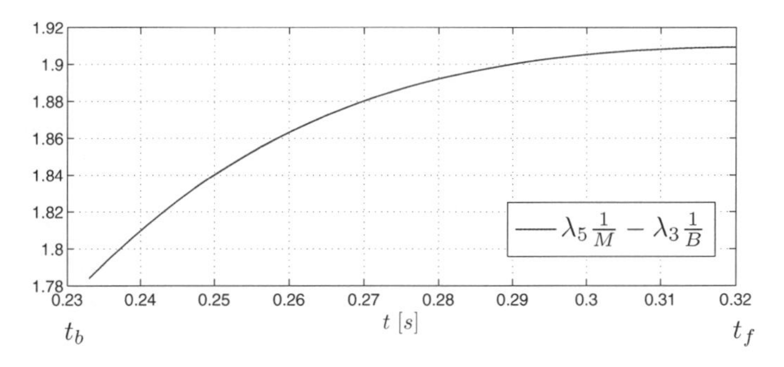

Fig. 10.26 Solution of the switching function with stiffness adjustment

function λ^* has to have positive sign in this interval. The system of differential equation for the adjoints is

$$\dot{\lambda}_1 = -\lambda_3 \frac{1}{B} K_P \tag{10.110}$$

$$\begin{aligned}\dot{\lambda}_2 = \lambda_3 \frac{1}{B}&((b_S - b_R)\exp(15(\varphi - \sigma)) \\ &-(a_S - a_R)\exp(15(\varphi - \sigma)) + K_P)\end{aligned} \tag{10.111}$$

$$\dot{\lambda}_3 = -\lambda_2 + \lambda_3 \frac{K_P}{B} \tag{10.112}$$

$$\begin{aligned}\dot{\lambda}_4 = &\left(\lambda_5 \frac{1}{M} + \lambda_3 \frac{1}{B}\right)((b_S - b_R)\exp(15(\varphi - \sigma)) \\ &+(a_S - a_R)\exp(15(\varphi - \sigma)))\end{aligned} \tag{10.113}$$

$$\dot{\lambda}_5 = -\lambda_4, \tag{10.114}$$

where $\varphi = x_2 - x_4$. With final values $\lambda^T(t_f) = [0\ 0\ 0\ 0\ 1]$ the problem can be formulated as final value problem and e.g. be solved with Runge-Kutta variants. Figure 10.26 depicts the solution of the switching function $\lambda^* = \lambda_5 \frac{1}{M} - \lambda_3 \frac{1}{B}$, showing the positive sign over the relevant time interval. Therefore, τ_J has to be maximized[12].

$$\tau_J = \frac{\mathrm{sgn}(\dot{\varphi})}{e^{15\sigma}} \left[(a_S - a_R)e^{15(x_2 - x_4)} - (b_S + b_R)e^{15(-x_2 + x_4)}\right] \tag{10.115}$$

The maximization of the elastic torque in turn necessitates the maximization of stiffness, respectively a minimization of σ at every time instant. Taking (10.108) into account the optimal stiffness trajectory is

$$\sigma^* = \begin{cases} 3; & \varphi \leq 3 \\ \varphi; & 3 < \varphi < 15. \end{cases} \qquad t_b \leq t \leq t_f \tag{10.116}$$

[12] Please note that for this case τ_J denotes the ideal elastic joint torque plus the friction model.

This means that the acceleration torque has to be sustained during relaxation as long as possible. From an energy point of view, the stiffness adjuster injects additional energy such that the joint maximally stores potential energy for a certain deflection. The potential energy that can be converted into kinetic energy is therefore maximized at the same time.

The according experimental verification is depicted in Fig. 10.24 and Fig. 10.25. For a moderate stiffness preset $\sigma = 9\,^o$ the achieved link velocity is 266 ^{o}/s, which is approximately 20 % higher than without adjustment. From Figure 10.25 (left) it can be observed that adjusting the stiffness according to (10.116) is not fully achieved due to too little dynamics of the stiffness motor[13]. Nonetheless, a significant velocity increase is observed here as well. Compared to the constant elasticity case the joint torque shows an increase from the moment of adjustment on, confirming the theoretical requirement to maximize the sustaining torque during relaxation phase.

In the next section the role of joint compliance for safety in Human-Robot Interaction is treated.

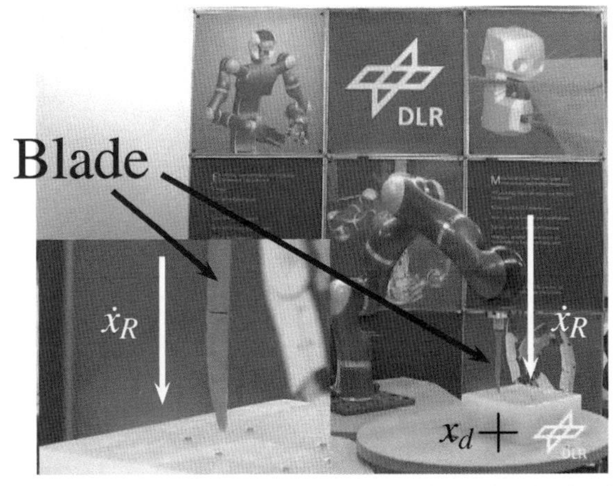

Fig. 10.27 The LWR-III equipped with a knife moves along a desired trajectory. The penetrated material is a silicone block. This experiment shows the benefit of intrinsic or controlled joint elasticity during impacts with sharp tools. The goal position x_d was ≈ 7 cm inside the silicone block.

10.8 Compliance as a Cornerstone of Safety?

As already discussed in Chapter 5, one can identify two immediate sources of possible human injury due to contact with a robot, namely sharp or blunt contacts. Furthermore, the contact stiffness plays a critical role since it defines the time constants of the collision. Furthermore, a reduction in joint stiffness cannot reduce the impact

[13] Please note that the stiffness adjuster is assumed to show ideal behavior for the simulation.

characteristics during rigid, fast blunt impacts for robots with similar link inertia to the LWR-III when reducing the stiffness from the quite high intrinsic values of the LWR-III to lower levels.

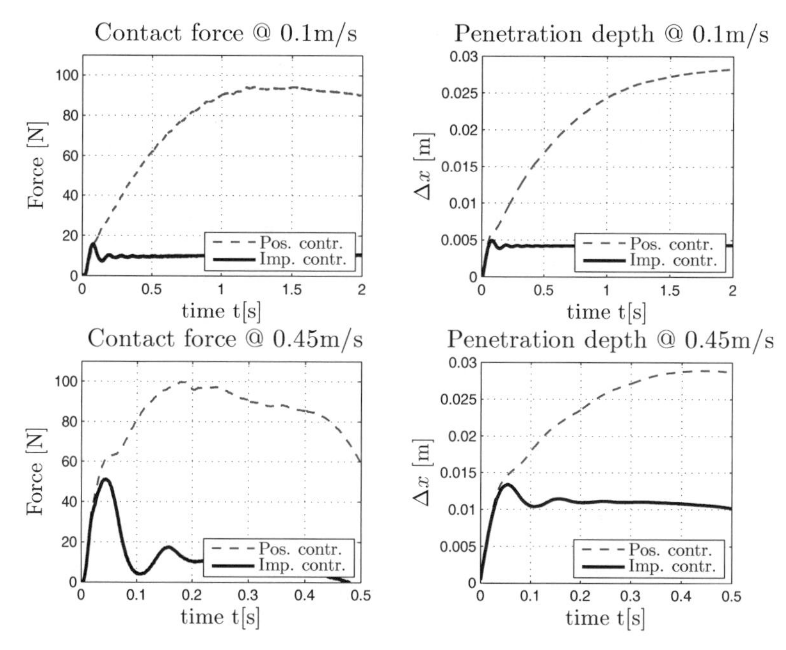

Fig. 10.28 Contact force and penetration depth for two different Cartesian velocities of 0.1 m/s and 0.45 m/s. Clearly, the benefit of (active) joint stiffness reduction is apparent. The force level can be decreased even below levels which would potentially harm a human, whereas in position control the force significantly exceeds this threshold. The goal position x_d was $\approx$ 7 cm inside the silicone block.

In case of injury caused e.g. with a slower motion, by sharp tools, the scenario changes drastically. In Chapter 6 the focus of possible injuries in Human-Robot Interaction was extended to various soft-tissue injuries as stab wounds and an extensive evaluation was given. The effect of the joint compliance in case that the robot has a potentially dangerous tool is discussed in the following.

10.8.1 Sharp Contact

As shown in Fig. 10.27 the LWR-III, in this case holding a knife, moves along a desired trajectory in position or joint impedance controlled mode, penetrating a silicone block. Although it is clear that the contact force will increase slower the lower the joint stiffness, it is not apparent what the maximum forces will be with such dangerous tools. According to [15] already contact forces of < 80 N are enough to penetrate the human skin and cause further injury with a knife in case of stabbing.

While moving in position control the joint stiffness was ≈ 6000 Nm/rad and during joint impedance control only 100 Nm/rad [14]. In Figure 10.28 the penetration force and depth during such a movement are visualized for these joint stiffnesses. It shows that with low joint stiffness the force and penetration depth increase much slower and for this particular trajectory one presumably could prevent damaging the human skin. The fundamental question regarding the point at which intrinsic joint stiffness is advantageous compared to actively controlled one is still to be answered. Generally speaking, this means to find the impact velocity and stiffness above which a controlled stiffness is no longer capable of realizing the required decoupling.

However, this experiment clearly shows the enormous benefit of actively controlled and intrinsic joint compliance: A possible collision detection and reaction scheme as presented in Chapter 3 gains valuable time for detection and reaction since the potential injury, correlating with penetration depth, increases significantly slower compared to the case of a stiff robot. However, in Chapter 6 it is also shown that injury can be prevented even in position control to a certain extent.

10.8.2 Blunt Contact

Usage of intrinsic joint compliance is mostly motivated by the achievement of motor and link inertia decoupling during human-robot impacts and therefore reducing collision danger by alleviating the impacting robot inertia. Furthermore, it was shown that the HIC, which is a criterion associated with resulting head acceleration, as well as similar head injury criteria could be reduced by introducing elasticity in the joint design. Some typical properties during blunt impacts and the role relevant mechanical parameters play are discussed now.

As pointed out in [4] a joint with relatively low reflected link inertia $M_x = 0.1$ kg is able to reduce the impact characteristics significantly if a contact stiffness of $K_H =$ 5 kN/m is assumed. Basically, the following conditions were assumed:

$$\text{stiff}: \; B_x >> M_x \quad M_x << M_H \quad K_{J,x} >> K_H \tag{10.117}$$

$$\text{compliant}: \; B_x >> M_x \quad M_x << M_H \quad K_{J,x} << K_H, \tag{10.118}$$

where $M_x, M_H, K_H \in \mathbb{R}^+$ are the reflected link inertia, head mass, and head stiffness. $B_x, K_{J,x} \in \mathbb{R}^+$ are the reflected motor inertia and joint stiffness, respectively. Similar to the work in [39] it was shown that a decrease in joint stiffness can significantly reduce the impact characteristics and thus is a powerful countermeasure against large contact forces. In [31] it was deduced that for the case of a 2-DoF planar intrinsically compliant robot, already slightly touching a rigid wall with its second link, the compliant mechanism can limit the maximum static force/torque

[14] The bandwidth of the controlled stiffness is high enough to emulate the behavior of a variable stiffness joint during an impact with soft material.

effectively if the motor torque is slowly increased. The corresponding conditions are

$$\text{stiff}: \ M_x \approx 0 \quad B_x << M_H \quad K_{J,x} < K_H \tag{10.119}$$

$$\text{compliant}: M_x \approx 0 \quad B_x << M_H \quad K_{J,x} << K_H \tag{10.120}$$

In the cited work fundamental insights into the aspects joint elasticity plays for safety at different impact conditions were given. It is clearly demonstrated that joint elasticity decouples the motor from the link. However, as was indicated in Chapter 5, it was observed that a reduction in joint stiffness cannot reduce the impact characteristics during rigid, fast blunt crash-test dummy impacts for the LWR-III. This was proven by measuring the decoupling of motor and link inertia via the integrated joint torque sensor and the additionally recorded external contact force. This is unexpected and shows that the compliance of the built in Harmonic Drive and the joint torque sensor is sufficient to decouple motor from link, making it unnecessary to further reduce joint stiffness for the given robot. There are two main aspects, which have to be considered to fully understand this result. On the one hand, the contact stiffness of the used crash-test dummy is significantly larger ($K_H \approx 10^6$ N/m) than the reflected elasticity of the LWR-III ($K_{J,x} \approx 10^5$ N/m). Furthermore, the reflected motor and link inertia of the LWR-III are $B_x \approx 13$ kg and $M_x \approx 4$ kg for the investigated configuration, i.e. in the order of magnitude of the head mass. The corresponding mass and stiffness relations are therefore:

$$B_x > M_x \quad M_x \approx M_H \quad K_{J,x} << K_H \tag{10.121}$$

This aspect is not unique to the LWR-III, but of more general characteristic. Consider the simplest two-link manipulator (q_1, q_2), having only point masses m_1, m_2 at the distal end of each link. The associated Operational Space mass matrix in body coordinates may be written as

$$M_x(\mathbf{q}) = \begin{bmatrix} m_2 + \frac{m_1}{\sin^2(q_2)} & 0 \\ 0 & m_2 \end{bmatrix} \tag{10.122}$$

The $x-$axis is pointing along the main axis of the second link. When considering the stretched out configuration and hitting the head in $y-$direction, the reflected inertia in this direction is simply m_2. Now it is assumed that the arm has human-like inertia properties with link weights $m_{1,2} \approx 2$ kg. This is an ambitious target weight for a full robot with similar torque capacities as the human as there is no manipulator yet available that possesses such desired properties. However, this would mean that the impact mass involved in the robot-human impact is $M_x = 2$ kg, i.e. $M_x \approx M_H$. Since the contact properties of human facial bones are also in similar range as the dummy head condition, (10.121) can be assumed to be realistic, since such a light arm would have at least similar flexibility in the joints as the LWR-III.

10.9 Blunt Impact Dynamics

In this section, simulation results are presented of impacting the QA-Joint (1-DoF) at different impact speeds and stiffness presets with the human head and abdomen. Furthermore, some theoretical insight are formed on the intrinsic properties of human-robot collisions with VIA joints. For the simulations following model is assumed

$$B\ddot{\theta} = \tau_m - \tau_J(\varphi,\sigma) \tag{10.123}$$

$$M\ddot{q} = \tau_J(\varphi,\sigma) - \tau_F - \tau_g - \tau_{\text{ext}}, \tag{10.124}$$

where $B,M \in \mathbb{R}^+$ are the motor and link side inertia. $\theta, q \in \mathbb{R}$ the motor and link side position, $\varphi = \theta - q$ the elastic deflection, and $\tau_m, \tau_J(\varphi,\sigma), \tau_F, \tau_g, \tau_{\text{ext}} \in \mathbb{R}$ the motor, elastic, friction, gravity, and external torque. $\sigma \in \mathbb{R}$ is the stiffness adjuster position. For sake of clarity $\tau_F = \tau_g = 0$ is assumed.

For the DLR QA-Joint the elastic joint torque is defined as

$$\tau_J = 40(e^{15((\theta-q)-\sigma)} - e^{15(-(\theta-q)-\sigma)}). \tag{10.125}$$

For details on the human models used in the impact simulations to generate τ_{ext} please refer to Chapter 4. Furthermore, the state feedback controller introduced in [2] and outlined in Sec. 10.3.5 is used for achieving good tracking performance of the QA-Joint. For all collision simulations a smooth trapezoidal velocity profile is commanded to hit the human body part at constant link velocity and without significant elastic deflection. In addition, the motor torque is assumed to be bounded.

10.9.1 Head Injury Criterion: Simulation

Figure 10.29 (left) shows the Head Injury Criterion for three different stiffness presets and various impact velocities. The tip impact velocity ranges up to 1.3 m/s and the stiffness preset is set to very low, medium, and high stiffness adjustment ($\sigma \in \{1,6,11\}\ ^o$). As already observed for rigid robots or for robots with moderate joint compliance as the LWR-III [14, 16, 28], Fig. 10.29 supports the statement of high impact velocity dependency of HIC also for a VIA joint. However, at the same time it becomes clear that an impact at such speeds with the QA-Joint is not harmful according to HIC. The HIC reaches maximum values of ≈ 10, representing a practically negligible injury probability [16]. Furthermore, as already predicted in Chapter 5, it cannot be confirmed that HIC significantly depends on joint stiffness. The curve is similar to the one for a relatively stiff joint (e.g. non-negligible joint elasticity due to Harmonic Drive and joint torque sensor). In other words, the joint inertias (motor and link) are decoupled for all stiffness presets already. Therefore, high stiffness of an intrinsically compliant joint is low enough to decouple motor and link inertia during a rigid impact compared to "industrial" robot rigidity. This is an unexpected result and significantly changes the knowledge about the role intrinsic joint compliance plays for safety. Basically, there is no need to demand more joint compliance than e.g. the intrinsic one of the LWR-III. This observation holds already for low reflected link inertias. One may say that for practically relevant inertias, joint elasticity in the range that is characteristic for intrinsically compliant joints does not add additional safety for head impacts except for such cases described in Sec. 10.8. However, please note that we refer to rigid impacts. As already outlined in Chapter 9 low intrinsic joint stiffness additionally increases the safety characteristics at low contact stiffness.

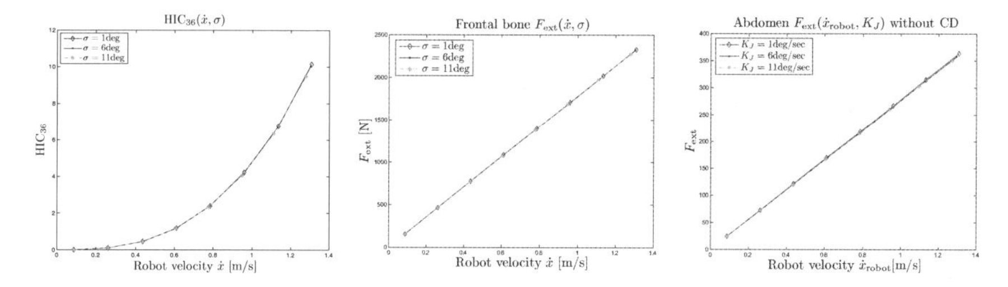

Fig. 10.29 Head Injury Criterion for different joint velocities and stiffness values (left). Frontal impact force for different joint velocities and stiffness values (middle). Abdominal impact force for different joint velocities and stiffness values (right).

The main reason for this effect is that the head stiffness is very high, at two orders of magnitude larger than the rigid joints of the LWR-III (see Sec. 10.8).

10.9.2 Frontal Impact Force

Figure 10.29 (middle) shows the impact force for the frontal bone, pointing out the linear relationship between peak force and impact velocity. This simulation also confirms that decoupling of motor from link inertia is present during all impacts. Even though the contact forces get large, they are still far below the corresponding fracture threshold value of 4 kN for the frontal bone. This leads to the conclusion that frontal fractures are very unlikely to occur.

10.9.2.1 Abdominal Impact Force

Figure 10.29 (right) shows the impact force for the abdomen, having similar behavior as the frontal impact force. For this simulation the mass of the human is considerably higher and the stiffness is lower by two orders of magnitude compared to the frontal skull area. However, also for this simulation the decoupling already applies due to the still lower joint stiffness compared to the human abdomen. Again, the occurring impact forces are significantly smaller than any critical value. Therefore, one can conclude that such an impact does not cause any harm to the human abdominal area by means of the force criterion. This states that a contact force of 2.5 kN must not be exceeded.

10.9.3 Maximum HIC for Compliant Joints

Intrinsic joint compliance is often considered to be the key to intrinsic safety [4, 11]. As argued in this part of the monograph, this statement needs some relevant extension, since there is clear evidence that under certain circumstances even the contrary may be concluded.

Consider the effect energy storage has on head injury again by means of HIC: An open-loop system is treated with respect to the link side position q. Furthermore, the already mentioned decoupling effect is assumed. According to [10] HIC can be expressed as

$$\mathrm{HIC} = 2\left(\frac{M_x}{(M_x+M_H)g}\right)^{\frac{5}{2}} \alpha^{-\frac{3}{2}}(\sin\alpha)^{-\frac{5}{2}} \left(\frac{M_x+M_H}{M_xM_H}\right)^{\frac{3}{4}} K_H^{\frac{3}{4}}||\dot{\mathbf{x}}_0||^{\frac{5}{2}} \tag{10.126}$$

when assuming a simple mass-spring model of the human head. $\dot{\mathbf{x}}_0$ is the Cartesian robot impact velocity. The constant α is

$$\alpha = \min(\alpha^*, \omega\Delta t_{\max}/2), \tag{10.127}$$

where α^* is the solution of

$$3\sin\alpha - 5\alpha\cos\alpha = 0 \tag{10.128}$$

in $]0,\pi/2]$. Its numerical solution is $\alpha = 1.0528$. ω is defined as

$$\omega := \sqrt{\frac{(M_x+M_H)K_H}{M_xM_H}}. \tag{10.129}$$

In Chapter 5 a more sophisticated nonlinear model was used for analysis which does not allow such a solution of HIC. However, the simple mass-spring model is sufficient for deducing some conclusions in the following. As one can see from (10.126), impact velocity affects HIC more than quadratically. When considering the infinite mass robot $M_x \to \infty$ in (10.126) it becomes clear that HIC saturates

$$\mathrm{HIC} = 2g^{-\frac{5}{2}}\alpha^{-\frac{3}{2}}(\sin\alpha)^{-\frac{5}{2}} M_H^{-\frac{3}{4}} K_H^{\frac{3}{4}}||\dot{\mathbf{x}}_0||^{\frac{5}{2}}. \tag{10.130}$$

Up to now this evaluation is for rigid robots with reflected inertia M (consisting for very rigid industrial robots of the motor and link inertia) as well as for decoupled compliant robots with link side reflected inertia M_x. As described in [19] it is possible to store a considerable amount of energy in the elastic mechanism of a VIA joint and use it for significant speedup of the link. Very high velocity could be achieved if one is able to apply bang-bang control if no elastic joint limits would be present. However, some oscillation cycles are still likely even if real-world constraints as e.g. maximum deflection are considered. Such motions would lead to very high and potentially life threatening HIC values in case of impact, see Fig. 10.30.

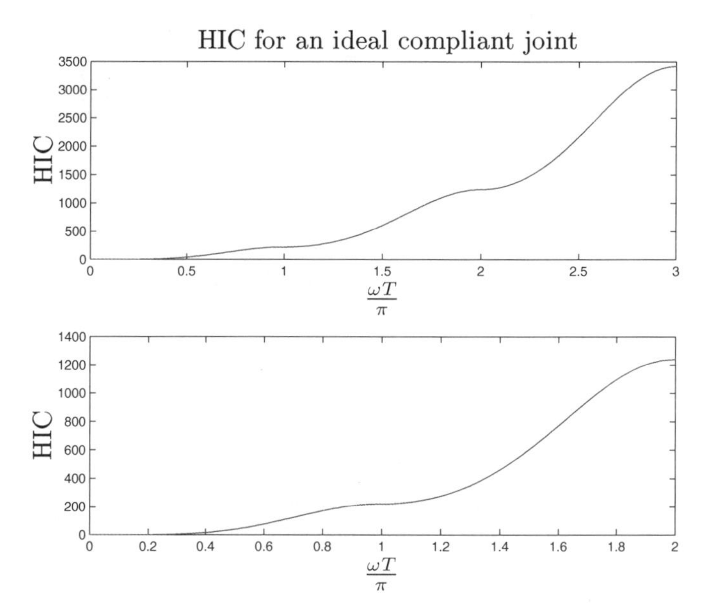

Fig. 10.30 Ideal HIC values for a compliant joint with constant joint stiffness and maximum motor velocity $\dot{\theta}_{\max} = 220\ ^o/\mathrm{s}$. The upper plot shows the HIC for a windup time of up to $\frac{\omega T}{\pi} = 3$ and the lower one for up to $\frac{\omega T}{\pi} = 1.5$.

In reality the limited elastic deflection $\varphi_{\max}$ is defining the maximum stored potential energy. This leads to a maximum link velocity, which is given by motor maximum velocity plus a term depending on the amount of the stored potential energy

$$\dot{q}_{\max} = \dot{\theta}_{\max} + \Delta\dot{q}_{\max}. \tag{10.131}$$

The velocity increase $\Delta\dot{q}_{\max}$ depends on the elastic energy.

$$\Delta\dot{q}_{\max} = \sqrt{\frac{2}{M} E_{\max}(\varphi, \sigma^*)}, \tag{10.132}$$

with $E_{\max}(\varphi, \sigma^*)$ being the maximum spring energy achievable by means of passive joint deflection. Constant stiffness preset σ^* is assumed for simplicity.

The maximum elastic energy for the QA-Joint is therefore

$$E_{max}(\varphi,\sigma^*) = \int_0^{\varphi_{max}} \tau_J \mathrm{d}\varphi \tag{10.133}$$
$$= 40\int_0^{\varphi_{max}} e^{15(\varphi-\sigma^*)}\mathrm{d}\varphi - 40\int_0^{\varphi_{max}} e^{15(-\varphi-\sigma^*)}\mathrm{d}\varphi \tag{10.134}$$
$$= \frac{8}{3}e^{-15\sigma^*}\left(e^{-15\varphi_{max}} + e^{15\varphi_{max}} - 2\right). \tag{10.135}$$

The corresponding velocity increase is

$$\Delta\dot{q}_{max} = \sqrt{\frac{2}{M}\frac{8}{3}e^{-15\sigma^*}\left(e^{-15\varphi_{max}} + e^{15\varphi_{max}} - 2\right)}, \tag{10.136}$$

which leads to a tip velocity $\dot{x} \in \mathbb{R}$ of

$$\dot{x}_{max} = l_M\left(\dot{\theta}_{max} + \sqrt{\frac{2}{M}\frac{8}{3}e^{-15\sigma^*}\left(e^{-15\varphi_{max}} + e^{15\varphi_{max}} - 2\right)}\right). \tag{10.137}$$

Inserting (10.137) in (10.126) leads to the maximum HIC for the QA-Joint. The maximum motor velocity of the QA-Joint is $\dot{\theta}_{max} = 220\,^o/\mathrm{s}$ and a reflected inertia of 3.1 kg (1 kg load and $M = 0.523\,\mathrm{kgm}^2$) is assumed. For a high stiffness preset value $\sigma^* = 15\,^o$ (lowest possible stiffness characteristic) the maximum elastic energy is ≈ 2.67 J, which leads to an increase in achievable link speed of ≈ 1.68. This in turn increases HIC by ≈ 3.6 (HIC $= 94.75$ in the rigid case[15], compared to HIC $= 348.73$ in the flexible case). This impression of a low value results out of the already very high given maximum motor velocity compared to the storable energy of the spring. If $\dot{\theta}_{max} = 80\,^o/\mathrm{s}$, a velocity increase of 2.9 could be achieved[16], leading to an HIC increase by 14.1.

To sum up, due to its ability to store potential energy and use it for achieving higher link speed, a compliant joint is in principle able to reach higher HIC values (for non-negligible link inertia) than its stiff counterpart[17]. However, this interpretation of safety level is also one-sided. If peak velocities are required only for a short period of time, intrinsic joint stiffness is an effective way to fulfill this with lighter robot design. In general, it is suggested to shift the focus of motivation for intrinsic joint compliance from achieving intrinsic safety of the human to utilizing compliance for joint protection and performance improvement. Similar to stiff joints the aspect of safety needs careful analysis of the particular design.

After this theoretical analysis on intrinsic impact properties of intrinsically compliant actuators some experimental analysis is presented in the following. For impact experiments are carried out with the QA-Joint and the dummy-dummy.

---

[15] The HIC was evaluated by (10.126).

[16] This is a slightly larger value than the one obtained for the full dynamic simulation/experiment in Sec. 10.7.5.

[17] Please note that joint compliance does not inherently come at the cost of higher joint weight. It is e.g. possible to make some structural parts compliant without increasing their weight and have the same energy storage effect.

10.9.4 Head Injury Criterion: Experiments

In the following experiment, the QA-Joint is equipped with an additional link side mass and let the joint collide with the dummy-dummy. The motor and link inertia are $B = 0.993374$ kgm^2, $M = 0.523808$ kgm^2 and the link length $l_M = 0.5$ m. The joint was commanded to move on a smoothened trapezoidal velocity profile and was controlled using the aforementioned state feedback controller. The measured acceleration was then used to calculate the resulting HIC values.

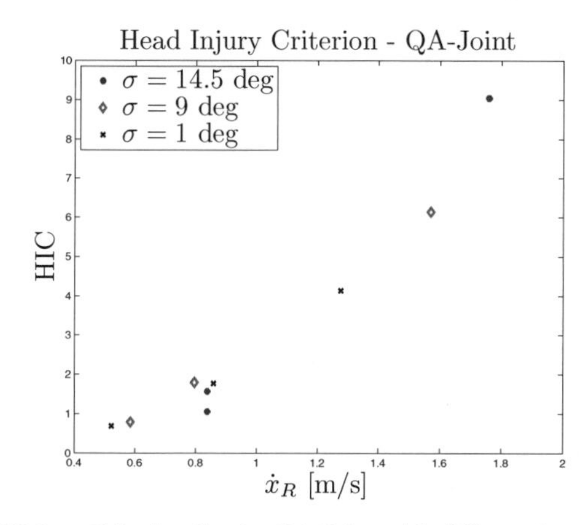

Fig. 10.31 Head Injury Criterion for the QA-Joint with different impact speeds and stiffness presets

In Figure 10.31 the experimental results for impacting the QA-Joint with the dummy-dummy are shown. They support the simulative predictions well. The calculated HIC values depend only on the link side velocity and not the stiffness preset at all. The "over quadratic" behavior [16] is clearly confirmed and the measured values are indicating very low injury by means of HIC. The impact velocities ranged up to 1.8 m/s, i.e. similar velocities as investigated for various other robots in recent work [17]. The joint shows considerably lower HIC values compared to them, which is mainly due to the lower reflected inertia of the test joint. Apart from evaluating the injury potential emanating from such a device, the robustness of the proposed control approach from [2] is tested at the same time. Although the impact results in large disturbance forces, it was not possible to destabilize the controlled joint.

Apart from the intrinsic properties of collisions, sophisticated collision detection and reaction schemes are needed in order to adequately react to external disturbances. This becomes especially important during sharp contact.

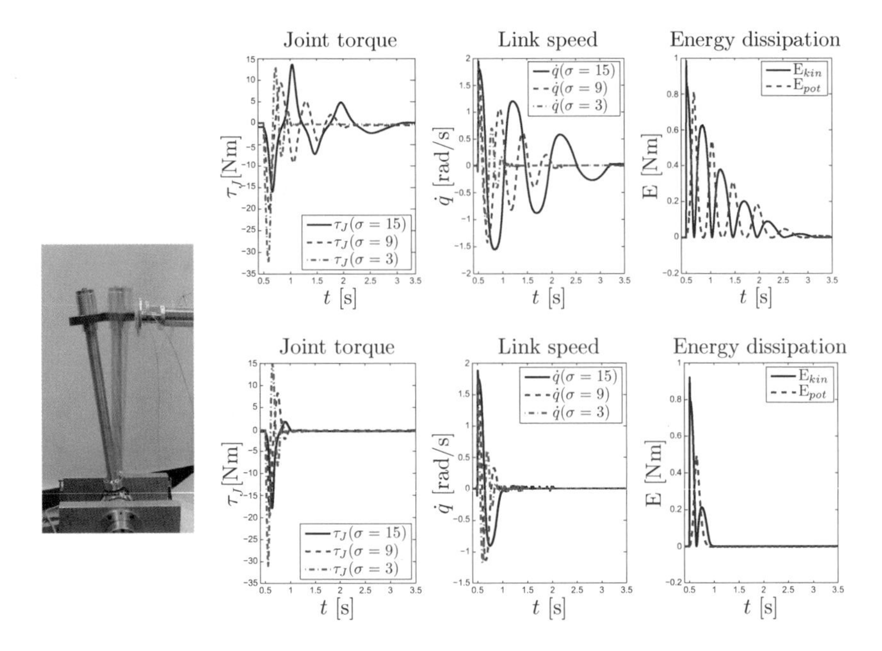

Fig. 10.32 Impact evolution without state feedback control (upper block) and with state feedback control (lower block). For readability the energies are only plotted for $\sigma = 15\,^o$.

10.9.5 Joint Protection and Control Performance

In order to show the shock resistance and control performance of the proposed joint design, impact drop tests were conducted with a rigid object acting on the link of the test joint, see Fig. 10.32 (left). Figure 10.32 (right upper and lower) shows the behavior for three different stiffness preset values covering the entire range of the mechanism. The upper row depicts the measurements with the joint in PD position control and the lower one with full state feedback control for vibration damping. The mass of the impactor is 4.2 kg and the impactor speed at the collision instant 1.07 m/s. The contact is rigid aluminum-brass, leading to very large collision forces of up to 5 kN, measured with a high bandwidth force sensor mounted on the impactor. The second column depicts the elastic joint torque, which oscillates strongly up to 3.5 s after the collision for simple motor side PD control. The state feedback controller diminishes these oscillations effectively. Similar observations can be drawn for link speed and energy dissipation. Apart from the control performance during these very high disturbance forces, the collision protection due to the elastic mechanism becomes apparent by taking a closer look at the joint torque. Even for the rigid stiffness preset, the maximum nominal joint torque of 40 Nm is not reached despite high impact forces. Furthermore, the large benefit of stiffness reduction can be observed. By setting the stiffness to the lowest preset, the impact joint torque can be almost halved.

After this analysis of mostly intrinsic impact properties of VIA joints, collision detection and reaction schemes are elaborated for intrinsically compliant devices. Although a collision detection and reaction will not reduce the impact dynamics during rigid blunt impacts, it is still important to prevent soft-tissue injury.

10.10 Collision Detection for VSA

In this section, we introduce two different collision detection methods for intrinsically compliant joints, based only on proprioceptive sensing and certain model knowledge. The first one is the straightforward extension of the method for flexible joint robots outlined in Chapter 3.3. It utilizes the measured joint torque and the known link side dynamics, whereas the second one relies on motor and link dynamics but does not require joint torque sensing. The estimation $\hat{r}$ is also a first order filtered version of τ_{ext}.

10.10.1 Generalized Link Side Momentum Observer

Similar to the collision detection method proposed for a flexible joint robot with joint torque sensing in Chapter 3, a momentum-based disturbance observer can be used, which uses the measured joint torque and the known rigid body dynamics for the VSA case as well. Instead of a designated joint torque sensor as the strain gauge-based ones in the LWR-III, the estimated joint torque $\hat{\tau}_J$ obtained from identification is utilized with a model-based joint torque sensor. The mathematical derivation being analogous to the one for constant joint elasticity, except for $\hat{\tau}_J \approx \tau_J = f(\theta, q, \sigma)$ is a possibly nonlinear relationship.

10.10.2 Generalized Joint Momentum Observer

The second method for collision detection is also based on momentum observation. However, in this case the momentum of both the motor and link inertia is monitored, and used for collision detection. The main characteristic of this scheme is that it does not require identification of τ_J. Note that such a scheme is sensitive to unknown friction torque.

The generalized momentum of the motor is defined as

$$p_1 = B\dot{\theta}. \tag{10.138}$$

Its dynamics can be written as

$$\dot{p}_1 = \tau_m - \tau_J. \tag{10.139}$$

In a similar fashion, the link side momentum

$$p_2 = M\dot{q} \tag{10.140}$$

leads to a reformulation of the rigid body dynamics

$$\dot{p}_2 = \tau_J - g(q) - \tau_{\text{ext}}. \tag{10.141}$$

Consider the momentum sum of the motor inertia B and link side inertia M

$$p = p_1 + p_2 = B\dot{\theta} + M\dot{q}. \tag{10.142}$$

Its timely evolution can be written as

$$\dot{p} = \dot{p}_1 + \dot{p}_2 \tag{10.143}$$
$$= \tau_m - \tau_J + \tau_J - g(q) - \tau_{\text{ext}} \tag{10.144}$$
$$= \tau_m - g(q) - \tau_{\text{ext}}. \tag{10.145}$$

The estimation $\hat{\dot{p}}$ of $\dot{p}$ is defined as

$$\hat{\dot{p}} := \tau_m - g(q) - \hat{r}, \tag{10.146}$$

i.e. substituting τ_{ext} in (10.145) with $\hat{r}$. Defining the weighted error dynamics as

$$\dot{\hat{r}} := K_0(\hat{\dot{p}} - \dot{p}) \tag{10.147}$$

this leads together with (10.142) and (10.146) to

$$\hat{r} = K_0 \int \tau_m - g(q) - \hat{r}\mathrm{d}t - (B\dot{\theta} + M\dot{q}). \tag{10.148}$$

From (10.143), (10.146), and (10.148) one can write in the Laplace domain

$$s\hat{r} = K_0(\tau_{\text{ext}} - \hat{r}), \tag{10.149}$$

which is equivalent to

$$\hat{r} = \frac{K_0}{K_0 + s}\tau_{\text{ext}}. \tag{10.150}$$

Therefore, a first-order filtered version of the real external torques is isolated with $1/K_0$ being the filter frequency. This signal can be directly used for collision detection and the appropriate reaction, taking into account full information about external forces. The extension to the N-DoF case is straightforward.

Next, the introduced collision detection schemes are analyzed during an abdominal impact with moderate joint stiffness.

10.10.3 Collision Detection and Reaction for the QA-Joint

Figure 10.33 (left) and Fig. 10.33 (middle) depict the impact behavior with the abdomen at 100 ^{o}/s with collision detection activated. As soon as the robot detects the collision, the desired trajectory is stopped abruptly, causing a jump in velocity. The joint stops its motion entirely after ≈ 300 ms.

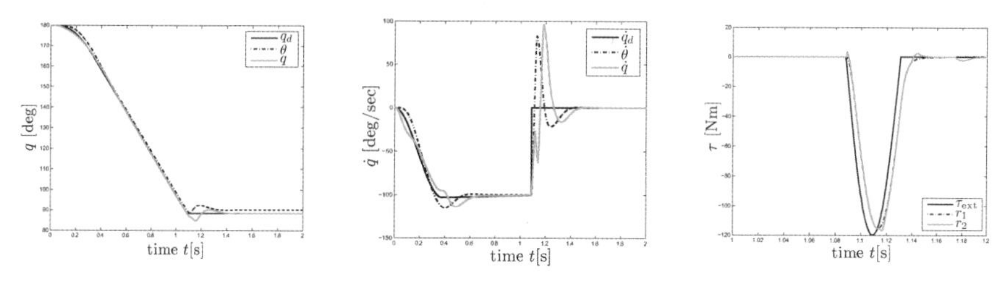

Fig. 10.33 Desired link position, motor position, and link position for collision detection and reaction (left). Desired link velocity, motor velocity, and link velocity for collision detection and reaction (middle). Real external torque, residual observer 1, and residual observer 2 (right). The reaction strategy is to simply stop the robot.

Figure 10.33 (right) depicts the real external torque τ_{ext} resulting from impact forces and its estimations r_1, r_2, which are given by the collision detection schemes presented earlier. The cutoff frequency $f_c = 1/K_0$ for both observers is chosen to be 250 Hz. Even though a small lag is therefore present in both cases, the proposed approaches show quick detection response. r_1 is characterized by some discontinuities in its behavior, which stem from the incorporation of the motor torque saturation in the simulation.

Figure 10.34 depicts the collision detection and reaction for a position-based strategy. Similar to the case of the LWR-III in [9, 18], r_i is used for implementing

$$q_d(t) = -\int K_A \, \hat{r}_i(t) \, \mathrm{d}t + q_{d,c}, \tag{10.151}$$

where $K_A \in \mathbb{R}^+$ is a gain factor and $q_d, q_{d,c}$ the desired position and the desired position at which the collision occurred. This enables the robot to retract from external collision sources and to show more reactive behavior than simply stopping the robot.

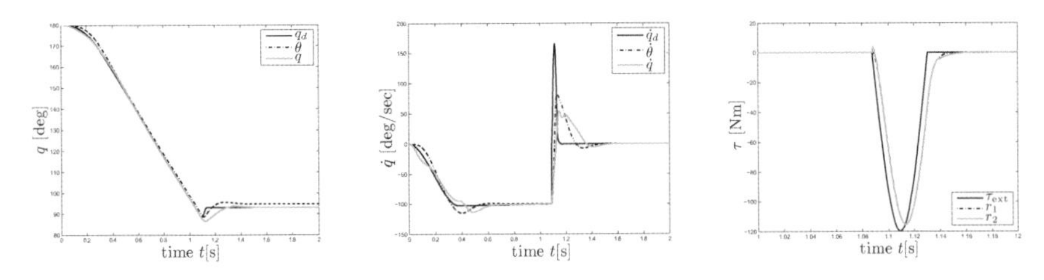

Fig. 10.34 Desired link position, motor position, and link position for collision detection and reaction (left). Desired link velocity, motor velocity, and link velocity for collision detection and reaction (middle). Real external torque, residual observer 1, and residual observer 2 (right). The reaction strategy is admittance-based.

10.10.4 Experimental Collision Detection Performance

Figure 10.35 depicts the result of the collision detection and reaction experiment. The upper plot visualizes the position of the motor and link side as well as the desired motor position. The lower one depicts the residual and the moment the collision detection activates. In this experiment, a simple stop is triggered as soon as an impact is observed. The robot collides at 30 ^{o}/s against the human arm and stops its motion as soon as the threshold 2 Nm for r_1 is exceeded. This value stems mainly from model uncertainties and noise in the order of 1 Nm. Compared to the very

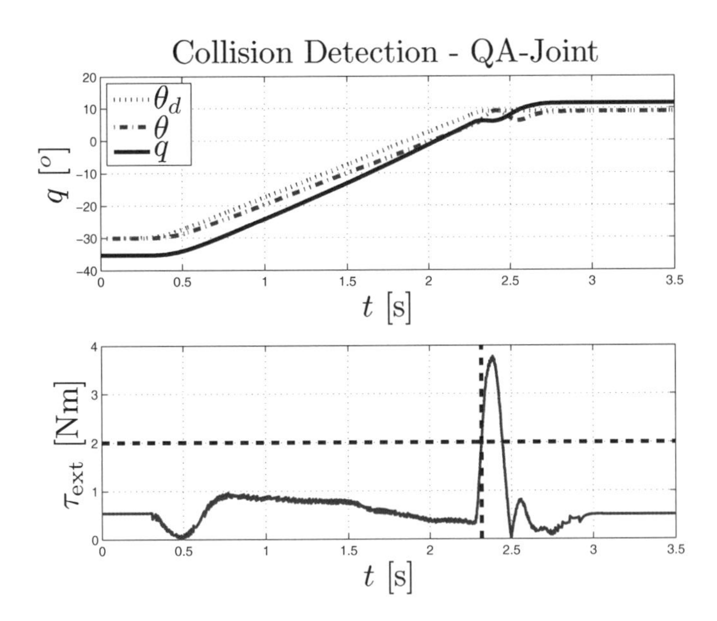

Fig. 10.35 Collision detection experiment with the QA-Joint

high collision speed in the simulations ($\dot{q}_{d,\max} = 100\ ^o/\mathrm{s}$), it is reasonable to chose $\dot{q}_{d,\max} = 20\ ^o/\mathrm{s}$ for the experiments in order to limit the load on the prototype. The sensitivity of the collision detection algorithm for the QA-Joint is comparable to the one for the LWR-III reported in [18].

10.11 Summary

In the first part of this chapter design considerations for intrinsically compliant joint designs were elaborated, which are believed to be important for building such joints. As a consequence of these investigations a novel concept could be realized that intends to cover the identified desired properties.

The theoretical basis for obtaining optimal control motions for VSA joints was elaborated. The effects of motor dynamics, deflection limits, and stiffness adjustment were analyzed. It was proven that these novel devices are capable of outstanding performance increase compared to classical actuation. The results were experimentally supported and this line of work will be extended in the future to the n-DoF case.

In the last part of the chapter the role intrinsic joint stiffness plays for safety in physical Human-Robot Interaction was analyzed. Collision detection schemes suited for such devices were presented and analyzed theoretically as well as experimentally. Insights concerning the inherently possible velocity increase were elaborated and this property was discussed in the context of intrinsic safety. It was shown that the initial motivation for such devices has to be revised when comparing with active compliance approaches. There are two major causes of potential injury, which are related to intrinsic joint stiffness. One of which is dominant for each class of the two designs:

- **Actively compliant robots:** For stiff impacts, motor and link inertia are already decoupled by the moderate joint compliance. However, for lower contact stiffness sophisticated soft-robotics algorithms are needed to realize compliance by software (otherwise the robot would be stiff due to its mechanical design).
- **Passively compliant robots:** The decoupling of link from motor inertia is always feasible. Nonetheless, it is possible to drive at very high speeds due to intrinsically very low joint stiffness, low damping, and the energy storage in the spring. In order to tackle that problem, effective vibration damping schemes for preventing oscillatory, energy storing and releasing motion, as well as the safe limitation of the maximum velocities by software are needed.

Apart from these two major aspects differentiating the inherent joint designs, it is crucial to develop collision detection schemes with high sensitivity for injury prevention for both joint classes. Clear advantages of intrinsic joint compliance, on the other hand, are significantly better joint protection during impacts with the environment, and the large speed performance increase that is possible.

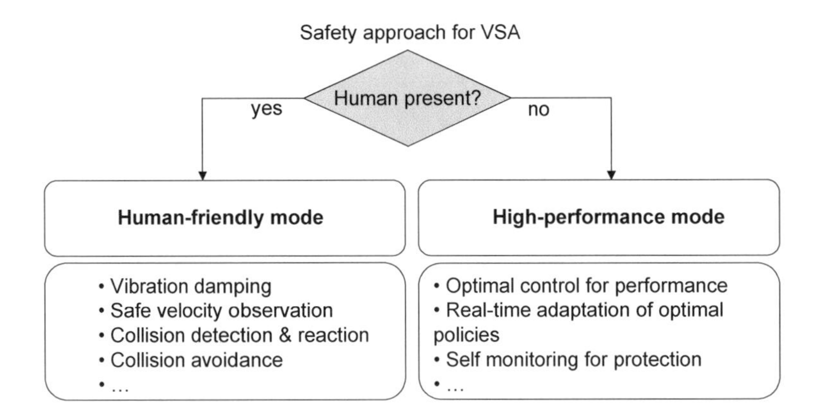

Fig. 10.36 Safety-oriented methodology for VIA control

To sum up, it is important to distinguish two modes for intrinsically compliant joints, see Fig. 10.36:

1. **human-friendly mode**
2. **high-performance mode**

The **human-friendly mode** focuses on providing intrinsically compliant behavior, while suppressing unwanted oscillations that may easily lead to high and therefore dangerous robot velocities. In order to provide high level of safety to the human, the stored elastic energy should be supervised, and the full toolbox of methodologies, ranging from collision avoidance, detection, and reaction to higher level fault modules should be embedded in the robot control. The **high-performance mode** on the other hand serves for the full exploitation of energy storage and release for carrying out high-speed motions. This differentiated view makes allowance for the desire of achieving human-like performance while providing safe behavior.

References

[1] Albu-Schäffer, A., Eiberger, O., Grebenstein, M., Haddadin, S., Ott, C., Wimböck, T., Wolf, S., Hirzinger, G.: Soft robotics: From torque feedback controlled lightweight robots to intrinsically compliant systems. IEEE Robotics and Automation Mag.: Special Issue on Adaptable Compliance/Variable Stiffness for Robotic Applications 15(3), 20–30 (2008)

[2] Albu-Schäffer, A., Wolf, S., Eiberger, O., Haddadin, S., Petit, F., Chalon, M.: Dynamic modeling and control of variable stiffness actuators. In: IEEE Int. Conf. on Robotics and Automation (ICRA 2010), Anchorage, Alaska, pp. 2155–2162 (2010)

[3] Benson, D.: A gauss pseudospectral transcription for optimal control (2005)

[4] Bicchi, A., Tonietti, G.: Fast and soft arm tactics: Dealing with the safety-performance trade-off in robot arms design and control. IEEE Robotics and Automation Mag. 11, 22–33 (2004)
[5] Bryson, A., Ho, Y.: Applied optimal control, rev. print. edn. Hemisphere Publ. Corp. (1975)
[6] Bulirsch, R., Stoer, J.: Einführung in die numerische Mathematik 2. Springer (1978) (German)
[7] Carl-Cranz-Gesellschaft: Optimierungsverfahren- Software und praktische Anwendungen (1981) (German)
[8] Chou, C., Hannaford, B.: Measurement and modeling of McKibben pneumatic artificial muscles. IEEE Transactions on Robotics and Automation (12), 90–102 (1996)
[9] De Luca, A., Albu-Schäffer, A., Haddadin, S., Hirzinger, G.: Collision detection and safe reaction with the DLR-III lightweight manipulator arm. In: IEEE/RSJ Int. Conf. on Intelligent Robots and Systems (IROS 2006), Beijing, China, pp. 1623–1630 (2006)
[10] Gao, D., Wampler, C.: On the use of the head injury criterion (HIC) to assess the danger of robot impacts. IEEE Robotics and Automation Mag. 16(4), 71–74 (2009)
[11] Edsinger, A.: Robot manipulation in human environments. Ph.D. thesis, Massachusetts Institute of Technology, Department of Electrical Engineering and Computer Science (2007)
[12] Garg, D., Patterson, M.A., Hager, W.W., Rao, A.V., Benson, D.A., Huntington, G.T.: A unified framework for the numerical solution of optimal control problems using pseudospectral methods. Automatica 46(11), 1843–1851 (2010)
[13] Grebenstein, M., van der Smagt, P.: Antagonism for a highly anthropomorphic hand-arm system. Advanded Robotics 22(1), 39–55 (2008)
[14] Haddadin, S.: Evaluation criteria and control structures for safe human-robot interaction. Master's thesis, Technical University of Munich (TUM) & German Aerospace Center (DLR) (2005)
[15] Haddadin, S., Albu-Schäffer, A., Hirzinger, G.: Safe physical human-robot interaction: Measurements, analysis & new insights. In: International Symposium on Robotics Research (ISRR 2007), Hiroshima, Japan, pp. 395–408 (2007)
[16] Haddadin, S., Albu-Schäffer, A., Hirzinger, G.: Safety evaluation of physical human-robot interaction via crash-testing. In: Robotics: Science and Systems Conference (RSS 2007), Atlanta, USA, pp. 217–224 (2007)
[17] Haddadin, S., Albu-Schäffer, A., Hirzinger, G.: Requirements for safe robots: Measurements, analysis & new insights. The Int. J. of Robotics Research 28(11-12), 1507–1527 (2009)
[18] Haddadin, S., Albu-Schäffer, A., Luca, A.D., Hirzinger, G.: Collision detection & reaction: A contribution to safe physical human-robot interaction. In: IEEE/RSJ Int. Conf. on Intelligent Robots and Systems (IROS 2008), Nice, France, pp. 3356–3363 (2008)
[19] Haddadin, S., Laue, T., Frese, U., Wolf, S., Albu-Schäffer, A., Hirzinger, G.: Kick it like a safe robot: Requirements for 2050. Robotics and Autonomous Systems 57, 761–775 (2009)
[20] Haddadin, S., Weis, M., Albu-Schaeffer, A., Wolf, S.: Optimal control for maximizing link velocity of robotic variable stiffness joints. In: Proceedings of IFAC 2011,World Congress, pp. 3175–3182 (2011)
[21] Hermann, M.: Numerik gewöhnlicher Differentialgleichungen: Anfangs- und Randwertprobleme. Oldenbourg, Müchen (2004) (German)

[22] Hurst, J., Chestnutt, J., Rizzi, A.: An actuator with physically variable stiffness for highly dynamic legged locomotion. In: IEEE Int. Conf. on Robotics and Automation (ICRA 2004), Barcelona, Spain, pp. 4662–4667 (2004)
[23] Kirk, D.: Optimal control theory. Prentice-Hall (1970)
[24] Lagarias, J., Reeds, J., Wright, M., Wright, P.: Convergence properties of the nealder-mead simplex method in low dimensions. SIAM Journal on Optimisation 9, 112–147 (1998)
[25] Liberzon, D.: Calculus of Variations and Optimal Control Theory: A Concise Introduction. Princeton University Press (2011), http://books.google.de/books?id=bXBJYgEACAAJ
[26] Migliore, S., Brown, E., DeWeerth, S.: Biologically inspired joint stiffness control. In: IEEE Int. Conf. on Robotics and Automation (ICRA2005), Barcelona, Spain (2005)
[27] Morita, T., Sugano, S.: Development of one-dof robot arm equipped with mechanical impedance adjuster. In: IEEE/RSJ Int. Conf. on Intelligent Robots and Systems (IROS 2006), Washington, DC, USA, p. 407 (1995)
[28] Oberer, S., Schraft, R.D.: Robot-dummy crash tests for robot safety assessment. In: IEEE Int. Conf. on Robotics and Automation (ICRA 2007), Rome, Italy, pp. 2934–2939 (2007)
[29] Oberle, H.: BNDSCO - A Program for the Numerical Solution of Optimal Control Problems (2001), Hamburger Beiträge zur angewandten Mathematik, Bericht 36, Reihe B (German)
[30] Papageorgiou, M.: Optimierung : statische, dynamische, stochastische Verfahren für die Anwendung, 2. erw. u. verb. aufl. edn. Oldenbourg (1996) (German)
[31] Park, J.J., Kim, H.S., Song, J.B.: Safe robot arm with safe joint mechanism using nonlinear spring system for collision safety. In: IEEE Int. Conf. on Robotics and Automation (ICRA 2009), Kobe, Japan, pp. 3371–3376 (2009)
[32] Pontrjagin, L.: Mathematische Theorie optimaler Prozesse, 2., verb. aufl. edn. Oldenbourg, München (1967) (German)
[33] Pratt, G., Williamson, M.: Series elastics actuators. In: IEEE/RSJ Int. Conf. on Intelligent Robots and Systems 1995 (IROS 1995), Victoria, Canada, pp. 399–406 (1995)
[34] Rao, A.V., Benson, D.A., Darby, C., Patterson, M.A., Francolin, C., Sanders, I., Huntington, G.T.: Algorithm 902: Gpops, a matlab software for solving multiple-phase optimal control problems using the gauss pseudospectral method. ACM Trans. Math. 37, 22:1–22:39 (2010), http://doi.acm.org/10.1145/1731022.1731032
[35] Schiavi, R., Grioli, G., Sen, S., Bicchi, A.: VSA-II: a novel prototype of variable stiffness actuator for safe and performing robots interacting with humans. In: IEEE Int. Conf. on Robotics and Automation (ICRA 2008), Pasadena, USA, pp. 2171–2176 (2008)
[36] Unbehauen, H.: Regelungstechnik III, Identifikation, Adaption, Optimierung, 6., durchges. Aufl. edn. Oldenbourg (1989) (German)
[37] Van Ham, R., Vanderborght, B., Van Damme, M., Verrelst, B., Lefeber, D.: MACCEPA, the mechanically adjustable compliance and controllable equilibrium position actuator: Design and implementation in a biped robot. Robotics and Autonomous Systems 55, 761–768 (2007)
[38] Wolf, S., Hirzinger, G.: A new variable stiffness design: Matching requirements of the next robot generation. In: IEEE Int. Conf. on Robotics and Automation (ICRA 2008), Pasadena, USA, pp. 1741–1746 (2008)
[39] Zinn, M., Khatib, O., Roth, B.: A new actuation approach for human friendly robot design. The Int. J. of Robotics Research 23, 379–398 (2004)

Chapter 11
Considerations for New Robot Standards

In pHRI there is the natural demand for a clear set of standards that provide a reliable basis on which manufacturers can rely on. The introduced ISO safety standard *ISO-10218* for direct human-robot collaboration suggests limitations of the speed, power, and force of the robot. However, these limitations do not concisely correspond to the risks and level of potential human injuries. Therefore they are often over conservative and/or in other situations not conservative enough, making a more elaborated safety model necessary. If a new standard proposal is not developed, severe risks for serious injuries remain on the one hand, and the standard will be over restrictive in many cases on the other hand, hampering the application of human-robot collaboration. Certainly, the knowledge of the injury mechanisms and the different ways that a robot can cause injuries will make the standards for human-robot collaboration more useful. Simultaneously, the work on a standard towards this direction will increase the awareness of the cause-effect chain in robot accidents and foster the certification of robot installations. For robot-related injuries in industrial sites, Fig. 11.1 lists the distributions of the "types of contact" and the human body parts affected in these events, according to studies of the German, Austrian and U.S. workers unions. This data has been drawn

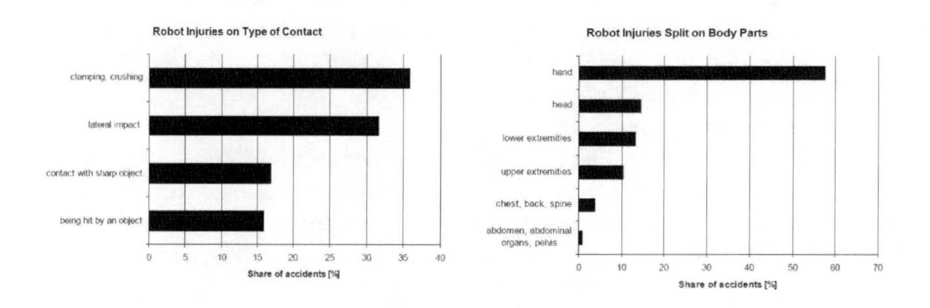

Fig. 11.1 Distribution of robot injuries on human body parts and type of contact

S. Haddadin: *Towards Safe Robots*,
Springer Tracts in Advanced Robotics 90,
DOI: 10.1007/978-3-642-40308-8_11, © Springer-Verlag Berlin Heidelberg 2014

from statistics on non-collaborative robot cells, so the contact events either arose during the installation phase, or were due to the manipulation of protective equipment. Nevertheless, they may serve as a first data set due to the lack of experience with collaborative scenarios. From these numbers it already becomes clear that clamping and lateral impacts should be investigated in much detail, and that relatively low-severity collisions with the hand are most frequent. However, this insight is due to its inherent nature not directly applicable to pHRI. In order to fully cover the threats that may arise from intensive interaction scenarios, a future standard based on the crashworthiness of the robot - assessing the consequences on the human health in case of a collision - should be proposed. This means that different types of impacts on various body parts must be examined with respect to different injury mechanisms. Furthermore, the safety capabilities of novel interaction schemes have to be incorporated. The full understanding of these effects would make it possible to formulate meaningful safety limits for robot operation.

In this monograph, first fundamental steps towards getting a systematic picture of human injury in robotics were made by conducting various impact studies in simulation and experiments. The investigations included blunt and sharp contacts for constrained and unconstrained situations. Furthermore, the beneficial effect of collision detection and reaction schemes was exhaustively analyzed. However, in order to obtain full understanding of the injury risks more simulations as well as soft-tissue and dummy tests are needed. For example extensive quasi-static and dynamic loading tests for different body parts and with varying shapes of the impacting robot surface are still to be carried out. The basis for a new safety standard could come from the evaluation of the injury mechanisms. Besides using available biomechanical data, further research is needed to obtain impact effects on bone structures, organs, and soft-tissues. Biomechanics experts are expected to contribute highly valuable expertise in relating injuries to criteria functions that are well suited for robotics. Apart from understanding the fundamental injury mechanisms of humans, it could also be important to take into account the pain tolerance of humans as examined in [31, 4]. This would be a strong, however, unmotivated constraint in performance if it is possible to significantly reduce the probability of collision and ensure a maximum amount of low injury probability.

In this chapter, various suggestions are contributed that could be useful for standardization efforts that are particularly focussing on pHRI. The chapter is organized as follows. Section 11.1 discusses the limitations of existing standards and requirements that could be useful for future, as well as existing ones. Section 11.2 outlines a synopsis of injuries in robotics based on the results of this monograph. Section 11.3 introduces a new injury scaling attempt, which extends the AIS to the demands of robotics. Section 11.4 gives a proposal on how future crash-testing experiments with crash-test dummies could be carried out for matching the requirements of robotics. Finally, Sec. 11.5 discusses open aspects and the next steps to be taken.

11.1 Limitations of Existing Safety Standards

The fact that the safety of Human-Robot Collaboration/Cooperation (HRC) has recently drawn so much interest and that research and development work has been intensified is attributable in part to the circumstance that industrial robot controllers have only recently been equipped with safety options, enabling safe monitoring of robot motion (e.g. KUKA's *Safe Operation and Safe Handling* and ABB's *SafeMove*). As a first step, this allows for routine co-existence and well-defined interaction between robots and humans. Nonetheless, strictly speaking, it does not allow for highly dynamic and unpredictable interaction. Complex interaction scenarios, however, will be a necessity in the future for the realization of numerous value-adding applications, in which industrial robots and humans work together or service robots are fulfilling tasks in human vicinity. As mentioned above, the robotics society is currently exhibiting increasing interest in deeper investigation of the specific injury risks and effects in human-robot collisions to enhance the safety of the systems and to apply this knowledge to propose well-founded limits for safe Human-Robot Interaction in terms of maximum velocity, impact forces, or transferred impact energy. Novel types of robots such as the LWR-III that are capable of fundamentally different interaction capabilities compared to classical industrial robots require new types of standards and limits and made it necessary to take a different view on the safety problem. In this line of thinking, cooperative research projects are also exploring safety requirements for Human-Robot Interaction. For example, the EC project PHRIENDS (IST-045359, http://www.phriends.eu) aimed at developing actively and passively compliant robots that can co-exist and cooperate with people. It seeked to enable physical Human-Robot Interaction that is both dependable and safe. Besides developing new ways of collision detection, this project also investigated different reaction strategies following unavoidable, undesired impacts of robots and humans.

The only standardized guideline, which focuses on Human-Robot Interaction up to now is the *ISO-10218* [16]. In this standard, new regulations are specified for the "collaborative operation-state in which robots work in direct cooperation with a human in a defined workspace". The presence of the human in this collaborative workspace requires one of the following conditions to be fulfilled:

- TCP/flange velocity ≤ 0.25 m/s,
- maximum dynamic power ≤ 80 W,
- maximum static force ≤ 150 N.

These requirements directly limit specific process characteristics of the industrial robot, not taking into account the real consequences for the physical well-being of the human in case of a system failure. This implies that the same maximum allowable process characteristics are to be applied to a lightweight compliant robot system as to a high payload standard industrial robot. This is an erroneous approach. The robot safety guidelines for collaborative robots are still set up independently of the robot size, structure, and available control strategy. Therefore, they are limiting the technological possibilities due to heavily over-restrictive over-the-board conditions. In addition, simple limitations on the ill-defined quantities "dynamic power" or static force are insufficient as design criteria. Consider for example the effect of a knife: if one compares a knife that is wielded against a human with 150 N to a frying pan applying the same amount of force to any body part it becomes obvious that a more careful specification of the contact event is required. However, it is not sufficient either to simply extend the measures to contact area or contact duration to estimate the resulting injuries. As discussed in the monograph, each body part responds rather differently to external dynamic influences. Furthermore, the guidelines for designing collaborative robot systems should make it possible to use the full power of existing technological methods for assuring the operators inviolacy while maximizing robot performance at the same time. Compliant control strategies and collision detection with appropriate reaction have been shown to be powerful methods to cope with uncertainties in the environment and provide entirely new ways of dealing with the safety problem. The research can only develop the methods to a certain level of maturity from which they have to be taken over by industry and development to a stage they are ready to go to market.

To sum up, current research activities on robot safety, into which existing biomechanical basics are incorporated, suggest the necessity to start with the human being as the central entity of any safety analysis in the context of collaborative robotic systems. This will enable a quantitative intrinsic safety evaluation to be introduced into the standardization. Furthermore, it is necessary to find a way for qualifying advanced control methods as safety features and make them a core demand for robots that are developed for pHRI applications. Novel standards need to take into consideration these findings and should to be open to novel ways of addressing safety.

Due to the fact that research within this area is still very new, a full set of appropriate criteria is not yet available and must be derived during the years to come. Nevertheless, a clear picture of where the standardization efforts have to go can already be sketched.

First, it is highly recommend to introduce a much finer granularity in the contact event phenomenology for the definition of future standards. A crucial action to be taken before introducing any relevant measure is, in contrast to *ISO-10218*, classifying and analyzing possible contact situations, as has been carried out in the rich existing biomechanical and forensic literature. A proposal is given in the next section. Furthermore, it is absolutely necessary to have different safety limits for each human body part and for all distinguishable contact situations as e.g. blunt and sharp contacts, meaning impacts with and without sharp tools or structural edges involved. From a biomechanical point of view, it is crucial to strictly differentiate them as the spectrum for potential injuries is highly complex and hardly integrable into a simple scheme. In order to provide the experience gained from the systematic analysis in this monograph to the standardization authorities, the author of the monograph is a DIN representative and committee member of the new ISO Working Group (WG) 7, Personal care robot.

In the next section, possible injuries in robotics are classified by means of the contact situation and contact properties. Furthermore, possible injuries are elaborated, worst-case factors, and appropriate injury measures.

11.2 Possible Injuries in Robotics: A Synopsis

Up to now, only isolated injury issues and mechanisms of robot safety were discussed and introduced in the robotics literature. In order to have an overview of the potential injury threats depending on the current state of the robot and the human, a classification of these mechanisms, governing factors of the particular process and possible injuries are proposed in Fig. 11.2. Physical contact can be divided into two fundamental subclasses: quasi-static and dynamic loading[1]. Fundamental differences in injury severity and mechanisms are observed as well if a human is (partially) constrained or not, leading to the second subdivision. For the quasi-static case it is differentiated between near-singular and non-singular clamping as already outlined. The last differentiation separates injuries caused by blunt contact from the ones induced by tools or sharp surface elements.

Each class of injury is characterized by possible injuries (PI), worst-case factors (WCF) and their worst-case range (WCR). WCF are the main contributors to the worst-case, such as maximum joint torque, the distance to singularity or the robot speed. The worst-case range indicates the maximum possible injury depending on the worst-case factors. In addition to the classification of injury mechanisms for each such class, suggestions for injury measures (IM) are given as well. They are specific injury measures which are appropriate, useful for the classification and measurement of injury potentially occurring during physical Human-Robot Interaction.

For example ① represents blunt clamping in the near-singular configuration, see Fig. 11.2. As already shown, even for low-inertia robots this situation can become very dangerous and is therefore a possible serious threat with almost any robot on a fixed base within a (partially) confined workspace. Possible injuries are fractures and secondary injuries e.g. caused by penetrating bone structures or an injured neck if the trunk is clamped but the head is free. This would mean that the robot pushes the head further while the trunk remains in its position. Another possible threat is shearing off a locally clamped human along an edge. Appropriate indices are the contact force and the Compression Criterion. ③ is the clamped blunt impact in non-singular configuration. The injury potential is defined by the maximum joint torque $\tau_{\max}$ and can range from no injury (as shown for the LWR-III) to severe injury or even death for high-inertia (and joint torque) robots. The robot stiffness does not contribute to the worst-case since a robot without collision detection would simply increase the motor torque to follow the desired trajectory. Therefore, robot stiffness only contributes to the detection mechanism by enlarging the detection time. Also, the contact force and the Compression Criterion are well suited to predict occurring injury. ⑧ denotes the classical free impact which was the first injury mechanism investigated in the robotics literature. This process is governed by the impact velocity and (up to a saturation value) by the robot mass. As shown in [13] and in Sec. 5.2, 5.3 even a robot of arbitrary mass cannot severely injure a human head by

[1] Only injuries for typical robot velocities are considered and no hypothetic extreme cases. As pointed out in previous sections and [13, 14] injury potential vastly increases with the impact velocity of the robot.

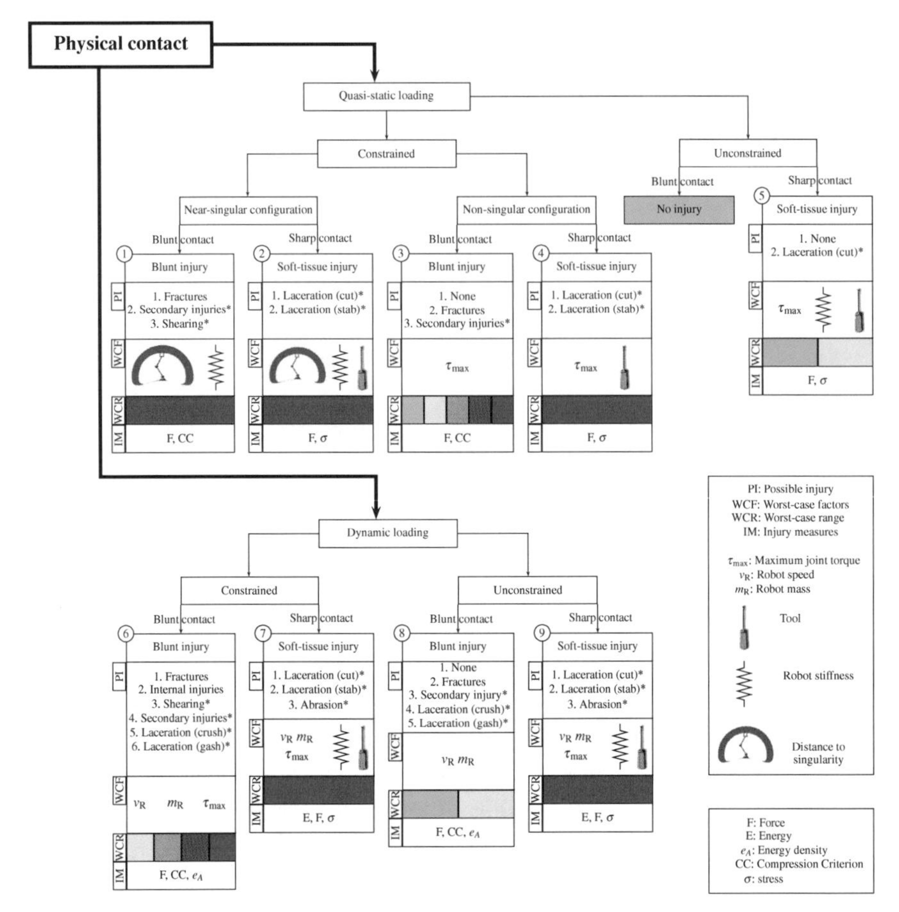

Fig. 11.2 Safety Tree showing possible injury (PI), major worst-case factors (WCF) and the possible worst-case range (WCR). * indicates still ongoing topics of research. Additionally, relevant injury criteria are given for the head, chest, and soft-tissue injuries.

means of impact related criteria from the automobile industry like HIC. However, fractures e.g. of facial bones are likely to occur but not all would be classified as a serious injury. Laceration by means of crushes and gashes are worth to be evaluated, especially with respect to service robotics. The contact force and the Compression Criterion are well suited severity criteria for this class and in order to evaluate lacerations the energy density has to be considered.

The preceding evaluation of injuries is intended as a worst-case analysis for the described contact cases. The next step is to ask which actions can be taken against each particular threat. ① to ⑤ can be handled by a collision detection and reaction as e.g. described in Chapter 3. Good countermeasures in case of ⑥ appears to be soft covering, lightweight design, and a fast and effective collision detection and reaction. ⑦ seems to be the most dangerous scenario one can think of and it needs special treatment. Safe robot speed to give the human enough time to react accordingly is indispensable. Secondly, an effective collision detection and a safe and carefully selected collision reaction have to be embedded. Similar countermeasures are appropriate for ⑨.

In this section, a classification attempt is described ending up with a Safety Tree which is intended to serve as a guideline of how to analyze potential threats during a Human-Robot Interaction scenario. By identifying all the possible physical contact situations for the particular application one does not need to address every theoretically possible injury sources but only the ones which are relevant in the particular context.

In this monograph, the Abbreviated Injury Scale is utilized for categorizing injury. Due to its coarse granularity, it is especially useful for qualifying injury of wide severity ranges. However, due to the special needs in robotics it is important to provide a more appropriate classification of injury, which especially subdivides low injury more specifically: the Extended Abbreviated Injury Scale[2].

11.3 Extended Abbreviated Injury Scale

Up to now, there is no injury classification system established in robotics in general and for low-severity soft-tissue injury in particular. Therefore, an extension of the Abbreviated Injury Scale is proposed, which additionally differentiates between injuries of lower severity, not relevant in automobile crash-testing but important to robotics. In forensic medicine and automobile crash-testing the AIS is an established injury classification tailored to injuries ranging from very low to lethal [2] (see the framed part of Tab. 11.1). Please note that at this point a generic classification of injury shall be defined, not a quantification for a particular body part, which is realized by severity indices.

Table 11.1 Definition of the Extended Abbreviated Injury Scale

		EAIS	SEVERITY	TYPE OF INJURY	INJURY EXAMPLE
Current industrial robots	Goal: service robots	0	None	None	-/-
		1.*A*	Minor	Superficial Injury, Class 1	Contusion (bruise)
		1.*B*	Minor	Superficial Injury, Class 2	Abrasion
		1.*C*	Minor	Superficial Injury, Class 3	Contusion (crush, severe bruise, hematoma)
		1.*D*	Minor	Superficial Injury, Class 4	Superficial/minor laceration (incised wound/cut)
	AIS	1.*E* ≡ 1	Minor	Superficial Injury, Class 5	Laceration, gash, superficial avulsion
		2	Moderate	Recoverable	Nerve contusion, linear fraction
		3	Serious	Possibly recoverable	Small brain contusion
		4	Severe	Not fully recoverable without care	Complex basal skull fracture
		5	Critical	Not fully recoverable with care	Diffuse axonal injury
		6	Fatal	Unsurvivable	Separation of brainstem

[2] In [4] the authors propose to reduce the limits in general to AIS 1 or ICD-10-GM2006.

Possible (future) applications of robots vary from classical industrial ones, such as car welding, to the household robot which will be part of everyday human life and environment. This necessitates the ability to classify injury with respect to the domain a robot operates in. In industrial applications, incorporating the use of heavy duty robots, severe injuries happened in the past as pointed out in [29]. This will remain so in the future. The Abbreviated Injury Scale covers the range of injuries already on a level of granularity, which is sufficient for such cases. For these classical industrial applications (no intentional interaction desired), injuries classified as minimally minor according to the AIS are probably inevitable if a constrained collision occurs, e.g. due to human failure or deactivation of safety devices. Within a typical domestic environment or in a scenario incorporating production assistants, however, such severe risks have to be avoided since active physical contact is desired and crucial at the same time. Future service/co-worker robots have to be designed such that even in the worst-case, at most superficial injuries can occur. In this context, a more granular severity classification is needed to capture the characteristics of the particular low severity injury. In order to take into consideration the entire range of applications, an extension of the AIS, fitting the needs of robotics seems reasonable. The lack of classification of non-severe injury according to their injury level makes it especially useful to define such a scale for cooperating service robots. Therefore, it is proposed to introduce five classes of minor injury, taking into account the order of injury severity for superficial injury. This reflects the fact that superficial avulsions are more severe than superficial lacerations, which in turn are more severe than abrasions or contusions. This gives the possibility to have a more precise description of occurring injury in low injury scenarios. This classification was developed in close cooperation with a biomechanics and forensic expert [18] and the proposed injury level classification is shown in Tab. 11.1. The lowest AIS category AIS = 1 is split up into five categories, representing superficial[3] injuries, rated according to their injury severity and indicated by examples. This leads to a more appropriate sub-classification of the potential injury a certain robot can cause.

In addition to the classification of injury, its pure description has to be agreed on as well for establishing a common taxonomy. Recent work in [15] exploits the so called AO-classification for describing injury objectively without classifying it into a severity category per se. In particular, the according soft-tissue injury definitions suitable to robotics are explained.

In the next section recommendations for standard blunt impact tests are given, which could serve as a basis for future standardized safety evaluation in robotics. In this sense, a first proposal for a set of standardized robot-dummy crash-tests is contributed.

[3] The officially used term **superficial** could be misleading in the context of robotics since even **superficial** is already unacceptable.

11.4 Standard Impacts

Based on the results with robot-dummy crash-testing, some recommendations with respect to a more standardized view on this topic are given in this section. If future robotic systems are going to act around humans and cooperate with them by physical means, a standardized crash-testing protocol will be needed to evaluate different robots on a meaningful and comparable basis. In this sense, this process is sought to be initiated by proposing *Standard Impact Phases* for the unconstrained impact, leading to a set of *Standard Impact Tests* for analyzing robot-human safety. As is explained later in the section, various standard blunt impact tests are proposed with

1. different impact directions,
2. sitting or standing dummy, and
3. defined secondary impact conditions.

11.4.1 Standard Impact Phases

In order to define standard impact tests one has to take into consideration the complexity of a collision process. It does not only consist of the immediate instance of interaction lasting only a few milliseconds, but a much more intricate process is related to it. This incorporates the behavior of the human body and its physical interaction with the robot and the environment. Establishing safety during head collisions is not only about determining the apparent head injury but also has to take into consideration all phases of a collision and the injury potential related to them. The following definition of major phases for the free unconstrained impact shows that this simplest case of a robot-human collision already consists of (minimally) five major phases, as can be deduced from the high-speed videos.

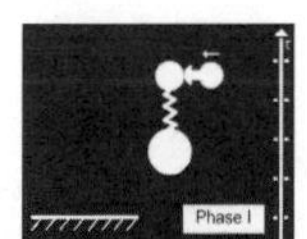

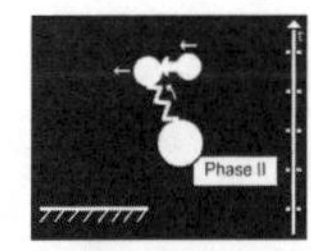

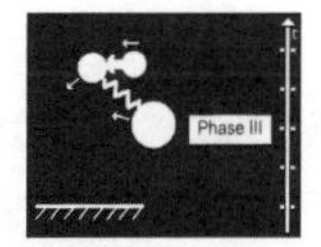

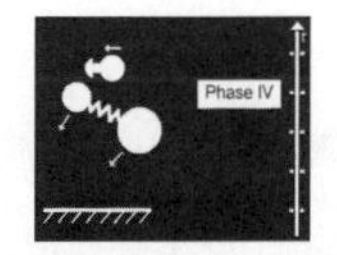

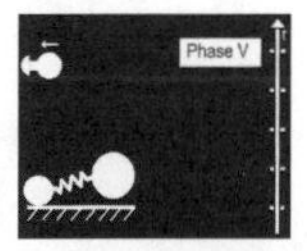

Fig. 11.3 Standard impact phases for an unconstrained robot-head impact. This can be applied to any single contact impact model, consisting of two bodies connected via a junction.

- **Phase I:** The short phase in which the direct impact between robot and head takes place.
- **Phase II:** The neck starts moving significantly due to the motion of the head.
- **Phase III:** The trunk begins to move significantly.
- **Phase IV:** The head loses contact with the robot and the entire body moves freely in space.

- **Phase V:** The body impinges on the ground usually first with the trunk and then with the head: The secondary impact occurs.

A pictogram visualizing these phases is shown in Fig. 11.3. Analogue to the head impact it is straightforward to define similar phases for the chest and other body parts. These standard phases are a good starting point to formulate standard impacts for robotics. A proposal is outlined in the following.

11.4.2 Standard Dummy Impact Tests

The following impact test proposal is a suitable starting point for a standardized set of blunt impacts tests. In this proposal, the evaluation of upper and lower extremities is excluded due to the fact that except for first experiments presented in Chapter 3 this is still an unresolved issue in robotics (also in biomechanics in general). In Section 11.5 the necessity of upper extremity injury is discussed in more detail and the state of the art is provided.

Table 11.2 Standard dummy impact tests

Configuration	Direction	Impact region	Barrier height
Sitting	Frontal	head, chest	$0 \ldots h_B^{\mathrm{T1}}$
*	Side	head, chest, abdomen, pelvis	$0 \ldots h_B^{\mathrm{T1}}$
*	Rear	head, chest	$0 \ldots h_B^{\mathrm{T1}}$
Standing	Frontal	head, chest	$0 \ldots h_B^{\mathrm{Leg}}$
*	Side	head, chest, abdomen, pelvis	$0 \ldots h_B^{\mathrm{Leg}}$
*	Rear	head, chest	$0 \ldots h_B^{\mathrm{Leg}}$

In order to consider the complexity of robot-human impacts it is suggested to first distinguish between a collision between a robot and A) a sitting dummy and B) a standing dummy. Furthermore, the major impact directions for collisions have to be covered, leading to the necessity of frontal, side, and rear impacts for which distinguished crash-test dummies exist. Then, the different impact locations are chosen according to the sensorial equipment of the particular dummy. The impact to the head should be directed normally towards the center of gravity of the head (partially adjustable with the head tilting angle φ_N) and the impacts at the other body parts have to act directly on the particular sensor. In addition to simply hitting the dummy in free space, varying barriers are proposed to evaluate the effect of constraints in the environment, see Tab. 11.2. For sitting configuration they should at most range to the trunk height of the dummy h_B^{T1} (T1 denotes the first thoracic vertebra) and for the standing configuration up to the leg height h_B^{Leg} of the dummy. The heights

$h_I^H, h_I^C, h_S, h_B, h_I^P$, and h_I^A have to be selected according to the specific dummy suited for the impact type, see Fig. 11.4. The aim of these tests is to provide a set of well defined testing setups, which allow not only to evaluate the direct impacts (Phase I) but also the subsequent motion (Phase II-IV) and even the secondary impact (Phase V). All following tests assume a hard base on which the secondary impact occurs. Therefore, the question about the consequences after the collision phase can be answered as well. In principle, arbitrary further situations can be imagined but this set of impact tests provides, similarly to automobile crash-testing, a clear evaluation of injury severity for blunt impacts. From high-speed recordings it becomes clear, which part of the recorded signals correlates to the particular impact phase and thus a separate analysis of each phase is possible. The main reason to distinguish between sitting and standing condition is, apart from the influence of partial constraints, a more detailed analysis of related secondary impacts. These will mainly depend on impact velocity and drop height.

The motion of the robot is commanded such that it moves at a constant velocity and all impact tests are to be carried out up to maximum velocity of the robot under the impact direction constraint. To quantify the effects of collision detection and reaction schemes for a robot, it is important to show under which conditions they contribute to increasing safety and where their limitations are. The analysis presented for the LWR-III can be seen as a first template.

It is clear that performing the entire set of measurements is an expensive and time consuming endeavor. However, the tests are related to different injury types, which do not obviously correlate. Therefore, it is reasonable to argue that they are mandatory in an incipient phase. If a subset of the tests captures all relevant aspects, a reduction of the full series extent will be done.

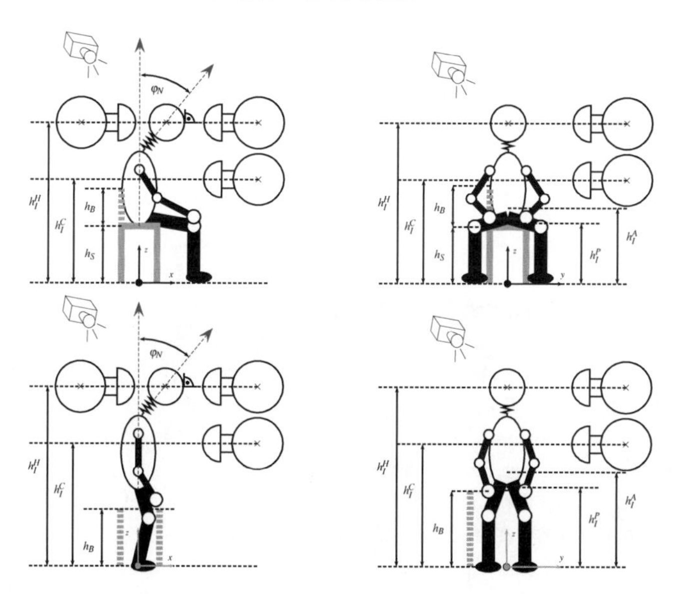

Fig. 11.4 Standard dummy configurations

11.4.2.1 The Standard Sitting Frontal and Rear Impact

In Figure 11.4 (upper left) the *frontal sitting* and *rear sitting* setup are shown. The dummy is sitting upright on a fixed object at height h_S and the head is adjusted such that the dummy is hit in normal direction against the head. The impact locations in this setup are the head and chest in the frontal case and the head only for rear impacts. The head is hit at h_I^H and the chest at h_I^C. In order to evaluate partial constraints the barrier height h_B is elevated until no further increase of injury severity is observed or the dummy is in danger to be destroyed.

11.4.2.2 The Standard Sitting Side Impact

In Figure 11.4 (upper right) the *side sitting* setup is depicted. The dummy is sitting upright on a fixed object at height h_S and $\varphi_N = 0^o$. The head is oriented horizontally such that the robot hits the dummy normal to the occiput. The impact locations tested in this setup are the head, the chest, the abdomen, and the pelvis. The head is hit at h_I^H, the chest at h_I^C, the abdomen at h_I^A, and the pelvis at h_I^P. In order to evaluate partial constraints the side barrier height h_B is elevated until no further increase of injury severity is observed or the dummy could be destroyed.

11.4.2.3 The Standard Standing Frontal and Rear Impact

In Figure 11.4 (lower left) the *frontal standing* and *rear standing* setup are shown. The dummy is standing upright and the head is adjusted such that the dummy is hit in normal direction against the head. The impact locations tested in this setup are the head and chest in the frontal case and the head for rear impacts. The head is hit at h_I^H and the chest at h_I^C. In order to evaluate partial constraints the barrier height h_B (in the back of the dummy for frontal impacts and in front of the dummy for rear impacts) is elevated until no further increase of injury severity is observed or the dummy could be destroyed.

11.4.2.4 The Standard Standing Side Impact

In Figure 11.4 (lower right) the *side standing* setup is depicted. The dummy is standing upright. The impact locations tested in this setup are the head, the chest, the abdomen, and the pelvis. The head is hit at h_I^H, the chest at h_I^C, the abdomen at h_I^A, and the pelvis at h_I^P. In order to evaluate partial constraints, the side barrier height h_B is elevated until no further increase of injury severity is observed or the dummy could be destroyed. Please note that in this test the barrier does only affect the lower extremities.

In order to carry out all these experiments, various testing devices are necessary. Therefore, a list of crash-test devices that are suitable in this sense is given.

Table 11.3 Dummies for standardized crash-testing in robotics

Impact test	Proposed dummy
Sitting frontal	Hybrid III 50th Percentile Male
Sitting side	EuroSID-1/EuroSID-2 (ES-2)
Sitting rear	BioRID-II
Standing front	Pedestrian **or** Hybrid III with standing support
Standing side	EuroSID-1/EuroSID-2 (ES-2) with standing support
Standing rear	BioRID-II with standing support

11.4.3 Crash-Test Dummies for Robot-Human Impacts

In Table 11.3 appropriate crash-test dummies for each of the proposed standard tests by biomechanical dimensioning are listed[4], which are tailored to the needs of the proposed impact tests. The first two for *Sitting frontal* and *Sitting side* are already established dummies in automobile crash-testing. The BioRID-II was designed for the rear impact assessment and is among other things especially designed for whiplash assessment. The Pedestrian can be used to simulate secondary impacts and their dependency on impact velocity and robot mass. As an alternative one could fix a Hybrid III in standing position and realize a simple release mechanism e.g. based on a light barrier to simulate standing during the impact.

11.4.4 Possible Extensions

In future the extension of the proposal to following body parts for the three impact directions is intended.

- Frontal impact
  - Knee, femur, pelvis
  - Lower leg
  - Upper Extremities
  - Cranium: mandible, maxilla, nasal,...
- Side impact
  - Cranium: temporal, parietal
- Rear impact
  - Spine
  - Cranium: parietal, occipital

[4] In this proposal, the appropriate human male dummies are listed. In future work other types as female and child versions should be included.

For the listed body parts distinct dummies exist, which will be used for detailed analysis in the future. The standardized evaluation of the face could be analyzed with face dummies as presented in [30, 21] and more detailed level aspects such as the eye with the new FOCUS (Facial and Ocular CoUntermeasure for Safety headform), developed by Denton [1]. Some of these tests are only relevant for pHRI-robots and not for large industrial robots. Apart from defining standardized blunt impact testing, it is necessary to get to a point at which soft-tissue injury can be evaluated in a standardized way as well. First evaluations in this direction were carried out in [12].

In the next section the impact of the monograph on industry and standardization is described. Furthermore, it is discussed how an injury analysis could be carried out in general, and which kind of further tests are necessary for this.

11.5 Impact of the Monograph and Next Steps

Recently, robots such as the LWR-III or the WAM arm were already commercialized and provide advanced control methods that allow close physical Human-Robot Interaction. Their entire design was driven by the desire to achieve high performance torque control and interaction. In case of the LWR-III significant parts of this

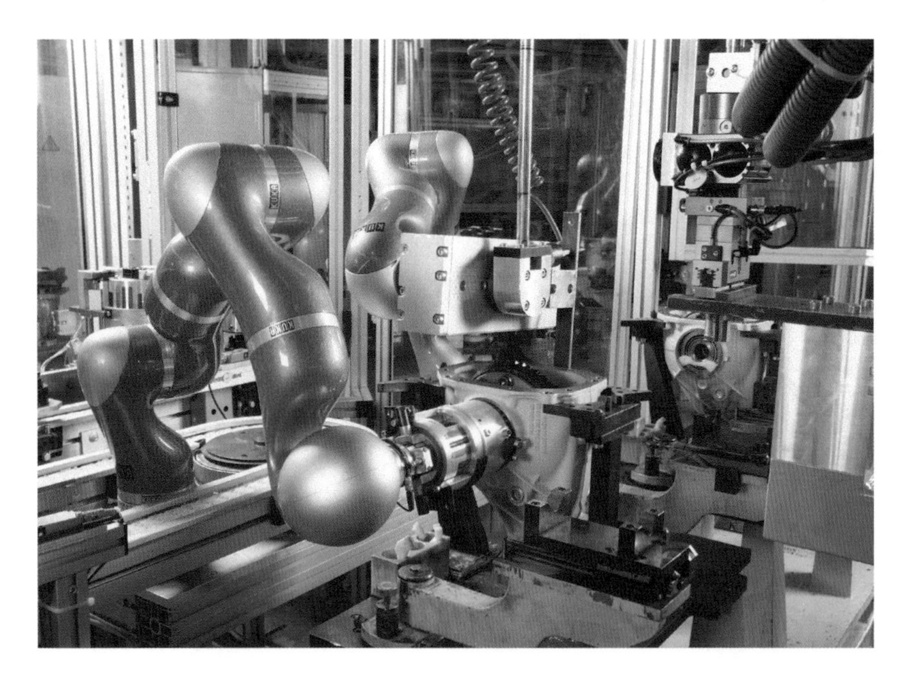

Fig. 11.5 LWR-III at Daimler-Benz

Fig. 11.6 Drop testing setup for analyzing soft-tissue injury during blunt and sharp contact

monograph were part of the technology transfer from DLR to KUKA. The collision detection and reaction methods as well as the trajectory scaling algorithm developed in this monograph are commercialized in the new KUKA Lightweight Robot [5, 20]. Apart from sensing collisions during operation, the collision detection is also used in the novel command *Trigger by contact* to switch between different controllers based on the estimation of external torques. This makes it possible to optimally combine the position accuracy of the state space controller with the adaptability of the impedance controller during physical interaction with the environment.

An interesting fact is provided by one of the first end customers using the KUKA Lightweight Robot, Daimler AG. They introduced the robot into an application for assembling rear-axle gear boxes and are currently analyzing further use in their car assembly processes. Generally, they state that the LWR-III is the next generation technology in production and that only the barrier free operation is the key for launching this innovative robot.

The fact that the work done in this monograph was also strongly noticed by the research community directly points toward the necessity to further investigate injury in robotics, find adequate countermeasures for limiting worst-case injury, and finally bring common research effort to the according standardization institution. As the monograph represents the first systematic evaluation of safety for pHRI it opens up entirely new research directions. Numerous awards and nominations at the most important robotics conferences and events were achieved. The impact of this monograph on the outcome of European projects such as SMErobot [27], PHRIENDS [7], and Viactors [3] was highly appreciated, as confirmed by the following quote from a Viactors review report.

> An important ingredient for the successful use of robots in a societal setting is their safety. DLR is the leading robotics group in safety in Human-Robot Interaction. [...] Safety definitions and procedures are of fundamental importance for Human-Robot Interaction in the design and control of robots that physically interact with humans.

However, despite the achievements made, there are still many open research questions to be answered. Furthermore, no agreement between researchers, standardization bodies, and industry has been achieved yet on which measures exactly will be used for defining robot safety in the future. The general approach in this monograph for understanding and rating injury is to analyze the cause-effect chain in a decreasing order of injury severity. It is important to first understand what exactly causes potentially lethal injury and prevent this by any means. One should therefore understand the effects of lower severity injuries and find according countermeasures. This kind of approach ensures that one will be able to determine exactly the possible worst-case injury for a given impact situation and therefore allow robots to exploit their maximum performance under the safety constraint. In order to systematically compare different robots and impact conditions it is necessary to build up various reproducible and verified dummies for fully analyzing the injury potential of a robot. For the verification of these testing devices, existing biomechanical data and insights should be used as much as possible. They form the fundamental basis to understand the dynamic behavior of humans and their response to mechanical inputs. In this line of thinking, other body parts than the ones in the monograph should also be considered. Body parts that play especially in industrial settings a major role are the upper extremities (especially the hand) as they are frequently injured in such settings. This leads to the necessity to understand hand injury, build hand dummies, and use them for verification of safety increasing control schemes. However, there are little experimental data or biomechanical understanding of hand injury, which would be valuable in robotics. Some work has been done on investigating the contact fracture mechanics of the outstretched hand and tolerance of human volunteers in crushing settings. [6, 25], e.g., analyzed the influence of impact stiffness on the impact force for outstretched hands. They investigated whether surface padding reduces the injury risk during down-fall accidents by comparing the experimental/simulation impact forces with fracture tolerance forces of 2.26 ± 1.0 [$\pm$SD] kN measured in [11], 1.6 ± 1.0 [$\pm$SD] kN measured in [28], and 1.8 ± 0.7 [$\pm$SD] kN measured in [23]. Concerning lower and upper arm there is some more extensive data to work with. According to [8] a 50% risk of elbow fracture, e.g., corresponds to forces as large as 1780 N. The particular loading tolerance of the humerus was investigated in [19, 9, 10, 26, 17] and for the lower arm in [24, 9, 10]. The impact experiments for the lower arm in [24], were conducted for impact speeds of 3.3 m/s and 7.6 m/s for both male and female subjects, respectively. The three point loading (elbow, wrist and impact force) caused linear to comminuted fracture. [9, 10] investigated impact speeds of 1.35 – 4.42 m/s for female subjects, where the lower arm was loaded in supination and pronation. For hand/finger injury [22] analyzed the tolerance of human subjects against crushing injury and pointed out the importance to differentiate between static and dynamic forces (as done in this monograph). The authors of the German *Institute for Occupational Safety and Health* defined admissible impact forces (in the sense of pain tolerance, not injury) and their analysis led also to the development of a testing device for simulating hand contact properties and measuring the contact forces. Recently, the same institute released a document listing pain tolerance values for forces and pressures of several body parts. However, in contrast to the work done in [22], it is not clear where this data originated from. This brief overview points towards the necessity of further large scale investigations to be done in the near future.

Furthermore, as discussed throughout this monograph and especially in Chapter 6, understanding of low severity blunt injuries as e.g. abrasions or hematoma, as well as sharp contact injury is still lacking to a large extent. In this line of thinking, it is crucial to analyze these two important classes of impact injury in more detail soon. For this purpose a drop test setup was developed (see Fig. 11.6) for analyzing the effect that different contact characteristics have on biological soft-tissue and use the outcome of these tests for further improving the safety characteristics of robots. Also the injury mechanisms of hands will be analyzed on a detailed level. The fundamental insight that is expected to be gained from these experiments is intended to build a basis for understanding the effect sharp tools mounted on a robot have, and how to design countermeasures to reduce their dangerous effect. The basis for this is given in Chapter 3.

11.6 Summary

Comparing the thresholds defined in *ISO-10218* with the results given in this monograph, it is clear that the listed requirements in *ISO-10218* are not based on biomechanical analysis. On the one hand, such an evaluation leads to much higher tolerance values for blunt impacts and on the other hand, to possibly lower ones for sharp contact. The intention of *ISO-10218* is to keep the velocity of the robot low in order to enable active avoidance of unintended contact by a human operator. If this is not possible, only very low exerted forces and power could avoid any kind of risk, i.e. *ISO-10218* is a conservative safety requirement. However, this appears to be an overly stringent restriction of robot performance for systems especially designed for pHRI applications. At the same time a well differentiated test standard is still missing. Particular tools and their corresponding injuries, which would demand even lower thresholds than currently required are not discussed in this standard. The definition of a a more sophisticated and differentiated basis to achieve an optimal safety-performance tradeoff is recommended. The qualification of new control and motion strategies is still a major unsolved issue, even though it seems crucial to equip robots using such methods. It is obviously evident that they contribute to a significant increase in safety and therefore have to be standardized to be well recognized as a human-friendly scheme.

For safety qualification and standardization purposes it is essential to define the level of possible injury in such granularity that the effect of safety enhancing methods can be quantified. Therefore, the AIS scale was extended such that a more differentiated rating of injuries is possible in the lower severity range, making it possible to generically classify occurring injury in robotics.

Due to the heavily varying conditions of human-robot contact, relevant injury mechanisms were classified, important factors governing each injury process, and the worst-case injury level derived from it. This classification should be considered as a basis for further investigations, as well as a roadmap pointing out open issues and the variety of possible injury mechanisms in physical Human-Robot Interaction.

Furthermore, a proposal on how future standardized blunt crash-testing could be formulated was given, similar to already established procedures in the automobile industry. The definition of such regulations makes it possible to compare different robots objectively and assess their qualification for Human-Robot Interaction. Finally, the discussed open issues point toward the most important questions to be answered and are intended to attract other researchers to this exciting new topic of robotics research.

References

[1] www.dentonatd.com
[2] AAAM: The Abbreviated Injury Scale (1990), Revision Update. Des Plaines/IL (1998)
[3] Systems emboying advances interaction behaviORS, V.V.I.A.: VIACTORS web site. Integrated project funded under the European Union's Sixth Framework Programme (FP7) (2009), http://www.viactors.org/
[4] BG/BGIA: Empfehlungen zur Gestaltung von Arbeitsplätzen mit kollaborierenden Robotern (German). Tech. rep. (2009)
[5] Bischoff, R., Kurth, J., Schreiber, G., Koeppe, R., Albu-Schäffer, A., Beyer, A., Eiberger, O., Haddadin, S., Stemmer, A., Grunwald, G., Hirzinger, G.: The kuka-dlr lightweight robot arm: a new reference platform for robotics research and manufacturing. In: International Symposium on Robotics (ISR 2010), Munich, Germany (2010)
[6] Chiu, J., Robinovitch, S.: Prediction of upper extremity impact forces during falls on the outstretched hand. Journal of Biomechanics 31(12), 1169–1176 (1998)
[7] PHRIENDS: Physical Human-Robot Interaction: Dependability and Safety: PHRIENDS web site. Integrated project funded under the European Union's Sixth Framework Programme (FP6) (2007), http://www.phriends.eu/

[8] Duma, S., Boggess, B., Crandall, J., MacMahon, C.: Fracture tolerance of the small female elbow joint in compression: the effect of load angle relative to the long axis of the forearm. Stapp Crash Journal 46, 195–210 (2002)
[9] Duma, S., Crandall, J., Hurwitz, S., Pilkey, W.: Small female upper extremity interaction with the deploying side air bag. SAE Paper No.983148, Proc. 42th Stapp Car Crash Conf., pp. 47–63 (1998)
[10] Duma, S., Schreiber, P., McMaster, J., Crandall, J., Bass, C., Pilkey, W.: Dynamic injury tolerances for long bones of the female upper extremity. In: International Research Council on Biomechanics of Injury (IRCOBI 1998), pp. 189–201 (1998)
[11] Frykna, G.: Fracture of the distal radius including sequelae-shoulder-hand-finger syndrome, disturbance in the distal radio-ulnar joint and impairment of nerve function; a clinical and experimental study. Acta Orthop Scand (1967)
[12] Haddadin, S., Albu-Schäffer, A., De Luca, A., Hirzinger, G.: Evaluation of collision detection and reaction for a human-friendly robot on biological tissue. In: IARP International Workshop on Technical challenges and for dependable robots in Human environments (IARP 2008), Pasadena, USA (2008), www.robotic.de/Sami.Haddadin
[13] Haddadin, S., Albu-Schäffer, A., Hirzinger, G.: Safety evaluation of physical human-robot interaction via crash-testing. In: Robotics: Science and Systems Conference (RSS 2007), Atlanta, USA, pp. 217–224 (2007)
[14] Haddadin, S., Albu-Schäffer, A., Hirzinger, G.: The role of the robot mass and velocity in physical human-robot interaction - part I: Unconstrained blunt impacts. In: IEEE Int. Conf. on Robotics and Automation (ICRA 2008), Pasadena, USA, pp. 1331–1338 (2008)
[15] Haddadin, S., Haddadin, S., Khoury, A., Rokahr, T., Parusel, S., Burgkart, R., Bicchi, A., Albu-Schäffer, A.: On making robots understand safety: Embedding injury knowledge into control. Int. J. of Robotics Research (IJRR 2012) 31(13), 1578–1602 (2012)
[16] ISO10218: Robots for industrial environments - Safety requirements - Part 1: Robot (2006)
[17] Jaffredo, A., Potier, P., Robin, S., Le Coz, J., Lassau, J.: Upper extremity interaction with side impact air bag. In: International Research Council on Biomechanics of Injury (IRCOBI 1998), pp. 485–495 (1998)
[18] Kallieris, D.: Personal communication (2007)
[19] Kallieris, D., Rizzetti, A., Mattern, R., Jost, S., Priemer, P., Unger, M.: Response and vulnerability of the upper arm through side air bag deployment. SAE Transactions 120, 143–152 (2004)
[20] KUKA, http://www.kuka-robotics.com
[21] Melvin, J., Little, W., Smrcka, J., Yonghau, Z., Salloum, M.: A biomechanical face for the Hybrid III dummy. In: Proceedings of the 39th Car Crash Conference (1995)
[22] Mewes, D., Mauser, F.: Safeguarding crushing points by limitation of forces. International Journal of Occupational Safety and Ergonomics 9(9), 177–191 (2003)
[23] Myers, E., Hecker, A., Rooks, D., Hipp, J., Hayes, W.: Geometric variables from DXA of the radius predict forearm fracture load in vitro. Calcif Tissue Int. 52, 199–204 (1993)
[24] Pintar, F., Yoganandan, N., Eppinger, R.: Response intolerance of the human forearm to impact loading. SAE Paper No.983149, Proc. 42th Stapp Car Crash Conf., pp. 65–74 (1998)
[25] Robinovitch, S., Chiu, J.: Surface stiffness affects impact force during a fall on the outstretched hand. Journal of Orthopaedic Research 16(3), 309–313 (1998)

[26] Schröder, G., Kallieris, D., Tschäsche, U., Scheunert, D., Schütz, J., Zobel, R.: Are sidebags dangerous in certain sitting positions? In: International Research Council on Biomechanics of Injury (IRCOBI 1998), pp. 477–483 (1998)

[27] SMErobot^{TM}: The European Robot Initiative for Strengthening the Competitiveness of SMEs in Manufacturing. Integrated project funded under the European Union's Sixth Framework Programme (FP6) (2005), `http://www.smerobot.org/`

[28] Spadaro, J., Werner, F., Brenner, R., Fortino, M., Fay, L., Edwards, W.: Cortical and trabecular bone contribute strength to the osteopenic distal radius. Journal of Orthopaedic Research 12, 211–218 (1994)

[29] United Auto Workers: Review of robot injuries - one of the best kept secrets. In: National Robot Safety Conference, Ypsilanti, USA (2004)

[30] Viano, D., Bir, C., Walilko, T., Sherman, D.: Ballistic impact to the forehead, zygoma, and mandible: Comparison of human and frangible dummy face biomechanics. The Journal of Trauma 56(6), 1305–1311 (2004)

[31] Yamada, Y., Hirasawa, Y., Huand, S., Umetani, Y.: Fail-safe human/robot contact in the safety space. In: IEEE Int. Workshop on Robot and Human Communication, pp. 59–64 (1996)

Chapter 12
Conclusion and Outlook

12.1 Conclusion

Achieving safe Human-Robot Interaction is one of the grand challenges of robotics. It is necessary to design systems that do not harm human beings during operation. However, due to the lack of real world applications for pHRI, there was very little research on how to assess, rate, and improve the safety of robots for tasks with direct human contact. Mostly, the term *safe* was used to label dependable robotic components, for which failure rate has to be minimized and reliability to be maximized. In this sense, the monograph gives the first large scale investigation of possible injuries a human would suffer from collisions with robots and elaborates the significant factors in this complex problem. For this standard equipment from automobile crash-testing was used, which has been applied over decades to rate the injury of humans in car crashes. However, the analysis is not only based on these well established methods and their applicability to robotics, but they were also extended to the needs in robotics. An analysis, grounded on a solid biomechanical basis, seems to be the only way to investigate the safety of robots, since it is not only a question of robot design alone, but to a major extent related to the physical effect a robot has on the human. Furthermore, it is not sufficient to rate the safety of a robot by simple dependability analysis, but the level of measurable physical harm has to be of primary concern. A major contribution of the monograph is that it gives general insights into the resulting impact dynamics for rigid blunt robot-human impacts. Furthermore, various injury measures for different human body parts are evaluated theoretically and experimentally with different robots of varying size. Apart from blunt collisions, also soft-tissue injuries caused by stabbing and cutting were investigated for the first time in robotics. For this purpose pig experiments were carried out to obtain quantitative measurements for injury assessment during sharp contact. Such investigations are necessary, since future robots will either be equipped with, or grasp sharp tools and objects in real-world applications. Generally, this part of the monograph gives fundamental insight into the influence mechanical design

S. Haddadin: *Towards Safe Robots*,
Springer Tracts in Advanced Robotics 90,
DOI: 10.1007/978-3-642-40308-8_12, © Springer-Verlag Berlin Heidelberg 2014

parameters as inertia, maximum velocity, or surface curvature have on the intrinsic safety properties of the robot.

In order to assemble a full image of injury mechanisms in human-centered robotics, an overview of possible injuries, a classification attempt, as well as related injury severity measures were developed. This was completely missing in the literature up to now, but is of high interest not only for industrial robots, but also for the safety standardization of service robots.

Apart from assessing possible injuries occurring during human-robot impacts, it is equally important to evaluate and rate the quality of robot control countermeasures for reducing or even preventing them. Primarily, a robot sharing its workspace with humans should be able to detect collisions quickly and to react safely in order to limit injuries due to physical contacts. In the absence of external sensing, relative motion between robot and human is not predictable and unexpected collisions may occur at any location along the robot arm. Efficient collision detection and reaction methods that use only proprioceptive robot sensors and provide also directional information for safe robot reaction after collisions were introduced and validated for this purpose. It was shown that the proposed methods are sufficiently powerful such that even sharp contact with a scalpel can be detected for a cutting motion and that the otherwise resulting very severe injury is entirely prevented.

Besides collision identification, isolation, and reaction two collision avoidance schemes with and without task preservation were developed. Both methods do not only consider proximity, but are also able to cope with contact forces at the same time. This makes it possible to address pre-collision, collision, and post-collision phase in a unified way, as showcased by different experiments at varying dynamic conditions. The overall approaches are able to cope with various kinds of disturbances in a safe and intuitive manner and provide diverse reaction patterns. An accompanying problem, arising from the variety of novel methods for human-friendly control is that the system complexity grows vastly. Therefore, a state-based control architecture was developed that is tailored for consistently, quickly, and safely activate the corresponding overall robot behavior in response to the current situation and according sensory input.

The underlying concept integrates the aforementioned novel capabilities of the robot into a safety-oriented approach, which enables its intuitive interfacing that is both reactive and flexible. The concept was also experimentally verified for a complex Co-Worker scenario. The developed architecture is currently in use in various applications, see Fig. 12.1. The first and second picture show an LWR-III billiard experiment [2] and an EMG-controlled LWR-III. The third image depicts the recent Braingate experiment [3] where the LWR-III is continuously controlled via a Brain-Machine-Interface that is attached invasively to the motor cortex of a tetraplegic person. The decoded neural data is used to command the robot, while several safety related behaviors are activated. The fourth image depicts the SAPHARI[1] setup, which is an experimental setup for evaluating safe and autonomous physical Human-Robot Interaction. At the recent trade fair AUTOMATICA 2010 the various complex

[1] Safe and Autonomous Physical HumAn-Robot Interaction.

Fig. 12.1 Several example setups using the developed safety oriented state-based robot control architecture. LWR-III controlled via Brain-Machine Interface, SAPHARI setup, Billiard playing and EMG-controlled robot.

interaction capabilities of the state-based controlled robots were showcased for different applications as e.g. interactive bin-picking.

Overall, the system provides a simple and intuitive access for robot task programming based on hybrid state machines, making task programming powerful, flexible, and efficient at the same time.

In addition to the safety investigations, software design contributions, and control schemes for achieving safe physical Human-Robot Interaction, the analysis was extended to Competitive Robotics. Safety problems were analyzed for situations in human-robot soccer, where human and robot are opponents and it was demanded that a robot may not be more dangerous than a human opponent. Furthermore, it was shown that intrinsically compliant joints are important for protecting the robot joint from external shock loads and are beneficial to store and release energy such that high link speeds can be achieved. At DLR, various joint prototypes and a full hand-arm system implementing such intrinsically compliant actuation were designed. In addition to showing passively compliant behavior, their stiffness characteristics can also be adjusted online. These novel actuation mechanisms allow different motion control schemes compared to classical stiff actuation. In order to optimally utilize their energy storage and release capabilities, results based on optimal control theory were derived regarding the optimal excitation of the joint elastic modes for reaching maximum link velocity. In particular, the appropriate timing of bang-bang control for the position motor was formulated and experimentally verified. Furthermore, it was shown that a similar bang-bang control for stiffness adjustment further maximizes the link velocity by optimally injecting additional energy into the actuation mechanism.

In the robotics literature, intrinsic joint compliance is mostly proposed for increasing safety due its inherently elastic behavior. However, in this monograph it was shown that this is only valid for specific cases. The already mentioned gain in link velocity can yield higher impact speeds and therefore more severe injury for unconstrained blunt impacts compared to stiff robots. This, in turn, necessitates effective controllers for vibration damping during motion, which utilizes the elastic energy storage mechanism only if needed. For further enhancing the safety of such systems, the methods for collision detection and reaction were extended to the variable stiffness case and their effectiveness was experimentally proven.

To sum up, this monograph made significant contributions to a variety of nowadays open research problems in human interactive robotics and has indeed opened up entirely new branches of robotics research. Developing the theoretical foundation and the experimental validation of various methods for collision avoidance, detection, and according reaction, as well as the development of a concept on how a human-friendly robotic Co-Worker can be designed from an architectural point of view form the foundation for bringing humans and robots closer to each other. All methods were experimentally verified on various robotic systems such as on the LWR-III, Justin, or the DLR Miro. The injury evaluation of robot-human impacts was the first in robotics to be carried out in such a systematic and extensive way. The experimental evaluation was particularly appreciated in the robotics community as a fundamental contribution for making robots safer. It contributed to the clarification of several misunderstandings and even errors present in the literature as a result of simulation only evaluation.

Finally, this monograph gives important insights into the safety and performance characteristics of Variable Impedance Actuators and how to optimally control them for achieving similar performance to humans.

The outcome of this work already found its way into commercialization and standardization. The developed collision detection and reaction schemes, the method of trajectory scaling, and partially also the methods for biomechanical safety analysis found their way to market products and into international standardization committees. Also, first industrial end users utilize the methods as an integral part of their applications.

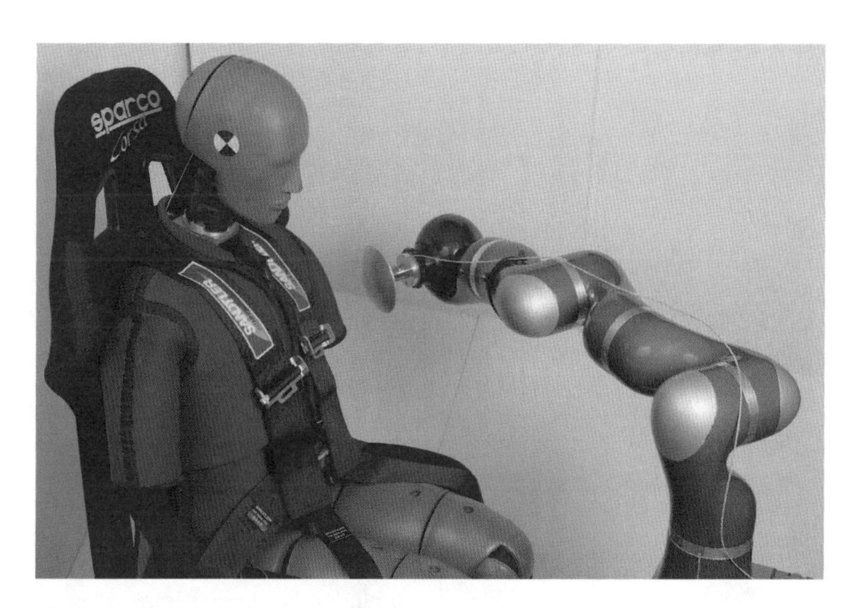

Fig. 12.2 Side impact dummy with LWR-III

Fig. 12.3 The multi-robot system SAPHARI

12.2 Outlook

Finally, some lines of research shall be presented that are directly related to the present monograph and were identified as the consequent extension of the presented work.

Certainly, the biomechanical safety investigations will be extended to other usually non-lethal injuries as abrasions and contusions, as well as the understanding of cutting and stabbing will be deepened. For this purpose a drop test setup was developed (see Fig. 11.6) for analyzing the effects that different contact characteristics have on biological soft-tissue and use the outcome of these tests for further improving the safety characteristics of robots. A long-term vision is to develop a catalog that classifies tools according to their potential injury level with the goal of providing standard guidelines for maximal robot speed, force, and possibly other relevant physical quantities, depending on the abovely mentioned geometric properties.

For carrying on the dummy crash-tests presented in this monograph, a side crash-test dummy was equipped with various impact sensors (see Fig. 12.2) and a testing suite was developed including automated robot crash-testing and evaluation software [1]. This shall lead to new insights into blunt robot-human impacts by carrying out various tests proposed in Chapter 11 and complete the picture given in this monograph.

Further experiments will focus on a detailed investigation of the effect of joint stiffness for improving safety and especially provide the experimental verification of developed methods on the new DLR hand-arm system. A focus in the future will be the extension of the optimal control results to full manipulators. For this, the effect multi-DoF dynamics have on the problem needs to be understood and it has to be investigated how the solutions for maximum link velocity in this monograph can be extended to other tasks.

In order to fully exploit the capabilities of such human-friendly robots as the LWR-III the proposed state based architecture will be further developed such that these complex devices can be controlled in an intuitive and abstract way also from high-level control and decision making processes. As a new experimental research platform SAPHARI was constructed, see Fig 12.3. This multi-robot system is equipped with various external sensors for workspace surveillance, human motion recognition, and object localization. It will enable the further development of methodologies towards flexible multi-robot systems that are capable of performing complex interaction scenarios with humans. For this, the unification of motion planning, interaction control, and collision detection/reaction will be further pursued. This is expected to finally lead to the versatility and dynamic behavior of a robot that is so crucially needed in pHRI. In order to accomplish this goal a unified way of treating motion and physical interaction needs to be found.

A further important aspect to be investigated in the future is the extension of the different methods for collision avoidance, detection, and reaction to the hand-arm system and biped systems as the DLR Biped.

To sum up, future research will be conducted in several fundamental areas, e.g., mechanisms of injury in humans cooperating with robots, further understanding and control of variable stiffness actuation systems, sensory integration for workspace monitoring and collision prevention, learning and understanding human motion, task-oriented and reactive motion planning, and safe control of Human-Robot Interaction forces.

Supporting the technological transfer of novel research results to industry and other trade branches has proven to be fruitful and successful for all parties over the last years. Therefore, the collaboration with robot manufactures such as KUKA and selected users in the risk estimation and identification process of Human-Robot Interaction systems will be continued.

Starting from the perspective of safety in robotics, the work evolved over the last years towards the much broader topic of pHRI. Recently, also novel aspects of cognitive Human-Robot Interaction, as e.g. the application of (industrial) design procedures for interaction processes in order to increase the robot's intuitiveness, usability, and feedback modalities, were investigated. On the long term the ultimate goal is this holistic view on HRI, which will potentially lead to a truly human-centered robot design from every perspective.

References

[1] Haddadin, S., Parusel, S., Vogel, J., Belder, R., Rokahr, T., Albu-Schäffer, A., Hirzinger, G.: Holistic design and analysis for the human-friendly robotic co-worker. In: IEEE/RSJ Int. Conf. on Intelligent Robots and Systems (IROS 2010), Taipeh, Taiwan, pp. 4735–4742 (2010)

[2] Parusel, S.: Playing billard with an anthropomorphic robot arm. Master's thesis, FH Kempten & German Aerospace Center (DLR) (2009)

[3] Vogel, J., Haddadin, S., Simeral, J.D., Stavisky, S.D., Bacher, D., Hochberg, L.R., Donoghue, J.P., van der Smagt, P.: Continuous control of the DLR Lightweight Robot III by a human with tetraplegia using the BrainGate2 neural interface system. In: International Symposium on Experimental Robotics (ISER 2010), Dehli, India (2010)

Appendix A
Braking Tests

In this appendix, the measurements of braking distances with the LWR-III, KR3-SI, KR6, and KR500 are given.

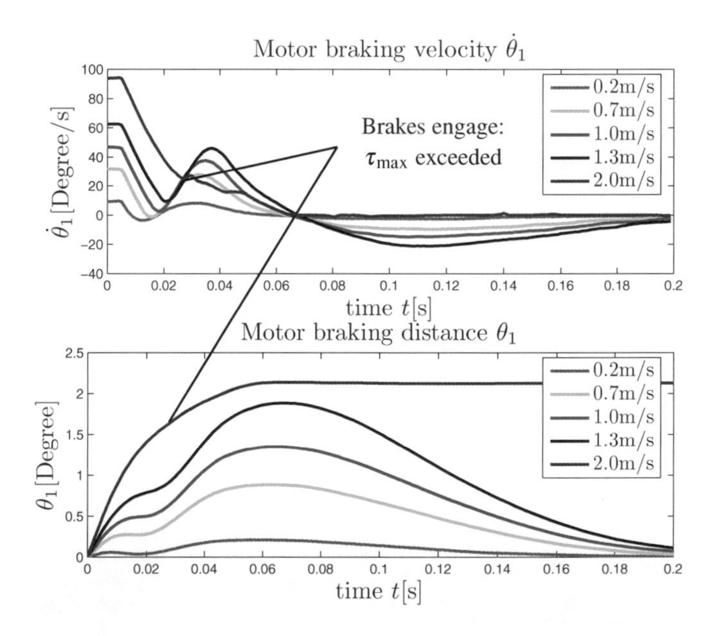

Fig. A.1 Braking behavior of the motor in axis 1 for the LWR-III at various velocities. At 2 m/s the maximal nominal joint torques are exceeded and a low-level safety feature causes the brakes to engage.

In Table 5.4 the braking distance of the LWR-III, resulting from impacting the mockup of a crash-test dummy head (dummy-dummy), illustrates the effect of external forces on braking distance. The robot's link side braking distance reduces by $> 1/3$ with the given additional impact forces. The motor braking behavior and

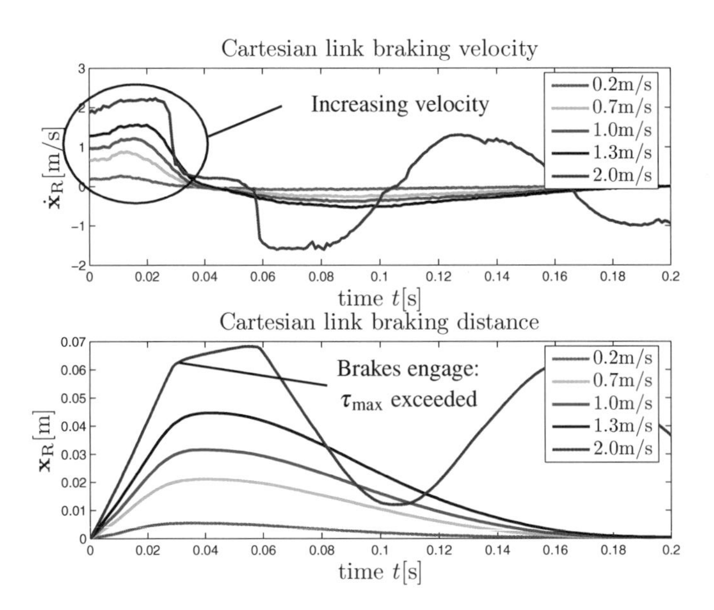

Fig. A.2 Cartesian, i.e. link side braking behavior of the LWR-III at various velocities. At 2 m/s the maximal nominal joint torques were exceeded, causing the robot to perform a low-level stop engaging the brakes. The stop time is the same for all velocities. $\dot{\mathbf{x}}_R$ possibly increases in the beginning due to pretension in the joint springs and the lack of constant velocity phase.

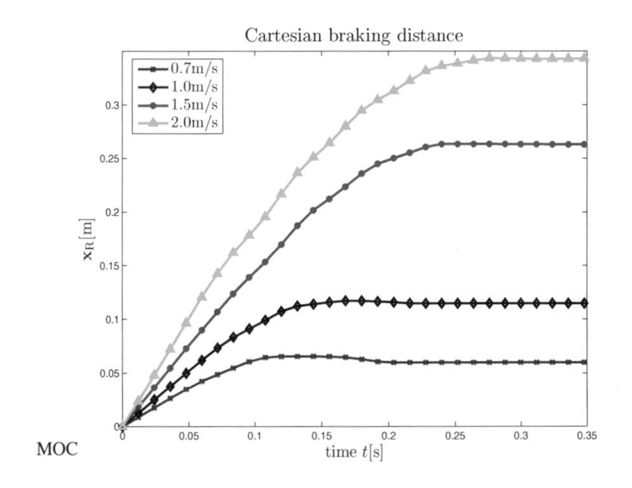

Fig. A.3 Category 0 stop with collision for the 54 kg KUKA KR3-SI at various impact velocities. The braking distance is almost 5× that of the LWR-III.

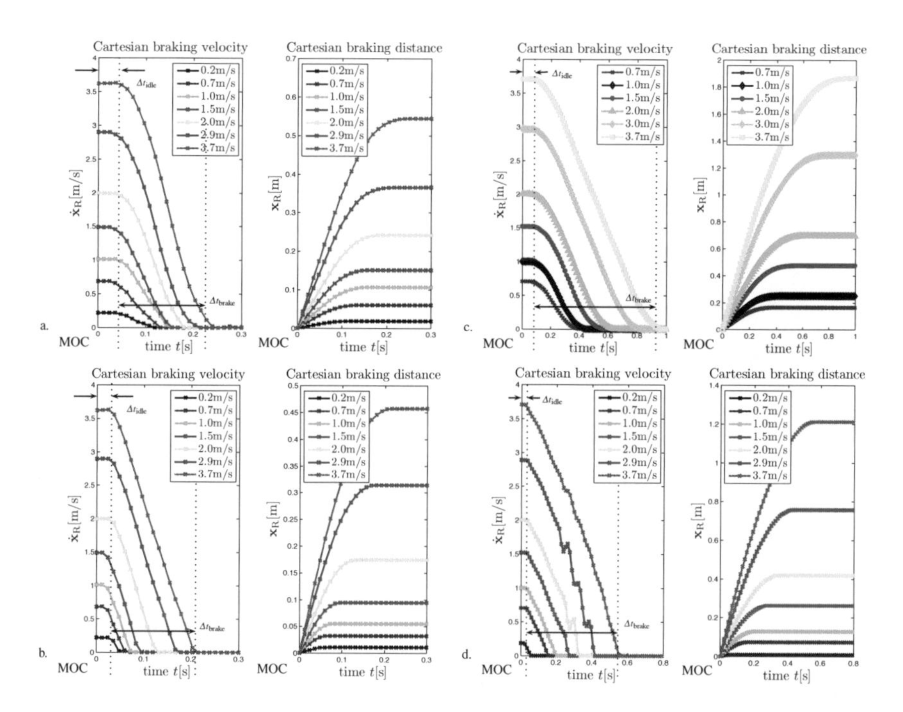

Fig. A.4 Category 1 stop (a.) and Category 0 stop (b.) with collision for the 235 kg KUKA KR6 at various velocities up to maximal TCP velocities possible with joint 1, i.e. $\dot{q}_1 = \dot{q}_1^{\max}$. The idle and braking time at 3.7 m/s are indicated (left column). Category 1 stop (a.) and Category 0 stop (b.) for the 2350 kg KUKA KR500 up to maximal TCP velocities possible with joint 1, i.e. $\dot{q}_1 = \dot{q}_1^{\max}$. The idle and braking time at 3.7 m/s are indicated (right column).

distance of the LWR-III can be extracted from Fig. A.1, where the measured curve in absence of a collision is plotted. The motor reacts 4 ms after the stop is initialized, whereas the link side is delayed due to the intrinsic joint elasticities, see Fig. A.2. The effect of increasing velocity is caused by the energy storage and release in the intrinsic joint spring[1].

Braking distance and velocity profiles of the industrial robots are given in Fig. A.3 and A.4, where the point of origin $t = 0$ s indicates the beginning of physical contact with the dummy-dummy (MOC=Moment of contact). Because of their high inertias the industrial robots were not noticeably influenced by the impact with the dummy-dummy. Therefore, the results are not differentiated in Tab. 5.4, unlike for the LWR-III. This was confirmed by braking tests without external disturbances.

[1] This effect is the same as already described in Chapter 9 and 10. Apparently, already the moderate stiffness of the LWR-III can be used to cause this effect.

Appendix B
Maximum Link Velocity for n_c Motor Cycles

It needs to be shown that the motor velocity switches when the angular deflection is zero, if the link velocity of an elastic joint is to be maximized with a limited number of motor cycles. For a bang-bang motor control the angular deflection can be described by the same differential equation as the joint's position q. Since $\dot{\theta} = \text{const.}$ holds between the switchings, from (10.63) follows:

$$M\ddot{\varphi} + \tau_J(\varphi) = 0,$$

where $\tau_J(\varphi)$ denotes the elastic joint torque. Mutliplying both sides with $\dot{\varphi}$ and integrating yields then:

$$\begin{aligned} \frac{1}{2}M\dot{\varphi}^2(t) - \frac{1}{2}M\dot{\varphi}^2(0) + \int_{\varphi(0)}^{\varphi(t)} \tau_J d\varphi &= 0 \\ \Rightarrow E_{kin_{rel}}(\dot{\varphi}) + E_{pot}(\varphi) &= E_{ges_{rel}}. \end{aligned}$$

The energy $E_{ges_{rel}}$ stays constant unless u_1 switches. When $u_1 = \dot{q} + \dot{\varphi}$ switches its sign, $\dot{\varphi}$ changes instantaneously as well, since $\dot{q}$ is continuous. Consequently, $E_{ges_{rel}}$ can only be changed by switching of u_1. In addition, the maximum link velocity depends on the maximum relative velocity $\dot{\varphi}_{max}$, since $\dot{q}_{max} = \dot{\varphi}_{max} + u_{1max}$[1]. This velocity is obtained when the total energy is transformed in to the kinetic energy so that $\dot{\varphi}_{max} = \sqrt{\frac{2E_{ges_{rel}}}{M}}$ holds. Clearly, maximizing $\dot{\varphi}_{max}$ is equivalent to maximize the energy $E_{ges_{rel}}$. We will next find the position where u_1 should switch, if $E_{ges_{rel}}$ is to be maximized.

Before the switching, the magnitude of $\dot{\varphi}$ can be found as a function of φ:

$$|\dot{\varphi}^-| = \sqrt{\frac{2\left(E^-_{ges_{rel}} - E_{pot}(\varphi)\right)}{M}},$$

[1] Note that the maximum values $\dot{q}_{max}$ and $\dot{\varphi}_{max}$ here consider only the number of the switchings allowed for the motor control. The angular deflection constraint is not being accounted for.

where $\dot{\varphi}^-$ and $E^-_{ges_{rel}}$ denote the time derivative of the angular deflection and the total energy before the switching. After the switching of u_1 at φ, $|\dot{\varphi}|$ will at most increase by $|\Delta\dot{\varphi}| = 2u_{1max}$:

$$|\dot{\varphi}^+| = \sqrt{\frac{2\left(E^-_{ges_{rel}} - E_{pot}(\varphi)\right)}{M}} + \Delta\dot{\varphi}.$$

The total energy after the switch $E^+_{ges_{rel}}$ will then take the form:

$$\begin{aligned} E^+_{ges_{rel}} &= \frac{1}{2}M(\dot{\varphi}^+)^2 + E_{pot}(\varphi) \\ &= E^-_{ges_{rel}} + M\Delta\dot{\varphi}\left(\frac{\Delta\dot{\varphi}}{2} + |\dot{\varphi}^-|\right). \end{aligned} \tag{B.1}$$

Obviously, the maximum value for $E^+_{ges_{rel}}$ will be obtained if $E_{pot}(\varphi)$ takes its minimum value at the time u_1 switches. This means that u_1 must switch, when the angular deflection φ is zero as claimed. $E^+_{ges_{rel}}$ and $\dot{\varphi}^+_{max} = \sqrt{\frac{2E^+_{ges_{rel}}}{M}}$ can now be computed from (B.1) with $\varphi = 0$:

$$E^+_{ges_{rel}} = \frac{1}{2}M(\dot{\varphi}^-_{max} + \Delta\dot{\varphi})^2, \tag{B.2}$$

$$\Rightarrow \dot{\varphi}^+_{max} = \dot{\varphi}^-_{max} + 2u_{1max} \tag{B.3}$$

where $\dot{\varphi}^-_{max} = \sqrt{\frac{2E^-_{ges_{rel}}}{M}}$ gives the maximum value for $\dot{\varphi}$, if no switching takes place. According to (B.3), the maximum value of $\dot{\varphi}$ can be increased at most by $2u_{1max}$ with one switching provided the switching takes place when $\varphi = 0$. Since the elastic joint we consider is initially at rest $|\dot{\varphi}(0)| = u_{1max}$, we can conclude that

$$\dot{q}_{max} = \dot{\theta}_{max} + \dot{\varphi}_{max} = 2n_c\dot{\theta}_{max},$$

holds for n_c motor cycles.

Printed by Printforce, the Netherlands